# LTE Air Interface Protocols

For a listing of recent titles in the
*Artech House Mobile Communications Library*,
turn to the back of this book.

# LTE Air Interface Protocols

Mohammad T. Kawser

ARTECH
HOUSE
BOSTON | LONDON
artechhouse.com

Library of Congress Cataloging-in-Publication Data
A catalog record for this book is available from the U.S. Library of Congress.

British Library Cataloguing in Publication Data
A catalog record for this book is available from the British Library.

ISBN-13: 978-1-60807-201-9

Cover design by Vicki Kane

© 2011 Artech House
685 Canton Street
Norwood MA 02062

The following figures are © 2010; 2011. 3GPP™ TSs and TRs and are the property of ARIB, ATIS, CCSA, ETSI, TTA and TTC who jointly own the copyright in them. They are subject to further modifications and are therefore provided to you "as is" for information purposes only. Further use is strictly prohibited. 3GPP TS sources are in parentheses: Figures 1.3, 1.4b, Table 23.10 (3GPP TS 23.401 V10.2.1),
Figures 1.4a, 1.6, 1.7, 2.2, (TS 36.300 V10.0.0), Figures 5.1, 5.2, 5.3, 5.4a,b, 5.5a,b, 5.7 (TS 36.211 version 10.1.0), Figures 6.1–6.13 (TS 36.321 V9.3.0), Figures 7.1, 7.2, 7.3, 7.4a–f, 7.5, 7.6a–c, 7.7a–c, 7.8 (TS 36.322 V9.2.0), Figures 8.1, 8.2, 8.5a,b, 8.6, 8.7a,b (TS 36.323 version 10.0.0), Figures 4.1, 24.3 (TS 36.331 V10.0.0)

The following figures and tables taken from 3GPP™ TS 23.203 v10.3.0, TS 23.401 v10.2.1, TS 36.101 v10.2.1, TS36.133 v9.3.0, TS 36.211 v10.1.0, TS 36.212 v9.2.0, TS 36.213 v10.0.1, TS 36.300 v10.0.0, TS 36.304 v10.0.0, TS 36.306 v 10.1.0, TS 36.321 v9.3.0, TS 36.322 v9.2.0, TS 36.323 v10.0.0, TS 36.331 v10.0.0 and further modified by the author of this book. 3GPP™ TSs are the property of ARIB, ATIS, CCSA, ETSI, TTA and TTC who jointly own the copyright in them: Figures 1.2, 1.5, 16.1, 24.1, 24.2

All rights reserved. Printed and bound in the United States of America. No part of this book may be reproduced or utilized in any form or by any means, electronic or mechanical, including photocopying, recording, or by any information storage and retrieval system, without permission in writing from the publisher.

All terms mentioned in this book that are known to be trademarks or service marks have been appropriately capitalized. Artech House cannot attest to the accuracy of this information. Use of a term in this book should not be regarded as affecting the validity of any trademark or service mark.

10 9 8 7 6 5 4 3 2 1

# Contents

Preface                                                                                                xiii

## CHAPTER 1
### Introduction                                                                                         1

1.1 Operating Bands                                                                                       2
    1.1.1 E-UTRA Absolute Radio Frequency Channel Number (EARFCN)                      4
1.2 Network Architecture                                                                                  5
    1.2.1 Evolved UMTS Terrestrial Radio Access Network (E-UTRAN)                       6
    1.2.2 Evolved Packet Core (EPC)                                                      8
    1.2.3 Tracking Area and MME Pool Area                                               10
1.3 Protocol Stack                                                                                       11
    1.3.1 Control Plane                                                                 11
    1.3.2 User Plane                                                                    13
1.4 Heterogeneous Network                                                                                13
    1.4.1 Home eNodeB                                                                   15
    1.4.2 Relay                                                                         17
1.5 UE Positioning                                                                                       18
    1.5.1 Network Architecture                                                          18
    1.5.2 Positioning Techniques                                                        19
1.6 Basic Definitions                                                                                    21
    1.6.1 Radio Signal Measurement                                                      21
    1.6.2 Identities in Network                                                         21
    1.6.3 Identities on UE Side                                                         22

## CHAPTER 2
### Information Bearer                                                                                   27

2.1 Channels and Signals                                                                                 27
    2.1.1 Logical Channels                                                              27
    2.1.2 Transport Channels                                                            28
    2.1.3 Physical Channels                                                             28
    2.1.4 Physical Signals                                                              29
2.2 Radio Bearer                                                                                         29
    2.2.1 Signaling Radio Bearer (SRB)                                                  29

|  |  |  |  |
|---|---|---|---|
| | 2.2.2 | Data Radio Bearer (DRB) | 31 |
| 2.3 | Mapping Among Channels and Bearers | | 32 |
| 2.4 | EPS Bearer | | 34 |
| | 2.4.1 | Service Data Flow (SDF) and Traffic Flow Template (TFT) | 35 |
| | 2.4.2 | QoS Profile | 36 |
| | 2.4.2.1 | Allocation and Retention Priority (ARP) | 36 |
| 2.5 | Piggybacking of Control Messages | | 39 |
| | 2.5.1 | Piggybacking of NAS Messages | 39 |
| | 2.5.2 | Piggybacking of RRC Messages | 40 |
| 2.6 | NAS Layer States | | 41 |
| | 2.6.1 | EPS Mobility Management (EMM) States | 41 |
| | 2.6.2 | EPS Session Management (ESM) States | 42 |

## CHAPTER 3

### Physical Layer Properties 43

|  |  |  |  |
|---|---|---|---|
| 3.1 | OFDMA for Downlink | | 43 |
| | 3.1.1 | Implications of OFDMA | 44 |
| | 3.1.2 | Transmission and Reception | 46 |
| 3.2 | Downlink Time-Frequency Resource Grid | | 48 |
| | 3.2.1 | Radio Frame Structure | 48 |
| | 3.2.2 | Resource Block (RB) | 51 |
| 3.3 | Downlink Physical Layer Process | | 53 |
| 3.4 | SC-FDMA for Uplink | | 54 |
| | 3.4.1 | SC-FDMA Transmission | 55 |
| | 3.4.2 | SC-FDMA Reception | 56 |
| 3.5 | Uplink Transmission Features | | 57 |

## CHAPTER 4

### System Information 59

|  |  |  |  |
|---|---|---|---|
| 4.1 | Mapping Among Layers | | 59 |
| 4.2 | Required System Information | | 60 |
| 4.3 | MIB Message | | 61 |
| 4.4 | SIB Type 1 Message | | 61 |
| 4.5 | SI Messages | | 62 |
| | 4.5.1 | Scheduling | 62 |
| | 4.5.2 | Modification | 63 |

## CHAPTER 5

### Physical Channels and Signals 67

|  |  |  |  |
|---|---|---|---|
| 5.1 | Downlink Physical Channels and Signals | | 67 |
| | 5.1.1 | Physical Broadcast Channel (PBCH) | 68 |
| | 5.1.2 | Physical Control Format Indicator Channel (PCFICH) | 68 |
| | 5.1.3 | Physical Downlink Control Channel (PDCCH) | 69 |
| | 5.1.4 | Physical Hybrid ARQ Indicator Channel (PHICH) | 77 |
| | 5.1.5 | Reference Signals | 80 |

| | | |
|---|---|---|
| 5.2 | Uplink Physical Channels and Signals | 87 |
| | 5.2.1 Uplink Reference Signals | 87 |
| | 5.2.2 Physical Uplink Control Channel (PUCCH) | 91 |
| | 5.2.3 Physical Random Access Channel (PRACH) | 94 |

## CHAPTER 6

### Medium Access Control — 101

| | | |
|---|---|---|
| 6.1 | MAC Layer Functions | 101 |
| 6.2 | MAC Protocol Data Unit (PDU) | 102 |
| | 6.2.1 Basic MAC PDU Format | 102 |
| | 6.2.2 Special Format for MAC PDU | 111 |

## CHAPTER 7

### Radio Link Control (RLC) — 117

| | | |
|---|---|---|
| 7.1 | Transparent Mode (TM) | 117 |
| | 7.1.1 TM Functions | 118 |
| | 7.1.2 TM PDU Format | 119 |
| 7.2 | Unacknowledged Mode (UM) | 119 |
| | 7.2.1 UM Functions | 119 |
| | 7.2.2 UM PDU Format | 122 |
| 7.3 | Acknowledged Mode (AM) | 125 |
| | 7.3.1 AM Functions | 125 |
| | 7.3.2 AM PDU Format | 130 |

## CHAPTER 8

### Packet Data Convergence Protocol (PDCP) — 137

| | | |
|---|---|---|
| 8.1 | PDCP Layer Functions | 137 |
| | 8.1.1 Processing the Data Packet | 137 |
| | 8.1.2 Security Function | 138 |
| | 8.1.3 Header Compression | 139 |
| | 8.1.4 Discard of Data Packets | 143 |
| | 8.1.5 Proper Delivery of Packets After Handover and Temporary Radio Link Failure | 144 |
| 8.2 | PDCP Protocol Data Unit (PDU) | 145 |
| | 8.2.1 PDCP Data PDU | 145 |
| | 8.2.2 PDCP Control PDU | 146 |

## CHAPTER 9

### Powering Up UE — 149

| | | |
|---|---|---|
| 9.1 | Cell Selection | 149 |
| | 9.1.1 Initial Cell Selection | 149 |
| | 9.1.2 Stored Information Cell Selection | 153 |
| 9.2 | Network Services | 153 |

## CHAPTER 10

### Synchronization — 155

10.1 Downlink Synchronization — 155
    10.1.1 Scheduling PSS and SSS — 155
    10.1.2 Physical Cell Identifier (PCI) — 156
    10.1.3 PSS Sequence — 156
    10.1.4 SSS Sequence — 157
10.2 Uplink Synchronization — 158
    10.2.1 Establishment of Timing Alignment — 159
    10.2.2 Maintenance of Timing Alignment — 161

## CHAPTER 11

### Random Access — 163

11.1 Configuration of Random Access — 163
11.2 Contention-Based Random Access — 164
    11.2.1 Transmission of the Random Access Preamble — 165
    11.2.2 Random Access Response (RAR) — 167
    11.2.3 Message 3 — 168
    11.2.4 Contention Resolution — 169
11.3 Noncontention-Based Random Access — 171
    11.3.1 Assignment of the Random Access Preamble — 172
    11.3.2 Transmission of the Random Access Preamble — 173
    11.3.3 RAR — 173

## CHAPTER 12

### NAS and AS Layer Security — 175

12.1 EPS Security Context — 176
12.2 EPS Authentication and Key Agreement (EPS AKA) — 176
12.3 Security in the NAS Layer — 178
12.4 Security in the AS Layer — 179
    12.4.1 Security Activation — 179
    12.4.2 COUNT — 180
    12.4.3 Ciphering — 181
    12.4.4 Integrity Protection — 181

## CHAPTER 13

### Paging — 183

13.1 Mapping Among Layers — 184
13.2 Indication of Paging Messages — 184
13.3 Default Paging Cycle — 185
    13.3.1 Allocation of the Paging Frame — 186
    13.3.1 Allocation of the Paging Occasion — 186
13.4 Paging Message — 186

## CHAPTER 14

### NAS Signaling Connection — 189

14.1 RRC Connection — 189
    14.1.1 RRC Connection Establishment — 190
    14.1.2 RRC Connection Release — 193
    14.1.3 Radio Link Failure (RLF) — 193
    14.1.4 RRC Connection Reestablishment — 195
    14.1.5 RRC Connection Reconfiguration — 198
14.2 MME Selection — 198
14.3 S1-MME Connection — 199
    14.3.1 S1 Setup — 199
    14.3.2 UE-Associated Logical S1-Connection Setup — 200
14.4 NAS Signaling Connection Release — 201

## CHAPTER 15

### Attach and Detach — 205

15.1 Initial Attach Procedure — 205
    15.1.1 NAS Signaling Connection Establishment and Attach Request — 206
    15.1.2 User Identification — 211
    15.1.3 Authentication and NAS Security Activation — 211
    15.1.4 ME Identity Check — 212
    15.1.5 Old EPS Bearer Context Delete — 212
    15.1.6 Location Update — 212
    15.1.7 PDN Connection Establishment — 213
15.2 Independent GUTI Reallocation — 223
15.3 Detach Procedure — 224
    15.3.1 UE-Initiated Detach Procedure — 225
    15.3.2 MME Initiated Detach Procedure — 227

## CHAPTER 16

### Discontinuous Reception — 231

16.1 DRX in the RRC_IDLE State — 231
16.2 DRX in the RRC_CONNECTED State — 231
    16.2.1 Configuration of the DRX Cycle — 232
    16.2.2 Continuous Reception and Resumption of DRX — 233
    16.2.3 HARQ During DRX — 233
    16.2.4 Active Time — 234

## CHAPTER 17

### Uplink Power Control — 235

17.1 Uplink Power Control Considerations — 235
17.2 Power Control on the PUSCH — 237
    17.2.1 Semistatically Configurable Terms — 238
    17.2.2 Dynamically Configurable Terms — 239

17.3 Power Control on PUCCH 244
    17.3.1 Semistatically Configurable Terms 244
    17.3.2 Dynamically Configurable Terms 245

## CHAPTER 18

### Multiple Antenna Techniques in the Downlink 249

18.1 Transmit Diversity: Transmission Mode 2 252
18.2 Spatial Multiplexing 255
    18.2.1 Single-User MIMO (SU-MIMO) 256
    18.2.2 Multiple Users MIMO (MU-MIMO): Transmission Mode 5 262
18.3 Beamforming: Transmission Mode 6 and 7 262
    18.3.1 Transmission Mode 7 263
    18.3.2 Transmission Mode 6 264

## CHAPTER 19

### Scheduling and Allocation of Resources 265

19.1 Scheduling Decision 265
    19.1.1 Information for the Scheduling Decision 266
    19.1.2 Considerations in Scheduling Decision 266
19.2 Resource Allocation Procedure 271
    19.2.1 Downlink Allocation 272
    19.2.2 Uplink Allocation 275
19.3 Apportionment of Resources Among Logical Channels 279

## CHAPTER 20

### Basic Activities During Data Transfer 283

20.1 Feedback from the UE 284
    20.1.1 Aperiodic CQI/PMI/RI Reporting 286
    20.1.2 Periodic CQI/PMI/RI Reporting 286
20.2 Link Adaptation 289
20.3 Hybrid Automatic Repeat Request (HARQ) 292
    20.3.1 HARQ Process 292
    20.3.2 Interrelationship Between HARQ and ARQ 296
    20.3.3 HARQ Information on the PDCCH 297
    20.3.4 Types of HARQ Retransmissions 297
    20.3.5 HARQ for the Downlink Transmission 298
    20.3.6 HARQ for the Uplink Transmission 299

## CHAPTER 21

### Data Transfer Session Setup 301

21.1 Service Request Procedure 302
    21.1.1 UE-Triggered Service Request 302
    21.1.2 Network-Triggered Service Request 305
21.2 Dedicated EPS Bearer Setup 306

|  |  | 21.2.1 | Network-Initiated Dedicated EPS Bearer Setup | 306 |
|---|---|---|---|---|
|  |  | 21.2.2 | UE-Requested Dedicated EPS Bearer Setup | 309 |
|  | 21.3 | Voice Services | | 310 |
|  |  | 21.3.1 | Voice Service Solutions | 310 |
|  |  | 21.3.2 | CSFB Procedure | 312 |
|  | 21.4 | Short Message Service (SMS) | | 318 |
|  |  | 21.4.1 | SMS Solutions | 318 |
|  |  | 21.4.2 | SMS over the SGs Procedure | 318 |

### CHAPTER 22

## Tracking Area Update — 323

| 22.1 | NAS Signaling Connection Establishment and TAU Request | 326 |
|---|---|---|
| 22.2 | UE Context Update | 327 |
| 22.3 | Bearer Update | 328 |
| 22.4 | Updating HSS | 329 |
| 22.5 | Tracking Area Update Accept | 330 |
| 22.6 | S1 Release | 331 |

### CHAPTER 23

## Change of Cell — 333

| 23.1 | Neighbor Cells | | 333 |
|---|---|---|---|
| 23.2 | Cell Reselection | | 334 |
|  | 23.2.1 | Measurement of Neighbor Cells | 336 |
|  | 23.2.2 | Cell Reselection Criteria | 339 |
| 23.3 | HANDOVER | | 343 |
|  | 23.3.1 | Types of Handover | 344 |
|  | 23.3.2 | Measurement and Reporting | 346 |
|  | 23.3.3 | Intra-E-UTRAN Handover Procedure | 362 |
|  | 23.3.4 | Inter-RAT Handover Procedure | 369 |

### CHAPTER 24

## Multimedia Broadcast Multicast Service — 381

| 24.1 | Multicell Transmission | | 381 |
|---|---|---|---|
| 24.2 | Network Architecture | | 382 |
| 24.3 | Mapping Among Layers | | 384 |
| 24.4 | SIB Type 13 | | 385 |
| 24.5 | MCCH | | 386 |
|  | 24.5.1 | Scheduling | 386 |
|  | 24.5.2 | MCS Configuration | 387 |
|  | 24.5.3 | Modification | 387 |
| 24.6 | MBMS Service and MBMS Session | | 389 |
|  | 24.6.1 | Session Start | 390 |
|  | 24.6.2 | Session Stop | 391 |
| 24.7 | PMCH | | 391 |

## CHAPTER 25

### Coordination Among Cells — 395

- 25.1 Intercell Interference Coordination (ICIC) — 395
  - 25.1.1 Exchange of Information to Support ICIC — 396
  - 25.1.2 Fractional Frequency Reuse (FFR) — 397
- 25.2 Load Balancing Among Cells — 399
- 25.3 Load Balancing and Rebalancing Among MMEs — 401
- 25.4 Self-Optimizing Networks (SON) — 402
  - 25.4.1 Self-Configuration (SC) — 402
  - 25.4.2 Self-Optimization (SO) — 403

## CHAPTER 26

### Public Warning System — 405

- 26.1 Earthquake and Tsunami Warning System (ETWS) — 405
- 26.2 Commercial Mobile Alert Service (CMAS) — 406
- 26.3 PWS Notification — 406
- 26.4 SIB Content for PWS — 407

### Appendix A: Major Information Elements (IEs) on RRC Messages — 409

- A.1 IEs on SIB Type 2 — 409
- A.2 IEs on RadioResourceConfigDedicated IE — 409
- A.3 IEs Used During Handover — 409

Acronyms — 413

Bibliography — 419

About the Author — 423

Index — 425

# Preface

The rapidly growing demand for bandwidth in cellular communication has impelled the telecommunication standardization authorities to introduce high performing and cost-effective platforms. The currently deployed technologies for high speed mobile wireless access are HSPA, EV-DO, and WiMAX. In order to provide solutions for next generation wireless broadband, the industries have demonstrated strong support for Long Term Evolution (LTE) specified by 3GPP. The first version of LTE is documented in Release 8 of the 3GPP specifications. Release 9 enhanced specifications for various features and also introduced several new features. By mid-2011, more than 200 operators in more than 80 countries were already known as proponents of LTE and more than 20 commercial LTE networks have already been launched around the world. Thus, it is widely anticipated that the majority share of cellular networks will be based on LTE/LTE-Advanced platform within a decade.

Since the advent of cellular communication, the overall technologies have been superseded several times by more capable ones. Whenever any new cellular technology came to light, it became an onerous job for the technical folks to get the knack. There was always a scarcity of tutorials to help teach the underlying technologies easily. This unnecessarily caused trouble and delayed the implementation as well.

This book delineates the air interface protocols of LTE in detail so the reader can learn them easily. The succinct and lucid descriptions using simple English will hopefully make the lives of thousands of people easier who work for real implementation. The reader is expected to find it easy as pie to learn as well as remember how LTE works! The book is expected to provide a great impetus to the implementation of LTE. The IEs in layer 3 messages are shown capitalizing the initial letters with a view to help the readers remember the IEs. In addition to Release 8, Release 9 has been covered to some extent. This edition of the book covers LTE FDD only and the next edition is expected to include LTE TDD.

Despite my best efforts, there may be inadvertent errors in the book and I apologize for them. The readers are welcome to send me any type of comments at mkawser@hotmail.com. Criticism will help me improve the content in subsequent editions of the book. On the other hand, the satisfaction of readers will encourage me.

# CHAPTER 1

# Introduction

Long Term Evolution (LTE) refers to the 3GPP Evolved UMTS Terrestrial Radio Access (E-UTRA) technology. Its first version is documented in Release 8 of 3GPP specifications. The advent of LTE is regarded as the latest upgrade in 3GPP technologies after GSM/GPRS/EDGE and UMTS/HSPA/HSPA+. Moreover, LTE, instead of UMB, has recently been chosen as the latest upgrade in 3GPP2 technologies after IS-95, CDMA 2000, and EVDO Rev 0/Rev A/Rev B. LTE is now considered the most promising cellular technology and expected to meet the growing demand of data rate. LTE is designed to achieve a peak data rate of 100 Mbps in downlink and 50 Mbps in uplink for 20-MHz bandwidth operation. However, the actual data rate is affected by radio link condition, the number of users in the cell, and various overhead. The overhead includes a guard band of the channel bandwidth, cyclic prefix, time windowing, and the control signaling at different layers of the protocol stack as shown in Appendix B. LTE can achieve a peak spectral efficiency of 5 bps/Hz in downlink and 2.5 bps/Hz in uplink using code rates exceeding 0.5. LTE uses optimized headers and control signaling in different layers with a view to attain a small difference between the physical layer and application layer throughputs. LTE supports QPSK, 16-QAM, and 64-QAM for data whereas HSPA supports QPSK and 16-QAM only. LTE uses an all-IP network and it supports advanced multiple antenna technologies. LTE uses optimized signaling. LTE offers enhanced support for end to end QoS. It also offers reduced complexity in UE and the network and, thus, allows reduced CAPEX and OPEX. The user plane latency is expected to be less than 5 ms for small IP packets. The user plane latency is critical, especially, for interactive applications (e.g., VoIP and gaming). LTE is optimized for user speed up to 15 km/h. But LTE supports high performance for user speed up to 120 km/h and it sustains mobility for user speed up to 350 km/h or even up to 500 km/h. The coverage is expected to be very good up to 5 km. The coverage can be good enough up to 30 km and it may range up to 100 km. However, LTE actually has a coverage issue because of its high data rate support. This makes the maximum allowable path loss (MAPL) for LTE less than other 3GPP/3GPP2 technologies.

The successor to LTE is LTE-Advanced, which is being standardized in Release 10. LTE-Advanced has the target peak data rate 1 Gbps in downlink and 500 Mbps in uplink. The target peak spectral efficiency is 30 bps/Hz in downlink and 15 bps/Hz in uplink.

## 1.1 Operating Bands

LTE supports both frequency division duplex (FDD) and time division duplex (TDD) modes allowing paired and unpaired spectrum respectively. LTE FDD and LTE TDD are often referred to as FD-LTE and TD-LTE, respectively. LTE FDD is more widely accepted by the industry and this book covers LTE FDD only. LTE FDD can be operated in both full- and half-duplex modes from the perspective of the user equipment (UE). In the case of half-duplex FDD, the UE transmits and receives over separate durations in uplink and downlink frequencies, respectively. The scheduler ensures that the UE does not need to transmit and receive simultaneously. This allows relaxed duplex-filter requirements at the UE. This is particularly useful when the spacing between uplink and downlink frequency ranges is small.

LTE can be operated on a channel bandwidth of 1.4, 3, 5, 10, 15, or 20 MHz. For the 1.4-MHz size, 22% of the channel bandwidth is used for the guard band, and for all channel bandwidths above 1.4 MHz, the guard band is 10% of the channel bandwidth. Table 1.1 shows the operating bands specified for LTE FDD and the different sizes of channel bandwidth that are permissible at these operating bands.

Operating bands 13 and 17 are the most promising in the United States for LTE FDD. In Europe, operating band 7 is the most promising for LTE FDD and frequencies around 2.6 GHz are also chosen for LTE TDD and WiMAX operation.

Release 10 supports channel bandwidth up to 100 MHz in order to achieve a very high data rate. It employs carrier aggregation to acquire the increased overall bandwidth while maintaining backward compatibility with earlier deployments. It supports aggregation of up to five carriers, each of which are 20 MHz wide, giving 100 MHz at the maximum. The aggregated carriers can be contiguous or noncontiguous.

A UE supports a few specific operating bands and it supports all the different sizes of channel bandwidth that is supported in the respective operating bands. The following points may be considered in the selection of channel bandwidth for deployment.

1. The channel bandwidths are scalable from 1.4 MHz up to 20 MHz among their different sizes. Such scalability is not available for UMTS. The use of OFDMA/SC-FDMA lends itself well to the scalability. A change in the FFT size and the number of subcarriers with fixed subcarrier spacing provides for changing the channel bandwidth. In every operating band, the consecutive sizes of channel bandwidth are permissible for the provision of scalability.
2. A larger channel bandwidth can support higher data rates and/or more users. A 5-MHz channel bandwidth may typically support around 200 VoIP calls.
3. A smaller channel bandwidth allows simpler spectrum refarming.
4. The transmit power required at the eNodeB is generally higher for larger channel bandwidth. For fixed power amplifier output power at the eNodeB, the transmit power spectral density (PSD) is inversely proportional to the channel bandwidth. For a particular throughput, channel coding gain increases with bandwidth. Therefore, if there are fixed power amplifier

## 1.1 Operating Bands

**Table 1.1** Operating Bands for LTE FDD

| Operating Band | Common Identity | Total Spectrum (MHz) | Uplink (MHz) | Downlink (MHz) | Channel BW (MHz) 1.4 | 3 | 5 | 10 | 15 | 20 |
|---|---|---|---|---|---|---|---|---|---|---|
| 1 | Europe/Asia/Australia IMT | 2 × 60 | 1,920–1,980 | 2,110–2,170 | | | ←—Supported—→ | | | |
| 2 | U.S. PCS | 2 × 60 | 1,850–1,910 | 1,930–1,990 | ←— | | | | —Supported—→ | |
| 3 | Europe/Asia DCS | 2 × 75 | 1,710–1,785 | 1,805–1,880 | | | ←— | | —Supported—→ | |
| 4 | U.S. AWS | 2 × 45 | 1,710–1,755 | 2,110–2,155 | ←— | | | | —Supported—→ | |
| 5 | U.S. 850 | 2 × 25 | 824–849 | 869–894 | ←— | | —Supported—→ | | | |
| 6 | Japan 800 | 2 × 10 | 830–840 | 875–885 | | | ←—Supported—→ | | | |
| 7 | Europe IMT Extension | 2 × 70 | 2,500–2,570 | 2,620–2,690 | | | ←— | | —Supported—→ | |
| 8 | Europe/Asia/Australia GSM | 2 × 35 | 880–915 | 925–960 | | ←—Supported—→ | | | | |
| 9 | Japan 1,700 | 2 × 35 | 1,749.9–1,784.9 | 1,844.9–1,879.9 | | | ←— | | —Supported—→ | |
| 10 | Extended AWS | 2 × 60 | 1,710–1,770 | 2,110–2,170 | | | ←— | | —Supported—→ | |
| 11 | Japan 1,500 | 2 × 20 | 1,427.9–1,447.9 | 1,475.9–1,495.9 | | | ←—Supported—→ | | | |
| 12 | U.S. Lower 700 | 2 × 18 | 698–716 | 728–746 | ←— | —Supported—→ | | | | |
| 13 | U.S. Upper 700 | 2 × 10 | 777–787 | 746–756 | | | ←—Supported—→ | | | |
| 14 | U.S. Upper 700 | 2 × 10 | 788–798 | 758–768 | | | ←—Supported—→ | | | |
| 17 | U.S. Lower 700 | 2 × 12 | 704–716 | 734–746 | | | ←—Supported—→ | | | |
| 18 | Japan 800 | 2 × 15 | 815–830 | 860–875 | | | | ←— | —Supported—→ | |
| 19 | Japan 800 | 2 × 15 | 830–845 | 875–890 | | | | ←— | —Supported—→ | |
| 20 | Europe 800 | 2 × 30 | 832–862 | 791–821 | | | ←— | | —Supported—→ | |
| 21 | Japan 1,500 Extended | 2 × 15 | 1,447.9–1,462.9 | 1,495.9–1,510.9 | | | ←—Supported—→ | | | |
| 23 | S-band [for Mobile Satellite Services (MSS)] | 2 × 20 | 2,000–2,020 | 2,180–2,200 | ←— | —Supported—→ | | | | |
| 24 | L-band [for Mobile Satellite Services (MSS)] | 2 × 34 | 1,626.5–1,660.5 | 1,525–1,559 | | | ←—Supported—→ | | | |
| 25 | U.S. PCS 1900 with G block | 2 × 65 | 1,850–1,915 | 1,930–1,995 | ←— | | —Supported—→ | | | |

output power and target throughput, then there is an optimum channel bandwidth.
5. The smaller channel bandwidths have limited guard bands, so their coexistence with other services in the same area is more challenging. The other services may include GSM, UMTS, CDMA, WiMAX, or even other LTE FDD/TDD. Channel bandwidths larger than 5 MHz have much better guard bands allowing simpler coexistence.
6. An operator may choose a small channel bandwidth in initial deployment. As the number of subscribers grows, the operator may opt for a larger channel bandwidth. This can help avoid increasing the number of cells with subscriber density. The increase in the number of subscribers increases the revenue for the operator and may pave the way for spending for the larger channel bandwidth.
7. Finally, the most important factor will most likely be the availability and the cost of acquiring a particular channel bandwidth.

In general, a lower frequency operating band may require slightly larger antennas, but nevertheless, it is more attractive for several reasons:

1. The free space path loss is smaller for lower frequency. The path loss depends on the number of wavelengths that the radio wave has traversed. If the frequency is doubled, making the wavelength half, then it offers the same free space path loss in half a distance.
2. The lower frequency offers less attenuation of the radio wave as it penetrates through objects. This improves signal quality in nonline-of-site (NLOS) link.
3. The lower frequency offers less attenuation by rain drops.
4. The lower frequency allows a higher angle of diffraction for the radio wave. Thus, the radio wave can bend more around obstacles. This improves signal quality in the NLOS link.
5. The lower frequency offers a smaller Doppler shift.
6. The lower frequency allows the radio wave to get better reflected by objects because less energy is absorbed. This improves signal quality in the NLOS link.
7. The lower frequency allows a simpler transmitter and receiver.

### 1.1.1 E-UTRA Absolute Radio Frequency Channel Number (EARFCN)

The E-UTRA absolute radio frequency channel number (EARFCN) is a number between 0 and 65,535 used to indicate the center carrier frequency of operation in uplink or downlink. The uplink EARFCN and downlink EARFCN indicate the uplink and downlink center carrier frequencies in an operating band, respectively. There is a specific range of values of uplink EARFCNs and downlink EARFCNs for each operating band. For all operating bands, the channel raster is 100 kHz (i.e., the EARFCNs correspond to center carrier frequencies with 100-kHz step size). The center carrier frequencies are integer multiples of 100 kHz as well.

## 1.2 Network Architecture

The Evolved Packet System (EPS) refers to the evolution of 3GPP UMTS radio access, packet core, and its integration to legacy 3GPP and non-3GPP networks (i.e., EPS refers to what is commonly termed the LTE network). EPS provides a data flow path with a particular QoS. The information transferred within EPS has two major types:

1. Control information;
2. User data.

Figure 1.1 shows the network architecture for EPS, which includes the following entities:

1. Radio access network, known as the Evolved UMTS Terrestrial Radio Access Network (E-UTRAN).
2. The nonradio aspects known as system architecture evolution (SAE), which includes the core network called the evolved packet core (EPC).

Two protocols are commonly used in EPS:

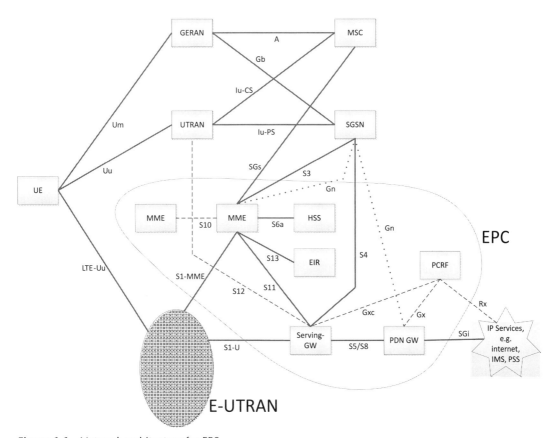

**Figure 1.1** Network architecture for EPS.

- *Stream Control Transmission Protocol (SCTP):* SCTP is commonly used protocol for the exchange of control messages within EPS. As opposed to conventional TCP, SCTP provides several streams within a connection. The SCTP stream represents a sequence of messages, whereas a TCP stream represents a sequence of bytes. An SCTP stream is unidirectional, so a pair of SCTP streams makes it bidirectional. Multiple pairs of SCTP streams are used. The key advantage of SCTP over TCP is that SCTP can overcome the head-of-line blocking problem. In TCP, the order of all packets is preserved. Therefore, if an earlier packet is lost, it is retransmitted and it arrives late; then a subsequent packet that reaches it duly has to wait. This is called a head-of-line blocking problem. This delay is unnecessary if the packets carry independent messages and this delay must be avoided if the messages are urgent. SCTP has multiple pairs of streams and data in different streams have no order constraints. Thus, head-of-line blocking does not occur. In addition, SCTP allows multihoming (i.e., reaching the destination using multiple network addresses). Furthermore, SCTP offers better transmission efficiency because of its message level framing characteristics, whereas TCP uses a byte stream.
- *GPRS Tunneling Protocol (GTP):* GTP is commonly used for packet routing within EPS. In this case, the IP packets for the UE are encapsulated and tunneled. The GTP tunnels are established to transfer control plane and user plane information. Thus, the GTP tunnels can be classified as GTP-C tunnels to transport control plane information and GTP-U tunnels to transport user plane information. A GTP tunnel, at each node, is identified by a tunnel end point identifier (TEID), an IP address, and a UDP port number. A transport bearer is identified by the GTP tunnel end points and the IP addresses (i.e., it is identified by the source TEID, destination TEID, source IP address, destination IP address, and UDP port numbers).

### 1.2.1 Evolved UMTS Terrestrial Radio Access Network (E-UTRAN)

The Evolved UMTS Terrestrial Radio Access Network (E-UTRAN) includes only the base station called eNodeB as shown in Figure 1.2. Unlike UMTS, there is no central node like the RNC and a flat hierarchy is used instead. Therefore, it is said to have a flat architecture. The flat architecture creates fewer nodes in the network and causes lower latency. The primary functions of the eNodeB are:

- Radio bearer control and radio admission control.
- Connection mobility control.
- Uplink and downlink radio resource scheduling and allocation.
- IP header compression and encryption of user data.
- Selection of an MME during UE attachment if it is needed. There is a NAS node selection function (NNSF) located in the eNodeB to determine which MME would be associated with a particular UE.
- Routing of the user plane data towards the serving gateway.
- Scheduling and transmission of paging messages.

## 1.2 Network Architecture

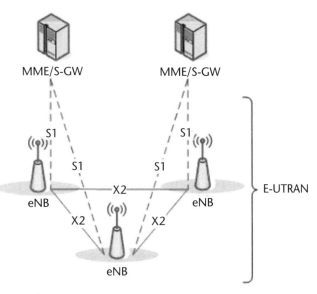

**Figure 1.2** E-UTRAN architecture.

- Scheduling and transmission of broadcast information and the Public Warning System (PWS) message.
- Measurement and reporting configuration.
- Handling a closed subscriber group (CSG).

#### 1.2.1.1 Interfaces with eNodeB

1. *S1 interface:* The eNodeB is connected with EPC via the S1 interface. An eNodeB can be connected with more than one MME/S-GW over multiple S1 interfaces. The S1 interface is split into two interfaces:
    - *S1-U:* S1-U is the part of the S1 interface for the user plane. S1-U connects the eNodeB with the Serving Gateway (S-GW). Separate GTP-U tunnels are established for each EPS bearer for each UE.
    - *S1-MME:* S1-MME is the part of the S1 interface for the control plane. S1-MME connects the eNodeB with the MME. As shown in Figure 1.3, the S1-AP protocol runs over the SCTP/IP on the S1 interface.

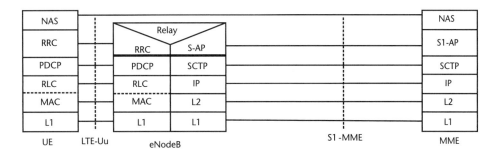

**Figure 1.3** Signaling protocol stack between the UE and the network.

2. *X2 interface*: The eNodeBs in a neighborhood are interconnected via X2 interfaces. The X2 interface is used for handover, load balancing, intercell interference coordination (ICIC), and so forth. The X2 interface is split into two interfaces:
   - X2-U: X2-U is the part of the X2 interface for the user plane.
   - X2-CP: X2-CP is the part of the X2 interface for the control plane.
3. LTE-Uu interface: The eNodeB communicates with the UE over an air interface known as the LTE-Uu interface.

### 1.2.2 Evolved Packet Core (EPC)

The Evolved Packet Core (EPC) includes the following major components:

1. *PDN gateway (P-GW):* The packet data network (PDN) gateway is the gateway to interface the network towards the PDN and thus provides connectivity to the IP backbone. The association between the UE and the PDN is known as the PDN connection. One IP address is used for a particular PDN connection. The primary functions of the PDN GW are:
   - IP address allocation for UEs;
   - Enforcement of the data rate of the UE based on APN-AMBR for non-GBR bearers and based on MBR for the GBR bearer;
   - Downlink packet filtering based on downlink TFT;
   - Uplink and downlink service level charging;
   - Serving as the mobility anchor during handover with the CDMA2000 network;
2. *Serving Gateway (S-GW):* The primary functions of the serving gateway are:
   - Routing and forwarding of IP packets;
   - Downlink packet buffering;
   - Serving as the local mobility anchor during handover between eNodeBs or handover with the GSM or UMTS network;
   - Uplink and downlink charging per UE, PDN, and QCI;
   - Initiation of a network-triggered service request procedure.
3. *Mobility Management Entity (MME):* The primary functions of the MME are:
   - NAS layer signaling;
   - NAS and AS layer security and authentication;
   - Management of the tracking area list;
   - Selection of the PDN GW and serving GW;
   - Selection of the MME for handover with the MME change;
   - Selection of the Serving GPRS Support Node (SGSN) selection for handover to the GSM or UMTS network;
   - Supporting Public Warning System (PWS) message transmission;
   - Roaming;
   - EPS bearer management;
   - CS fallback functions.
4. *Home Subscription Server (HSS):* The HSS stores the subscription data for the subscribers of an operator. It stores a long-term key used for

authentication. The HSS is updated with the information of the currently serving MME of the UE. It also stores the maximum bit rate restrictions for the UE and the APN, QoS information for different EPS bearers, and the visitor PLMNs in which roaming is allowed. The HSS typically includes an Authentication Center (AuC), which authenticates the UEs. The HSS is located in the home network of the subscriber and is connected to the MME via an S6a interface. The HSS uses Diameter protocol at the S6a interface instead of SS7. The Diameter protocol supports the transfer of subscription and authentication information in order to authorize the access of a user.

5. *Policy Control and Charging Rules Function (PCRF):* The Policy Control Enforcement Function (PCEF) is located in the P-GW. Each service data flow corresponds to a defined IP packet flow through the PCEF. The Policy and Charging Control (PCC) rule is set up for each service data flow depending on the subscription of the user. The PCC rule consists of a set of information to identify the service data flow and provide parameters for policy control and charging control of the service data flow. PCC includes:
   - *Packet Flow–based Charging:* This performs charging for each service data flow.
   - *Policy Control:* This includes QoS control, QoS signaling, and so forth.
   
   PCRF is used to provide PCEF with the PCC rule for policy and charging control. Thus, PCRF makes PCC-related decisions.

6. *Equipment Identity Register (EIR):* The EIR keeps records about UEs in the network and may maintain three lists of IMEIs:
   - *White list:* The white list contains the list of IMEIs of UEs that are permitted for use.
   - *Black list:* The black list contains the list of IMEIs of UEs that are permitted for use and so need to be barred.
   - *Gray list:* There may be a gray list that contains the list of IMEIs of UEs that are permitted for use but need to be tracked by the network for evaluation or other purposes.

The vendor may opt for developing multiple EPC entities in one box. The vendors may develop MME and S-GW in one box or S-GW and P-GW in one box or even all MME, S-GW, and P-GW in one box. Such integration can help reduce manufacturing costs because of fewer nodes and interfaces. It can also reduce latency in signaling between the entities. However, it increases complicacy and reduces scalability and flexibility. Thus, integration in a single box is more likely to be used during early deployment of the network.

#### 1.2.2.1 Interfaces in EPC

1. *S5/S8 interface:* The Serving Gateway (S-GW) is connected with the PDN Gateway (P-GW) via S5 interface when the UE is not roaming and via the S8 interface when the UE is roaming in the visitor PLMN (VPLMN). The S5/S8 interface uses the GPRS Tunneling Protocol (GTP) or the Proxy Mobile IP (PMIP). If GTP is used, then one GTP-C tunnel is established per

PDN connection for each UE. Also, separate GTP-U tunnels are established for each EPS bearer for each UE.
2. *S6a interface:* It connects MME with HSS. It uses the Diameter protocol. It uses SCTP or TCP. It supports subscription and security control between MME and HSS. The MME uses this interface to indicate the UE location. Then the HSS uses this interface to send subscription and security data for the UE.
3. *S10 interface:* It interconnects two MMEs. It supports mobility between MMEs. It uses the GTP-C tunnel.
4. *S11 interface:* It connects the MME with the S-GW. One GTP-C tunnel is established per PDN connection for each UE. It supports mobility and EPS bearer management.
5. *S13 interface:* It connects MME with EIR.
6. *Gx interface:* It connects P-GW with PCRF.
7. *Gxc interface:* It optionally exists to connect S-GW with PCRF.
8. *SGi interface:* It connects P-GW with external packet data network. Alternatively, it can connect the P-GW with an internal packet data network, for example, for IMS services.
9. *SGs interface:* It connects MME with MSC. It uses Gs protocols and supports mobility management and paging procedures between EPS and CS domain.
10. *S3 interface:* It connects MME with SGSN. It carries control information using GTP-C tunnel to support switching between LTE and UMTS technologies and Circuit-Switched Fallback (CSFB).
11. *S4 interface:* It connects S-GW with SGSN. It carries control information using the GTP-C tunnel and it carries user data using the GTP-U tunnel.
12. *S12 interface:* The S12 interface can optionally exist between the S-GW and Radio Network Controller (RNC) to support direct tunneling. It carries user data using the GTP-U tunnel. When the S12 interface exists, the S4 interface carries only control information, and thus, S-GW has no connection with the SGSN on a user plane.
13. *Gn interface:* Instead of S3, S4, and S12 interfaces, alternatively, the Gn interface can be used to connect the EPC with the SGSN. In this case, both the MME and P-GW are connected to the SGSN using Gn interfaces. Thus, the legacy 3GPP interfaces are used. From UTRAN or GERAN, the P-GW acts like a Gateway GPRS Support Node (GGSN) and the MME acts like a SGSN.

### 1.2.3 Tracking Area and MME Pool Area

A particular geographical area is split into a number of tracking areas in order to facilitate tracking of the UE. A tracking area covers a number of complete cells served by a number of eNodeBs. The Tracking Area Identity (TAI) is a globally unique identifier of a tracking area. A TAI list provides a list of tracking areas within which the UE can move around without performing a tracking area updating procedure. The tracking areas in the TAI list are served by the same MME area. However, in the TAI list, the MME does not have to include all tracking areas served by the

MME area. If the TAI list covers too many tracking areas (i.e., too large of an area), then the MME will need to page over all these tracking areas in order to initiate communication with the UE. On the other hand, if the TAI list covers too small of an area, it will trigger a high number of tracking area update procedures. Also, there can be ping-pong between two such areas (i.e., the UE moves between the two areas back and forth frequently, triggering unnecessary tracking area update procedures). The MME selects a number of tracking areas to include in the TAI list considering the above trade-off. Coordination with the geographical boundaries between tracking areas and location/routing areas of UTRAN/GERAN is expected, as will be explained in Section 21.3.2.

A number of MMEs can be connected by a mesh network to a set of eNodeBs and serve a particular geographical area. These MMEs are then called an MME pool and the area served is called an MME pool area or an MME area. An MME pool area is a collection of complete tracking areas. MME pool areas may overlap each other. The same MME usually maintains the UE context as long as the UE is located within a particular MME pool area. An eNodeB in an MME pool area may be served by multiple MMEs. In this case, the S1 interface is known as the S1-flex. The S1-flex allows load sharing among MMEs and also allows holding a connection if there is a failure in the interconnection with EPC.

## 1.3 Protocol Stack

The protocol stack has two types:

1. Control plane for the transfer of control information;
2. User plane for the transfer of user data.

The protocol stacks among various entities are shown for the control plane and user plane in the following sections.

### 1.3.1 Control Plane

Figure 1.4 shows the signaling protocol stack within differnt components of the network. It includes the following layers:

1. *Access Stratum (AS):* The AS layer includes the following layers. The AS protocols run between the UE and the eNodeB.
    - *Layer 3:* Radio Resource Control (RRC) layer. The connection between the RRC layers of the UE and the eNodeB is referred to as the *RRC connection*. The RRC connection is explained in Section 14.1. The UE is in the RRC_IDLE state when there is no RRC connection between the UE and the eNodeB. Conversely, the UE is in the RRC_CONNECTED state if the RRC connection has been established. A message generated by the RRC layer is called the *RRC message* or *layer 3 message*. The RRC layer performs the following major functions:
        - The RRC layer at the eNodeB broadcasts system information including ETWS and CMAS notification. This will be explained in Chapters 4 and 26.

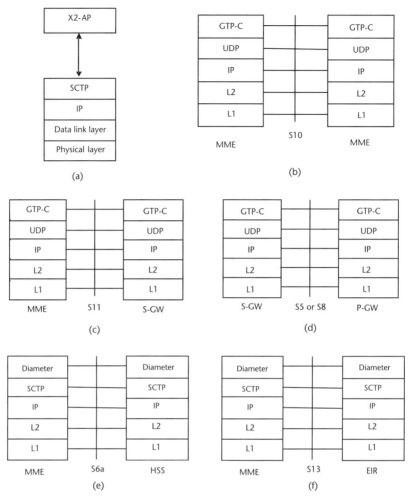

**Figure 1.4** Protocol stack (a) between two eNodeBs, (b) between two MMEs, (c) between MME and S-GW, (d) between S-GW and P-GW, (e) between MME and HSS, and (f) between MME and EIR.

- The RRC layer at the eNodeB performs paging. This will be explained in Chapter 13.
- The RRC layer at the eNodeB performs radio resource scheduling. This will be explained in Chapter 19.
- The RRC layer at the eNodeB performs security activation in AS layer. This will be explained in Section 12.4.
- The RRC layer at the eNodeB configures layer 1 and layer 2 parameters. This will be explained in different sections.
- The RRC layer at the eNodeB configures the measurement procedure of the UE including measurement gaps and conducts handover. This will be explained in Chapter 23.
- The RRC layer at the UE performs cell selection. This will be explained in Section 9.1.
- The RRC layer at the UE performs cell reselection. This will be explained in Section 23.2.

- The RRC layer at the eNodeB and the UE performs the establishment, modification, and release of the RRC connection. This will be explained in Section 14.1.
- The RRC layer at the UE performs recovery from radio link failure. This will be explained in Section 14.1.
  - Layer 2: This layer includes the following sublayers:
    - *Packet Data Convergence Protocol (PDCP):* The functions of PDCP sublayer will be explained in Chapter 8.
    - *Radio Link Control (RLC):* The functions of RLC sublayer will be explained in Chapter 7.
    - *Medium Access Control (MAC):* The functions of the MAC sublayer will be explained in Chapter 6.
  - Layer 1 (physical layer): The operation of the physical layer will be explained in Chapter 5.
2. *Non Access Stratum (NAS):* The NAS layer is located above the AS layer. The NAS protocols run between the UE and the MME. A NAS signaling connection is defined between the UE and the MME. As explained in Chapter 14, the NAS signaling connection consists of an RRC connection over an LTE-Uu interface and a UE-associated logical S1-connection over the S1-MME connection. A message generated by the NAS layer is called the *NAS message*. The NAS layer includes the following major protocols:
   - *EPS Mobility Management (EMM):* The EMM protocols are used for attach, detach, and tracking area update procedures. They are also used for UE identification, UE authentication, and security mode control. In addition, they support EMM connection-related services (e.g., service request, paging). The EMM connection-related protocols are often referred to as EPS Connection Management (ECM) protocols.
   - *EPS session management (ESM):* The ESM protocols are used for activation, deactivation, and modification of default or dedicated EPS bearers.

### 1.3.2 User Plane

Figure 1.5 shows the user plane protocol stack between the UE and the EPC. In the case of UTRAN, NodeB implements up to the MAC layer and a separate node in network, namely, Radio Network Controller (RNC) implements form the RLC layer.

The X2-U interface uses the same protocol stack as S1-U does. The Transport Network Layer (TNL) includes GTP on top of UDP/IP.

## 1.4 Heterogeneous Network

The eNodeBs can be used like typical base stations with transmit power more than 40 dBm and antenna gain between 12 and 15 dBi. Then a homogeneous network is deployed and the cells served are known as macro cells. The macro cells do not provide the most desirable performance in every scenario. For example, the indoor coverage can be very difficult in some cases. Besides, the user distribution and

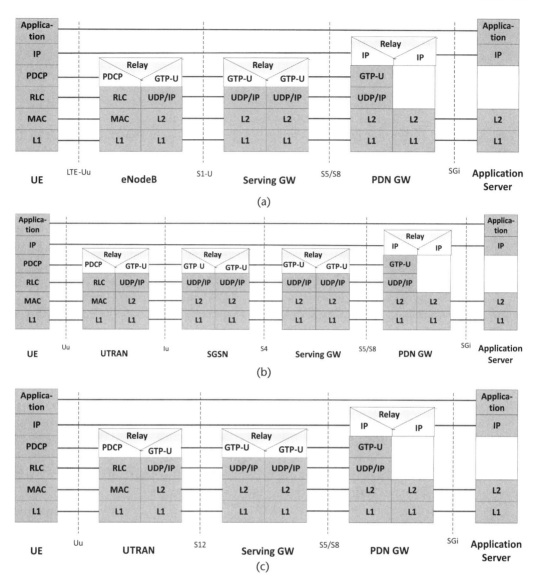

**Figure 1.5** User plane protocol stack (a) between UE and EPC via E-UTRAN, (b) between UE and EPC via UTRAN when direct tunneling using an S12 interface is not available, and (c) between UE and EPC via UTRAN when direct tunneling using an S12 interface is available.

propagation environment can vary significantly with time and the macro cells deployment may not be able to cope with the variation satisfactorily. The macro cells may even need to be split. The acquisition of a site for large base stations may also be difficult. Thus, the Home eNodeB and relay function, explained in the following sections, can be better alternative solutions in certain cases. They offer extension of the reach of coverage with very little additional backhaul expenses. Another type of deployment is pico cells where the eNodeB has lower transmit power and antenna gain. Its typical transmit power is between 23 dBm and 30 dBm with antenna gain up to 5 dBi. The pico cells are suitable for urban areas with a highly dense population. Among these different types of deployment, only the Home eNodeB or femto cell may restrict access to the users by using a closed subscriber group.

With the availability of these different cell levels, the heterogeneous network is introduced in Release 10. It uses simultaneous deployment of eNodeBs with different levels of capacities (i.e., it supports coexistence of macro cells, pico cells, femto cells, and/or relay nodes in the same spectrum). It uses low power nodes overlaid by a macro cell. The initial network deployment may include only macro cells. Thereafter, pico cells, femto cells, and/or relay nodes may be added with time based on subsequent requirements. The addition of low power nodes can enhance coverage and capacity greatly. It improves data rate for particular users who would experience a poor data rate otherwise. The low power nodes are smaller base stations with lower antenna gain compared to macro cells, so the site acquisition for their addition can also be simpler. The macro cells may attempt to keep supporting high speed users in order to reduce the number of handovers.

### 1.4.1 Home eNodeB

The service coverage may be limited or unavailable inside the home, small business area, and so forth. In such cases, an access point can be set up at customer premises and it functions as a small base station. The access point is called Home eNodeB (HeNB) and such Home eNodeB cell is traditionally known as a femtocell. The Home eNodeB can provide service indoors where the coverage would normally be limited or unavailable. Also, a number of Home eNodeBs can be installed in a campus or large business environment. The transmit power of Home eNodeB is typically less than 23 dBm. Both cell reselection in the RRC_IDLE state and handover in the RRC_CONNECTED state can take place between a Home eNodeB cell and other cells and between Home eNodeB cells. There should be very careful planning in deployment of Home eNodeBs as it would create significant interference to macro cells otherwise.

The Home eNodeB is assigned a human readable and user friendly name. The length of the Home eNodeB name does not take more than 48 bytes. When the eNodeB is a Home eNodeB, it transmits SIB type 9, which contains the Home eNodeB name. SIB type 9 does not include any other information. SIB type 9 is a system information message. System information messages are explained in Chapter 4.

#### 1.4.1.1 Home eNodeB Gateway

A Home eNodeB Gateway (HeNB GW) may be used to connect a large number of Home eNodeBs with the EPC using the S1 interface as shown in Figure 1.6. The Home eNodeB Gateway is placed at the operator's premises.

The HeNB GW appears as a Home eNodeB to the MME and the HeNB GW appears as MME to the HeNB. The HeNB GW is connected with MME via the S1-MME interface. The S1-U interface between the Home eNodeB and MME can be connected directly or via the HeNB GW.

#### 1.4.1.2 Access Mode

The Home eNodeB is configured for operation in any of the following modes.

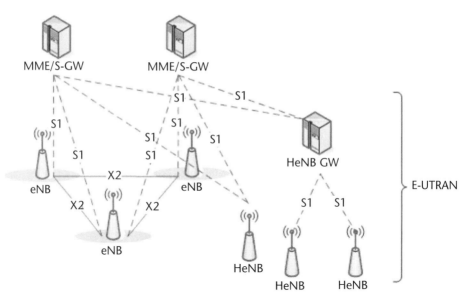

**Figure 1.6** Network architecture for Home eNodeB.

1. *Open access mode:* The Home eNodeB operates like an ordinary eNodeB. It provides services to all subscribers of any PLMN if there is proper roaming agreement.
2. *Closed access mode:* The Home eNodeB provides services only the users who belong to a closed subscriber group (CSG) and it controls the access of users. The HeNB is typically operated in this closed access mode.
3. *Hybrid access mode:* The Home eNodeB provides services to its associated CSG members preferentially. In addition, it provides services to subscribers of any PLMN that is not allowed to access the CSG, if there is proper roaming agreement.

Thus, the HeNB typically serves a CSG cell. The CSG holds a CSG identity, which is unique within the PLMN. The UEs that are the members of the CSG carry the same CSG identity. The CSG may include a number of Home eNodeBs located in a campus or business environment. When the UE switches to a Home eNodeB within a CSG, it will prefer to stay on the same CSG instead of going back to the external network. The HeNB broadcasts a CSG-Indication field on SIB type 1, which is a system information message. If this CSG-Indication field is true, then the cell is a CSG cell. In this case, SIB type 1 also includes a CSG-Identity field, which depicts the CSG identity of the cell. A UE is only allowed to access the CSG cell if the CSG identity stored in the UE matches with this CSG identity. Thus, only the UEs who are the members of the associated CSG can access the CSG cell. SIB type 4 includes the range of Physical Cell Identifiers (PCIs) reserved for a CSG cell. The PCI is explained in Section 10.1.2. This range of PCIs can be used within 24 hours from the reception of SIB type 4 although other information in SIB type 4 remains valid for only 3 hours as shown in Section 4.5.2.

The CSG selection by the UE can be either automatic or manual. Both automatic and manual CSG selections use the CSG identity for selection. In case of manual selection, the user can initiate CSG selection. The manual selection additionally uses the Home eNodeB name for selection because this name is easy to use. The UE performs a scan for available CSGs and finds out CSG identities and Home eNodeB names for the available CSGs. Then the UE displays Home eNodeB names and PLMN names for the available CSGs. If the Home eNodeB name is not available, the UE displays the CSG identity instead. Each Home eNodeB name or CSG identity belongs to one of the following lists and the UE also indicates its list placement:

- Allowed CSG list: A user-controlled list of allowed CSG identities;
- Operator CSG list: An operator-controlled list of allowed CSG identities;
- Neither the allowed CSG list nor the operator CSG list.

The user manually selects a CSG from the available CSGs and then the UE attempts to camp on it. The UE reselects any of the available cells within the CSG chosen. All CSG identities on the allowed CSG list and the operator CSG list have the same priority in case of cell reselection.

### 1.4.2 Relay

The relay function or multihop transmission is introduced in Release 10. It can be used to improve coverage with high data rate or to improve cell-edge throughput. It can also be used to provide coverage in new areas especially as a temporary deployment. In the case of a relay, an eNodeB is set up without any direct connection with EPC and it is rather served by another eNodeB being wirelessly connected as shown in Figure 1.7. The serving eNodeB is called Donor eNodeB (DeNB) and the eNodeB operated via relay from the DeNB is called the Relay Node (RN). The radio interface between DeNB and the RN is called the Un interface. The relay node can use an inband or outband connection while the inband connection means the Un interface uses frequencies, which is a part of what is used for LTE-Uu interface. The DeNB provides the functionality of the connection of RN with other eNodeBs, MME, and

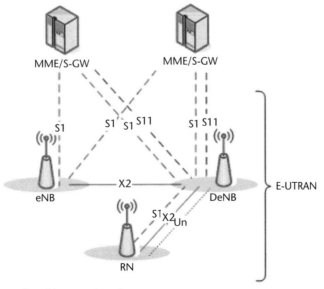

**Figure 1.7** Network architecture for relay.

S-GWs by proxy. Thus, the DeNB appears as an MME, an eNodeB, and an S-GW to the RN. The DeNB configures the physical layer of the relay node and updates it with a change in system information by sending an RN reconfiguration message to the relay node. Only single hop relay is supported. So, the relay operation cannot propagate (i.e., the relay node should not act as a DeNB in order to serve another relay node). The relay nodes typically operate at a transmit power less than macro cells and the transmit power may be below 30 dBm. Intercell handover is not supported for relay nodes. If a UE operating under main eNodeB enters the coverage area of a relay node, it may cause significant interference with the relay node, so there must be very careful planning in deployment of relay nodes.

## 1.5 UE Positioning

The geographic position of the UE may be determined for various purposes. It can be employed for various Location-Based Services (LBSs), for example, navigation, friend finding, location aware advertising, city guides, local service search, fleet management, billing, and so forth. Also, the network may use the UE position for radio resource management.

The Global Navigation Satellite System (GNSS) is the widely satellite-based technology to determine geographic position. However, GNSS exhibits limited performance because of poor visibility of satellites in many cases. This may happen, for example, in narrow streets with high buildings on the two sides or inside buildings. Besides, signal propagation delays or multipath fading can gives rise to errors in case of GNSS. The UE should receive a much stronger signal from eNodeB compared to GNSS; thus, the LTE signal can be used to support UE positioning. Therefore, LTE caters to the UE positioning function in order to determine the position of the UE and possibly the velocity of the UE as well. It may be noted that LTE assists in terrestrial position determination whereas GNSS is a three-dimensional positioning method.

### 1.5.1 Network Architecture

The network architecture for the purpose of UE positioning is shown in Figure 1.8. The LTE Positioning Protocol (LPP) is defined between a location server and the UE. The location server is defined as an enhanced serving mobile location center (E-SMLC) when communicated over the C-plane. The location server is defined as SUPL location platform (SLP) when communicated over U-plane. Secure User Plane Location (SUPL) is a location service and positioning protocol that works on the user plane. SUPL may be used for authentication, charging, triggering of location reporting to SLP, and so forth. Besides supporting protocol, LPPa operates between eNodeB and E-SMLC.

LPP supports exchange of information between the UE and the E-SMLC. This information includes UE positioning capabilities, data that assists in measurement or position computation, measurement reports, and estimated UE position. Thus, the LPP message may carry any of the following messages.

- Request capabilities;

## 1.5 UE Positioning

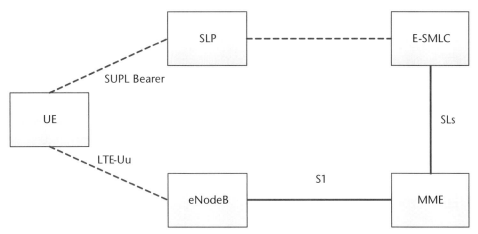

**Figure 1.8** Network architecture for UE positioning.

- Provide capabilities;
- Request assistance data;
- Provide assistance data;
- Request location information;
- Provide location information.

### 1.5.2 Positioning Techniques

LTE supports the following UE positioning techniques. A combination of multiple positioning techniques from this list can also be employed:

- Observed Time Difference of Arrival (OTDOA);
- Assisted-GNSS (A-GNSS);
- Enhanced Cell-ID (E-CID).

The UE positioning techniques, in general, allows the UE or the eNodeB measure signal from the eNodeB. In case of A-GNSS, the UE also measures signal from GNSS. Then the position of the UE and, optionally, the velocity of the UE are computed based on the measurements. This computation may be made by the UE or by the E-SMLC in case of A-GNSS (i.e., A-GNSS can be either UE based or E-SMLC based. The other two methods are always E-SMLC based (i.e., the E-SMLC always performs the computation). The E-SMLC–based methods are always assisted by the UE and are called UE assisted. If the E-SMLC performs the positioning computation, it may send the results to the UE. Similarly, if the UE performs the positioning computation, it may send the results to the E-SMLC. The UE positioning techniques are explained here:

1. *Observed Time Difference of Arrival (OTDOA):* The UE measures the timing of the signals received from multiple eNodeBs and observes their time differences. The UE may use positioning reference signal as the signal from the eNodeB. The positioning reference signal is explained in Section

5.1.5.4. The UE position is computed based on these time differences and the geographical coordinates of the eNodeBs.

The E-SMLC provides the UE with PCI and ECGI of the cells, which may be used for measurements. The E-SMLC also provides the UE with timing of the candidate cells relative to the serving eNodeB. The UE measures the time difference of signals received from multiple eNodeBs. Then the UE sends the measurement results and PCI and ECGI of the measured cells to the E-SMLC. The E-SMLC performs the positioning computation. Each OTDOA measurement describes a hyperbolic locus of constant difference along which the UE may be located. The E-SMLC determines the position of the UE as the intersection of two such hyperbolic loci.

2. *Assisted-GNSS (A-GNSS):* The UE has to be GNSS measurement capable. The UE receives both GNSS signals from the satellite and signals from the eNodeB for positioning computation. The GNSS signals used can be GPS and its modernization, Galileo, GLONASS, Quasi-Zenith Satellite Systems (QZSS), or Satellite-Based Augmentation Systems (SBAS) including WAAS, EGNOS, MSAS, and GAGAN. More than one type of GNSS signals can be used separately or in combination with others to determine the UE position more accurately and reliably. This UE positioning computation may be made by the UE or by the E-SMLC (i.e., A-GNSS can be either UE-based or E-SMLC-based).

   - *E-SMLC-based or UE-assisted:* The eNodeB sends information to the UE in order to assist GNSS measurements. The UE performs GNSS measurements and sends these measurements to the E-SMLC. The E-SMLC performs the positioning computation.
   - *UE-based:* The eNodeB sends information to the UE in order to assist both GNSS measurements and positioning computation. The UE performs GNSS measurements. The UE may also measure signals from the eNodeB. The UE performs the positioning computation.

   Alternatively, the UE can operate in autonomous mode. In this case, the UE does not receive any assistance from the eNodeB. The UE simply performs GNSS measurements and positioning computation.

   The information to assist GNSS may include reference time, visible satellite list, satellite signal Doppler, code phase, Doppler and code phase search windows, and so forth. The information to assist UE position calculation may include reference time, reference position, satellite ephemeris, clock corrections, and so forth.

3. *Enhanced Cell-ID (E-CID):* The E-CID method performs estimation of the position of the UE using its serving eNodeB and cell. The E-SMLC can acquire the information of the serving eNodeB and cell through paging, tracking area update, and so forth. Then the UE and the eNodeB perform additional measurements in order to improve the UE location estimate. The UE may measure E-UTRA Carrier RSSI, RSRP, RSRQ, and the time difference between transmission and reception. The eNodeB may measure the angle of arrival (AoA) of the signal from the UE and the time difference between transmission and reception. These measurements along with the PCI and ECGI of the measured cells are reported to the E-SMLC. The E-SMLC performs the positioning computation.

## 1.6 Basic Definitions

### 1.6.1 Radio Signal Measurement

Reference Signal Received Power (RSRP) is a metric of the downlink cell-specific reference signal strength. The UE calculates RSRP as the linear average over the power contributions of the resource elements that carry cell-specific reference signals within the considered measurement frequency bandwidth. The power per resource element is determined from the energy received during the useful part of the symbol, excluding the cyclic prefix. This measured signal strength involves path loss, antenna gain, log-normal shadowing, and fast fading, which are averaged over all the reference symbols within the measurement bandwidth. This averaging over the reference symbols is done at layer 1 and so it is called L1 filtering.

The cell-specific reference signals, $\mathbf{R}_0$ (on antenna port, 0) are used for RSRP determination. The UE may use cell-specific reference signals, $\mathbf{R}_1$ (on antenna port 1) in addition to $\mathbf{R}_0$, if the UE can reliably detect that reference signals from the second antenna port is transmitted. The measurement frequency bandwidth and the measurement period used to determine RSRP is left up to the UE implementation but the measurement accuracy requirements must be fulfilled.

Reference Signal Received Quality (RSRQ) is a metric of downlink cell-specific reference signal quality that takes the intended signal strength as well as interference and noise into account. The UE determines RSRP and E-UTRA Carrier RSSI over the same set of resource blocks within the considered measurement bandwidth. The UE calculates E-UTRA Carrier RSSI as the linear average of the total received power from all sources, including cochannel serving and nonserving cells, adjacent channel interference, and thermal noise within the considered measurement bandwidth. Then, the UE calculates RSRQ as the ratio between N × RSRP and E-UTRA Carrier RSSI, where N is the number of resource blocks of the measurement bandwidth. RSRP and RSRQ are used as metrics to rank different candidate cells in the neighborhood.

Energy Per Resource Element (EPRE) represents the energy used for transmission per resource element in the downlink. It considers the average energy taken over all constellation points for the particular modulation scheme used. The energy is taken prior to the insertion of cyclic prefix.

### 1.6.2 Identities in Network

The Public Land Mobile Network (PLMN) code of a particular network owned by an operator is constructed from the Mobile Country Code (MCC) and Mobile Network Code (MNC). The MCC identifies the country where the network is located. Thus, every country is assigned a unique MCC. A few countries are allocated multiple MCCs. The MCC is three decimal digits long. Each network in a country is assigned a particular MNC. The MNC is two or three decimal digits long depending on the value of MCC. A mixture of two and three digits length for MNC is not recommended with the same value of MCC. The PLMN code, also known as MCC/MNC tuple, is a globally unique identifier of a network owned by an operator.

The Access Point Name (APN) identifies the packet data network (PDN) for connection with the UE. More than one APN may be used to connect a UE to internet so that different application traffic is routed via different APNs. The APN follows the DNS naming conventions.

- *Tracking area code (TAC):* The TAC uniquely identifies a tracking area within its PLMN.
- *Tracking area identity (TAI):* The TAI is a globally unique identifier of a tracking area. TAI is constructed from MCC, MNC, and TAC.
- *MME group ID (MMEGI):* The MMEGI is a 16 bits long unique identifier of a MME group within the PLMN. The most significant bit of MMEGI is always 1.
- *MME code (MMEC):* The MMEC is an 8-bit-long unique identifier of an MME within the MME group.
- *MME identifier (MMEI):* The MMEI is constructed from MME Group ID (MMEGI) and MME Code (MMEC). Thus, MMEI identifies a MME uniquely within a PLMN.
- *Globally unique MME identity (GUMMEI):* The GUMMEI is constructed from the Mobile Country Code (MCC), Mobile Network Code (MNC), and the MMEI. Thus, GUMMEI is a globally unique identifier of a MME.
- *eNodeB identifier (eNB ID):* The eNB ID uniquely identifies an eNodeB within its PLMN. The eNB ID is 20 bits long.
- *Global eNodeB identifier (Global eNB ID):* The Global eNB ID is globally unique identifier of an eNodeB. Global eNB ID is constructed from MCC, MNC, and eNB ID.
- *E-UTRAN cell identifier (ECI):* The ECI uniquely identifies a cell within its PLMN. ECI is 28 bits long. The most significant 20 bits of the ECI is equal to the eNB ID of the eNodeB that serves this cell. Thus, each cell under an eNodeB possesses a unique value in the least significant 8 bits of the ECI.
- *E-UTRAN cell global identifier (ECGI):* The ECGI is a globally unique identifier of a cell. ECGI is constructed from MCC, MNC, and ECI.
- *Physical cell identifier (PCI):* The PCI is explained in Section 10.1.2.

### 1.6.3 Identities on UE Side

Each individual piece of MS equipment is assigned a unique number called the International Mobile Station Equipment Identity (IMEI). The IMEI cannot be changed and it resists tampering.

The Universal Subscriber Identity Module (USIM) is the application that stores information related to subscription and authentication of the user. It also manages functions related to subscription and authentication. The USIM typically runs in a removable card in the UE called Universal Integrated Circuit Card (UICC). In common speech, the UICC is referred to as the USIM or the SIM.

The International Mobile Subscriber Identity (IMSI) is a globally unique identifier of a user and is associated with his subscription. It is stored in the USIM. It is not more than 15 decimal digits long. It is constructed from the MCC, MNC, and the Mobile Subscriber Identification Number (MSIN). MSIN uniquely identifies a subscriber within a PLMN.

The PLMN code part of the IMSI is known as the Home Public Land Mobile Network (HPLMN). More than one PLMN codes may be permissible for a user and the UE should select one of them for services. In order to allow provision for selection from multiple PLMN codes, the USIM of the UE can store a list of PLMN codes known as Equivalent Home PLMN (EHPLMNs). This EHPLMN list includes the EHPLMNs with priority order. This EHPLMN list may or may not include HPLMN.

The visited public land mobile network (VPLMN) is a PLMN different from HPLMN or EHPLMN. When the UE operates on VPLMN, it is considered to be roaming. If an EHPLMN list exists but it does not include the PLMN code derived from the IMSI, then that PLMN code is treated as a VPLMN instead of HPLMN.

M-TMSI is a 32-bit-long temporary identity of a UE allocated by the MME. It can be expressed as hexadecimal.

The globally unique temporary UE identity (GUTI) is constructed from the GUMMEI and the M-TMSI. Thus, GUTI is a globally unique temporary identifier of the UE. The MME allocates a dedicated GUTI to the UE during an initial attach procedure. Later, the MME can reallocate the GUTI during a tracking area update procedure or even at other times independently. The allocation/reallocation of the GUTI to the UE and this is referred to as *GUTI reallocation*. The GUTI reallocation is actually used to allocate a new GUTI, a new TAI list, or both to the UE. Section 15.1.7.4 will explain a GUTI reallocation during an initial attach procedure, Section 22.5 will explain a GUTI reallocation during a tracking area update procedure, and Section 15.2 will explain an independent GUTI reallocation.

The SAE TMSI (S-TMSI) is a short form to identify the UE over the radio link during initial communication. S-TMSI is constructed from MMEC and M-TMSI. Thus, S-TMSI is unique for a MME group. S-TMSI is also unique in a tracking area. S-TMSI is 40 bits long. It may be noted that S-TMSI is a subset of GUTI; thus, when the UE is allocated a new GUTI, the UE also becomes aware of its new S-TMSI.

The Cell Radio Network Temporary Identifier (C-RNTI) is a unique identification of the UE at the cell level during the RRC_CONNECTED state. The C-RNTI identifies the RRC connection. Thus, C-RNTI is commonly used by the eNodeB to identify the UE in the RRC_CONNECTED state. The eNodeB sends indications using PDCCH by having its CRC scrambled by the C-RNTI of the target UE so that the UE can identify that the PDCCH instance is intended for itself. The value of C-RNTI can be a value between 0001 and FFF3 in hexadecimal.

*Note:* Since IMSI is a permanent identifier, an eavesdropper may be able to know it. Therefore, for the sake of confidentiality, attempts are typically made to avoid inclusion of IMSI in over-the-air messages and GUTI or S-TMSI is used instead as the UE identity. However, as shown in Sections 13.4, 15.1.1, and 15.3.2, IMSI may have to be transmitted over the air sometimes. S-TMSI is used only during initial communication and so it is used in RRCConnectionRequest message and Paging message. GUTI is basically used as the UE identity between the EPC and the UE. On the other hand, the eNodeB basically uses C-RNTI as the UE identity in the RRC_CONNECTED state. Other RNTIs are used by the eNodeB for specific purposes.

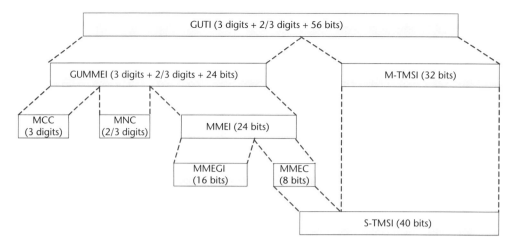

**Figure 1.9** Construction of various UE identities.

- *Semipersistent Scheduling Cell Radio Network Temporary Identifier (SPS C-RNTI):* As shown in Section 19.2, the PDCCH has its CRC scrambled by SPS C-RNTI when semipersistent scheduling is used. Also, the C-RNTI is used for dynamic scheduling. Thus, the UE can differentiate between dynamic scheduling and semipersistent scheduling. The value of SPS C-RNTI can be a value between 0001 and FFF3 in hexadecimals.
- *Temporary C-RNTI:* As shown in Section 11.2, the temporary C-RNTI is used as a temporary identity of the UE during random access procedure. The value of temporary C-RNTI can be a value between 0001 and FFF3 in hexadecimals.
- *TPC-PUCCH-RNTI:* As shown in Section 17.3.2.3, the eNodeB may send the TPC command for power control in PUCCH using PDCCH with DCI format 3 or 3A; in this case, the CRC of PDCCH instance is scrambled by TPC-PUCCH-RNTI. The UEs, which are assigned the same TPC-PUCCH-RNTI, will decode the PDCCH instance and apply the TPC command. The value of TPC-PUCCH-RNTI can be a value between 0001 and FFF3 in hexadecimals.
- *TPC-PUSCH-RNTI:* As shown in Section 17.2.2.3, the eNodeB may send TPC command for power control in PUSCH using PDCCH with DCI format 3 or 3A; in this case, the CRC of PDCCH instance is scrambled by TPC-PUSCH-RNTI. The UEs, which are assigned the same TPC-PUSCH-RNTI, will decode the PDCCH instance and apply the TPC command. The value of TPC-PUSCH-RNTI can be a value between 0001 and FFF3 in hexadecimals.
- *Paging Radio Network Temporary Identifier (P-RNTI):* As shown in Section 13.2, the eNodeB sends a notification for the presence of a paging message using PDCCH and the particular PDCCH instance is identified by having its CRC scrambled by P-RNTI. The P-RNTI has a fixed value, which is FFFE in hexadecimals.

- *System Information Radio Network Temporary Identifier (SI-RNTI):* As shown in Chapter 4, the eNodeB sends a notification for the presence of a system information message using PDCCH, and the particular PDCCH instance is identified by having its CRC scrambled by SI-RNTI. The SI-RNTI has a fixed value, which is FFFF in hexadecimals.

- *Random Access Radio Network Temporary Identifier (RA-RNTI):* As shown in Chapter 11, the eNodeB sends notification for presence of a random access response (RAR) message using PDCCH and the PDCCH instance is identified by having its CRC scrambled by random access RNTI (RA-RNTI). The value of RA-RNTI can be a value between 0001 and 003C in hexadecimals. This range of values is also allowed for C-RNTI, SPS C-RNTI, Temporary C-RNTI, TPC-PUCCH-RNTI, and TPC-PUSCH-RNTI. However, the values used for RA-RNTI are not used for other types of RNTIs in the same cell.

- *MBMS Radio Network Temporary Identifier (M-RNTI):* As shown in Section 24.5.3, the eNodeB sends notification for a change in MCCH using PDCCH and the PDCCH instance is identified by having its CRC scrambled by MBMS RNTI (M-RNTI). The M-RNTI has a fixed value, which is FFFD in hexadecimals.

- *MBMS Service ID and Temporary Mobile Group Identity (TMGI):* MBMS Service ID is a 3-octet-long identity of the UE, which is used for group paging in MBMS service. TMGI is the IE that carries the MBMS Service ID. The TMGI IE has a minimum length of 5 octets. The octets 3–5 contains the MBMS Service ID. The TMGI IE can also optionally include MCC and MNC representing the PLMN; in this case, the TMGI IE becomes 8 octets long.

- *UE categories:* Depending on the uplink and downlink capabilities of the UE, five UE categories are defined as shown in Table 1.2. Separate UE categories are not defined for uplink and downlink capabilities like HSPA. The UE categories are often referred to as UE classes. All UE categories support 20 MHz. A UE of a higher category should be more expensive; thus, different UE categories can satisfy different market demands. In this initial market, Category 3 UEs are getting most availability. Category 4 UEs are also becoming available. The UE categories 6, 7, and 8 are introduced in Release 10 and they support carrier aggregation. The UE informs the network of its category. Some UE capability information is not defined by the UE category, and such information can be separately signaled. This is shown in Section 15.1.7.4.

- *UE power class:* Depending on maximum output power capability of the UE for any transmission bandwidth within the channel bandwidth, four power classes are defined. They are namely, Power Class 1, Power Class 2, Power Class 3, and Power Class 4. Power Class 3 UEs are expected to be available initially. The maximum output power for Power Class 3 UEs is set to 23 dBm with ±2 dB tolerance.

**Table 1.2** Capabilities of Different UE Categories

| UE Category | Reception on DL-SCH | | | | | Transmission on UL-SCH | | | | Total Layer 2 Buffer Size for UL and DL (kB) |
|---|---|---|---|---|---|---|---|---|---|---|
| | Maximum Transport Block Size (bits) | Maximum Transport Block Bits Within a TTI (1 ms) | Maximum Number of Layers for Spatial Multiplexing | Maximum Number of PDCP SDUs per TTI | Maximum Data Rate (Mbps) | Maximum Transport Block Size (bits) | Maximum Transport Block Bits Within a TTI (1 ms) | Highest Modulation Scheme | Maximum Data Rate (Mbps) | |
| Category 1 | 10,296 | 10,296 | 1 | 10 | 10 | 5,160 | 5,160 | 16-QAM | 5 | 150 |
| Category 2 | 51,024 | 51,024 | 2 | 10 | 51 | 25,456 | 25,456 | 16-QAM | 25 | 700 |
| Category 3 | 75,376 | 102,048 | 2 | 20 | 102 | 51,024 | 51,024 | 16-QAM | 51 | 1,400 |
| Category 4 | 75,376 | 150,752 | 2 | 30 | 150 | 51,024 | 51,024 | 16-QAM | 51 | 1,900 |
| Category 5 | 149,776 | 299,552 | 4 | 50 | 300 | 75,376 | 75,376 | 64-QAM | 75 | 3,500 |
| Category 6 | 149,776 (4 layers) 75,376 (2 layers) | 301,504 | 2 or 4 | — | 300 | 51,024 | 51,024 | 16-QAM | 50 | 3,300 |
| Category 7 | 149,776 (4 layers) 75,376 (2 layers) | 301,504 | 2 or 4 | — | 300 | 51,024 | 102,048 | 16-QAM | 100 | 3,800 |
| Category 8 | 299,856 | 2,998,560 | 8 | — | 3,000 | 149,776 | 1,497,760 | 64-QAM | 1,500 | 42,200 |

# CHAPTER 2
# Information Bearer

The information between the UE and the network can be either control information or user data. Different types of information require different types of treatment in transmission in terms of scheduling, priority, power control, modulation, scrambling, coding, rate matching, multiple antenna techniques, and so forth. Also, the flow of the information via different layers in a protocol stack needs to be organized. Therefore, different bearers and channels are defined where each bearer or channel carries certain types of control information or user data between certain layers. The various bearers and channels are explained in this chapter.

## 2.1 Channels and Signals

### 2.1.1 Logical Channels

The logical channels describe what data are transferred over the radio interface. The Service Access Point (SAP) between the MAC sublayer and the RLC sublayer provides the logical channel. The following logical channels are available.

#### 2.1.1.1 Downlink Logical Channels

Control channels carry control-plane information:

- Broadcast Control Channel (BCCH);
- Paging Control Channel (PCCH);
- Common Control Channel (CCCH);
- Multicast Control Channel (MCCH);
- Dedicated Control Channel (DCCH).

Traffic channels carry user-plane data:

- Dedicated Traffic Channel (DTCH);
- Multicast Traffic Channel (MTCH).

### 2.1.1.2 Uplink Logical Channels

Control channels carry control-plane information:

- Common Control Channel (CCCH);
- Dedicated Control Channel (DCCH).

Traffic channels carry user-plane data on the Dedicated Traffic Channel (DTCH).

## 2.1.2 Transport Channels

The transport channels describe how and with what characteristics data are transferred over the radio interface. The service access point (SAP) between the physical layer and the MAC sublayer provides the transport channel. The following transport channels are available.

### 2.1.2.1 Downlink Transport Channels

1. Broadcast Channel (BCH);
2. Downlink Shared Channel (downlink-SCH);
3. Paging Channel (PCH);
4. Multicast Channel (MCH).

### 2.1.2.2 Uplink Transport Channels

1. Uplink Shared Channel (uplink-SCH);
2. Random Access Channel (RACH).

## 2.1.3 Physical Channels

The physical channels at layer 1 provide air interface support for different transport channels. The following physical channels are available.

### 2.1.3.1 Downlink Physical Channels

1. Physical Broadcast Channel (PBCH);
2. Physical Control Format Indicator Channel (PCFICH);
3. Physical Downlink Control Channel (PDCCH);
4. Physical Hybrid ARQ Indicator Channel (PHICH);
   - Physical Downlink Shared Channel (PDSCH);
5. Physical Multicast Channel (PMCH).

#### 2.1.3.2 Uplink Physical Channels

1. Physical Uplink Shared Channel (PUSCH);
2. Physical Uplink Control Channel (PUCCH);
3. Physical Random Access Cchannel (PRACH).

### 2.1.4 Physical Signals

The physical signals carry signals for layer 1 and do not carry any upper layer information. The following physical signals are available.

#### 2.1.4.1 Downlink Physical Signals

The downlink physical signals include reference signals and synchronization signals as shown here:

1. *Reference Signal (RS)*: There are five types of RSs:
   - Cell-specific reference signal;
   - UE-specific reference signal;
   - MBSFN-specific reference signal;
   - Positioning reference signals;
   - Channel-State Information (CSI) reference signals.
2. *Synchronization signal*: There are two types of synchronization signals:

   - Primary Synchronization Signal (PSS);
   - Secondary Synchronization Signal (SSS).

#### 2.1.4.2 Uplink Physical Signals

1. Demodulation Reference Signal (DM RS);
2. Sounding Reference Signal (SRS).

## 2.2 Radio Bearer

A radio bearer transports control signaling or user data packets like a traffic flow with common QoS treatment between the UE and the eNodeB. The radio bearer can be classified as follows:

1. Signaling Radio Bearer (SRB);
2. Data Radio Bearer (DRB).

### 2.2.1 Signaling Radio Bearer (SRB)

The radio bearers that transport RRC messages and NAS messages on the control plane are known as Signaling Radio Bearers (SRBs). The SRBs are of the following types.

#### 2.2.1.1 SRB0

SRB0 transports RRC messages using the CCCH logical channel. SRB0 is used during RRC connection establishment procedure. The RRC connection establishment involves setting up SRB1; thereafter, SRB1 is used for further signaling messages. Neither ciphering nor integrity protection is used for SRB0. SRB0 carries the following RRC messages.

- RRCConnectionRequest message;
- RRCConnectionSetup message;
- RRCConnectionReject message;
- RRCConnectionReestablishmentRequest message;
- RRCConnectionReestablishment message;
- RRCConnectionReestablishmentReject message.

#### 2.2.1.2 SRB1

SRB1 transports RRC messages using DCCH logical channel. The RRC messages may or may not have piggybacked NAS messages. SRB1 includes messages that are ciphered, not ciphered, integrity protected, and not integrity protected. SRB1 is used for security activation. In general, SRB1 carries the RRC messages in RRC_CONNECTED state. SRB1 carries the following RRC messages.

- RRCConnectionSetupComplete message;
- RRCConnectionReconfiguration message;
- RRCConnectionReconfigurationComplete message;
- RRCConnectionReestablishment message;
- RRCConnectionReestablishmentComplete;
- RRCConnectionRelease message;
- SecurityModeCommand message;
- SecurityModeComplete message;
- SecurityModeFailure message;
- MeasurementReport message;
- UECapabilityEnquiry message;
- UECapabilityInformation message;
- CounterCheck message;
- CounterCheckResponse message;
- ULHandoverPreparationTransfer message;
- MobilityFromEUTRACommand message;
- HandoverFromEUTRAPreparationRequest message;
- UEInformationRequest;

## 2.2 Radio Bearer

- UEInformationResponse;
- CSFBParametersResponseCDMA2000 message;
- CSFBParametersRequestCDMA2000 message.

The following RRC messages have piggybacked NAS messages with no RRC layer signaling information. These messages are generally carried by SRB2, but they can be carried by SRB1 only if SRB2 is not yet established.

- DLInformationTransfer message;
- ULInformationTransfer message.

SRB1 is identified by its radio bearer identity (RB ID), which is called SRB-Identity. The value of SRB-Identity is set to 1 for SRB1. The eNodeB sends the RRCConnectionSetup message to the UE to establish SRB1. The eNodeB sends RRCConnectionReestablishment message to the UE in order to reestablish the SRB1, resolving contention. The RRCConnectionSetup message and RRCConnectionReestablishment message include RadioResourceConfigDedicated IE, which contains the SRB-Identity assigned.

### 2.2.1.3 SRB2

SRB2 uses the DCCH logical channel. SRB1 and SRB2 are mapped onto two different DCCH logical channels. SRB2 has a lower priority than SRB1. Both ciphering and integrity protection are used for SRB2 because SRB2 is always established after security activation.

As mentioned above, SRB2 transports the following RRC messages; they can be carried by SRB1 instead only if SRB2 is not yet established. If SRB2 is suspended, then these messages are not transmitted until SRB2 is established again.

- DLInformationTransfer message;
- ULInformationTransfer message.

SRB2 can also transport UEInformationResponse message. This message can also be carried by SRB1.

The value of SRB-Identity is set to 2 for SRB2. The eNodeB sends the RRCConnectionReconfiguration message to the UE to establish SRB2. This message includes RadioResourceConfigDedicated IE, which contains the SRB-Identity assigned.

### 2.2.2 Data Radio Bearer (DRB)

The radio bearers, which transport user plane data using DTCH logical channel, are known as data radio bearers (DRBs). Ciphering is used for DRBs but integrity protection is not used. DRBs are always established after security activation. A UE can have a maximum of eight DRBs established for each PDN. A DRB is uniquely identified by its radio bearer identity (RB ID), which is called DRB-Identity. The eNodeB sends RRCConnectionReconfiguration message to the UE in order to establish a DRB. This message includes RadioResourceConfigDedicated IE, which contains the DRB-Identity assigned. The DRB-Identity can be any integer up to 32.

## 2.3  Mapping Among Channels and Bearers

On the transmitting side, a layer receives the data packet as a service data unit (SDU), performs necessary processing, and generates a protocol data unit (PDU). Then the layer sends the PDU to its next lower layer and the lower layer receives it as its own SDU. Thus, the PDU of a layer becomes the SDU of the next lower layer. It occurs in the reverse way on the receiving side. A layer receives the data packet as a PDU, performs necessary processing, generates SDU, and sends the SDU to its next upper layer. Figure 2.1(a, b) shows mapping among radio bearers, logical channels, transport channels, and physical channels through different layers in downlink and uplink, respectively.

The mapping of radio bearers and channels are explained below for the transmitting side. Evidently, this mapping is much simpler and more straightforward compared to the mapping for HSPA.

- SRB0 carries RRC PDUs. There is one-to-one mapping between SRB0 and CCCH. RRC PDUs are directly mapped on RLC SDUs without using PDCP layer. The RLC layer uses the transparent mode (TM).
- SRB1 and SRB2 carry RRC PDUs. There is one-to-one mapping between SRB1 and a DCCH; also, there is one-to-one mapping between SRB2 and another DCCH. RRC PDUsare mapped on PDCP SDUs. The RLC layer uses the acknowledged mode (AM) for both SRB1 and SRB2.
- DRBs carry IP packets. There is one-to-one mapping between a DRB and a DTCH. The IP packets are mapped on PDCP SDUs. The RLC layer uses the acknowledged mode (AM) or unacknowledged mode (UM) for DRBs.
- The PDCP layer receives PDCP SDUs from radio bearers, performs necessary processing, generates PDCP PDUs, and sends them to the RLC layer.
- PDCP PDUs and RRC PDUs are treated as RLC SDUs. The RLC layer receives RLC SDUs, performs necessary processing, generates RLC PDUs, and delivers RLC PDUs onto logical channels. As shown in Section 7.2.1.3, it may perform segmentation and/or concatenation of RLC SDUs in order to fit in RLC PDUs.
- The MAC layer treats RLC PDUs from different logical channels as MAC SDUs, performs necessary processing, generates MAC PDUs, and delivers MAC PDUs onto transport channels, DL-SCH, or UL-SCH as transport blocks (TBs). The MAC layer multiplexes the MAC SDUs in order to generate MAC PDUs and deliver them onto the transport channels. A MAC PDU is fitted into a transport block and the transport block does not contain more than one MAC PDU. As shown in Section 6.2, there may be one or more MAC control elements in the MAC PDU. Also, padding bits are added at the end of the MAC PDU, if needed, to complete the transport block size.

## 2.3 Mapping Among Channels and Bearers

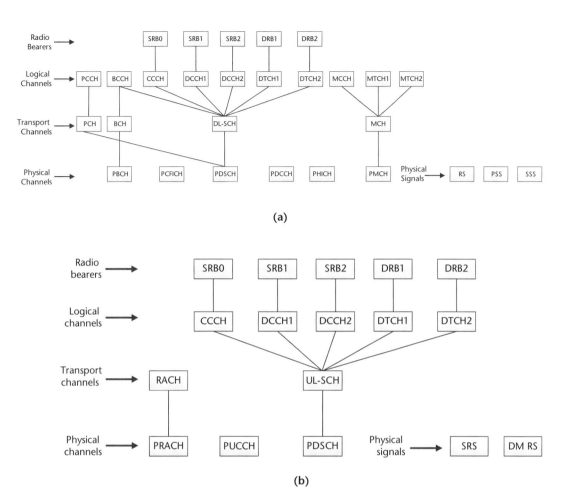

**Figure 2.1** (a) Mapping among bearers and channels in downlink. (b) Mapping among radio bearers and channels in uplink.

- BCCH and PCCH logical channels have special mapping as explained in Section 4.1 and Section 13.1, respectively.
- The mapping of MBMS-related information is shown in Section 24.3.
- The transport block is encoded independently and it is delivered to physical layer. It is then called *codeword*. The physical layer performs necessary processing and transmits codewords over physical channels.

In addition to the bearers and channels described earlier, a few other bearers are defined within EPS to carry user plane data. Mapping among all these bearers for the user data are explained here and it is schematically shown in Figure 2.2.

- *EPS bearer:* An EPS bearer is established between the UE and the PDN GW to carry user data. The MME allocates an EPS bearer identity, which uniquely identifies an EPS bearer for one UE. A UE can have a maximum of 11 EPS bearers for each PDN. The EPS bearer is explained below in Section 2.4.

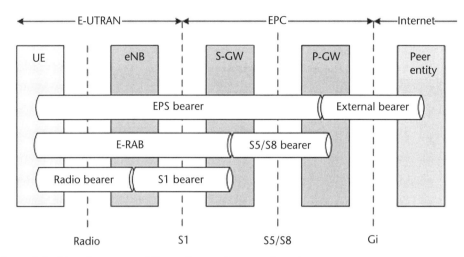

**Figure 2.2** Mapping among different bearers for user plane data.

- *E-UTRAN Radio Access Bearer (E-RAB):* An E-RAB transports the packets of an EPS bearer between the UE and the serving GW. There is one-to-one mapping between an EPS bearer and an E-RAB. An E-RAB is uniquely identified by an E-RAB ID whose value is the same as the EPS bearer identity value. The E-RAB ID remains unique for the UE in RRC_IDLE state when the UE-associated logical S1-connection is released.
- *S5/S8 bearer:* An S5/S8 bearer transports the packets of an EPS bearer between a serving GW and a PDN GW.
- *S1 bearer:* An S1 bearer transports the packets of an EPS bearer between an eNodeB and a serving GW. The serving GW stores a one-to-one mapping between an S1 bearer and an S5/S8 bearer.
- *Data Radio Bearer (DRB):* A DRB transports user plane data. The eNodeB stores a one-to-one mapping between the DRB and the S1 bearer. Thus, there is one-to-one mapping between an EPS bearer and a DRB.
- *Logical channel:* There is one-to-one mapping between a DRB and a DTCH logical channel. There are individual RLC buffers for different logical channels.
- *Transport channel:* All DTCH logical channels are multiplexed onto the DL-SCH transport channel for downlink. Similarly, all DTCH logical channels are multiplexed onto the UL-SCH transport channel for uplink.
- *Physical channel:* The DL-SCH is mapped on PDSCH for downlink and the UL-SCH is mapped on PUSCH for uplink.

## 2.4 EPS Bearer

An EPS bearer is established between the UE and the PDN GW to carry user data. The EPS bearer carries traffic flows that receive a common QoS treatment between the UE and the PDN GW. All traffic mapped on the same EPS bearer receive the same treatment in terms of scheduling policy, queue management policy, rate shaping policy, RLC configuration, and so forth. Providing a different bearer level packet forwarding treatment requires separate EPS bearers.

A default EPS bearer is established between the UE and the PDN GW and this enables always-on IP connectivity to the UE. The default EPS bearer remains established throughout the lifetime of the PDN connection. A UE is allowed to have connections with multiple PDNs. The default EPS bearers are created on per PDN basis. Connectivity with multiple PDNs may be used, for example, to have one PDN for connection with an IP multimedia subsystem (IMS) network and another PDN for connection with the Internet.

Additional EPS bearers can also be established to the same PDN; if established, they are referred to as dedicated EPS bearers. A UE can have 11 EPS bearers established at the maximum per PDN. It is the same IP address that is used for the default EPS bearer and any dedicated EPS bearer established towards the same PDN.

The following identities are used for an EPS bearer.

- *EPS Bearer Identity (EBI):* The MME allocates an EPS bearer identity, which uniquely identifies an EPS bearer for the UE. Every EPS session management (ESM) message contains the EPS bearer identity.

   The EBI is 4 bits long. It may be noted that NSAPI, which identifies a PDP context in 2G/3G, is also 4 bits long. Thus, EPS bearer identity can have 16 values. Five of these 16 values are reserved, leaving 11 valid. The UE can have 11 EPS bearers established at the maximum per PDN.

- *Linked Bearer Identity (LBI):* The LBI is the EBI of the default EPS bearer for a particular PDN connection. LBI is used during the establishment of dedicated EPS bearers with a view to indicate the default EPS bearer for a PDN connection.

### 2.4.1 Service Data Flow (SDF) and Traffic Flow Template (TFT)

An EPS bearer consists of one or more service data flows (SDF). Each SDF is intended for a particular category of traffic and may obtain different service level QoS within an EPS bearer. Each SDF is characterized by a Traffic Flow Template (TFT). The TFT contains a set of packet filters in order to allow only the traffic that the SDF is intended for. The packet filters have component values depending on IP address, port number, protocol information, traffic class type, security parameter, and so forth.

Uplink TFT or downlink TFT binds an SDF to an EPS bearer in uplink or downlink direction, respectively. The uplink TFT is applied by the UE and the downlink TFT is applied by the PDN GW. Individual TFTs are specified to filter IP packets for each of the EPS bearers. If multiple SDFs are multiplexed onto the same EPS bearer, then multiple packet filters are specified in uplink TFT or downlink TFT. The traffic flow aggregate contains the aggregate of all packet filters that determine the traffic mapping to a particular EPS bearer.

## 2.4.2 QoS Profile

The EPS bearer QoS profile characterizes an EPS bearer; it includes the following parameters:

- QoS class identifier (QCI);
- Maximum bit rate (MBR);
- Guaranteed bit rate (GBR);
- Allocation and retention priority (ARP).

All SDFs pertaining to the same EPS bearer have the same value for QCI and ARP. The EPS bearer QoS profile may be defined per SDF or per SDF aggregate at the service level in the network. Thus, each SDF may obtain different service level QoS within an EPS bearer. For the same IP-Connectivity Access Network (IP-CAN) session, multiple SDFs with the same QCI and ARP can be treated as a single traffic aggregate, which is referred to as an *SDF aggregate*. An SDF is a special case of an SDF aggregate.

In order to notify the EPS bearer QoS profile, the MME sends EPS quality of service IE to the UE when it establishes or modifies an EPS bearer. So, the MME includes the EPS quality of service IE when it sends the following NAS messages to the UE.

- Activate Default EPS Bearer Context Request message;
- Activate Dedicated EPS Bearer Context Request message;
- Modify EPS Bearer Context Request message.

The EPS Quality of Service IE includes QCI, GBR, and MBR out of the parameters of EPS bearer QoS profile. QCI, GBR, and MBR can impact packet forwarding treatment, such as scheduling and rate control. ARP, the other parameter of EPS bearer QoS profile, is not included in EPS Quality of Service IE. ARP is not sent to the UE and it is rather used within the network.

The EPS bearer QoS profile parameters are explained in the following sections.

### 2.4.2.1 Allocation and Retention Priority (ARP)

The Allocation and Retention Priority (ARP) is used during establishment or modification of EPS bearers and it has no bearing on how the packets should be forwarded after the establishment of the EPS bearer. The ARP contains the following information.

- *Priority level*: The value of this field ranges between 1 and 15 with 1 as the highest level of priority. It defines the relative importance of a resource request. Network uses this priority level to decide if a bearer establishment or modification request can be accepted when there are limited resources. This is typically required for GBR bearers. ARP can also be used to decide if existing bearers should be released during resource limitations.

- *Preemption capability:* It is a flag that can be set to either yes or no. It defines if a bearer with a lower ARP priority value should be dropped to make resources available for the requested bearer.
- Preemption vulnerability: It is a flag that can be set to either yes or no. It defines if the bearer can lose resources in order to make resources available for a preemption capable bearer with a higher ARP priority value.

#### 2.4.2.2 GBR Bearer and Non-GBR Bearer

The EPS bearer can be categorized as a GBR bearer and non-GBR bearer based on the QoS profile. The default EPS bearer is always made a non-GBR bearer. A dedicated bearer can be either a GBR or a non-GBR bearer.

*Guaranteed Bit Rate (GBR) Bearers*

An EPS bearer is a GBR bearer if it is expected to provide a particular bit rate known as GBR. Each GBR bearer is also expected to provide a maximum bit rate known as maximum bit rate (MBR). MBR is required to be equal to GBR in Release 8. In case of GBR bearers, congestion related packet drops are not generally expected when the data rate is lower than or equal to GBR. The GBR and MBR have uplink and downlink components.

The GBR SDF aggregates are typically authorized on demand, which requires dynamic policy and charging control. The eNodeB performs uplink scheduling for the UE and controls uplink bearer level rate based on MBR. The PDN GW enforces the downlink rate of the UE based on the accumulated MBRs of the aggregate of SDFs.

*Non-GBR Bearers*

An EPS bearer with no reserved or guaranteed bit rate resources is known as a non-GBR bearer. The following parameters are defined for non-GBR bearers.

- *APN Aggregate Maximum Bit Rate (APN-AMBR):* For the UE, APN-AMBR is the maximum allowed aggregate bit rate that can be expected to be provided across all non-GBR bearers from all PDN connections with the same APN. Total traffic from non-GBR bearers of the UE may not exceed the APN-AMBR from the same APN. All non-GBR bearers of the UE from the same APN share the whole APN-AMBR and any non-GBR bearer may use the entire APN-AMBR when other non-GBR bearers do not carry any traffic. The APN-AMBR of the UE depends on its subscription parameter stored in the HSS.
- *UE Aggregate Maximum Bit Rate (UE-AMBR):* The MME adds APN-AMBR of all active APNs for a UE and sets it to UE-AMBR of the UE. Thus, total traffic from all non-GBR bearers of a UE may not exceed the UE-AMBR. All non-GBR bearers from the same APN share the whole APN-AMBR and any non-GBR bearer may use the entire APN-AMBR when other non-GBR bearers do not carry any traffic. The UE-AMBR of a UE depends on its subscription parameter stored in the HSS.

Congestion-related packet drops are not unexpected especially in case of traffic load peak or poor coverage area for non-GBR bearers. A non-GBR SDF aggregate may be preauthorized through static policy and charging control. The APN-AMBR and UE-AMBR have uplink and downlink components. The eNodeB performs uplink and downlink scheduling for a UE and thus, controls data rate based on UE-AMBR. The PDN GW enforces uplink and downlink rate of the UE based on its APN-AMBR. The UE enforces uplink rate based on its APN-AMBR.

### 2.4.2.3 QoS Class Identifier (QCI)

The QoS Class Identifier (QCI) assigned to an EPS bearer can have values from 1 through 9. In order to attain appropriate treatment in packet forwarding, the eNodeB can configure few settings based on the value of QCI, such as, scheduling weights, admission thresholds, queue management thresholds, link layer protocol configuration, and so forth. The characteristics for any particular QCI are shown in Table 2.1. Every QCI is associated with a priority level where priority level 1 indicates the highest priority. The packet delay budget (PDB) defines an upper limit for packet delay between the UE and the Policy and Charging Enforcement Function (PCEF). The delay should not exceed PDB for a certain QCI in the case of 98% of the packets that have not been dropped due to congestion. Scheduling between different SDF aggregates is performed primarily based on the PDB. However, if the PDB of certain SDF aggregates cannot be met, then the scheduler will attempt to meet the PDB of SDF aggregates that have higher priority while allowing more delay for SDF aggregates that have lower priority. PDB can also be used for setting target HARQ operating points and other link layer configurations. The packet

**Table 2.1** Characteristics for Different QCI Values

| GBR/ Non-GBR | QCI | Priority | Packet Delay Budget (PDB) | Packet Error Loss Rate (PELR) | Application |
|---|---|---|---|---|---|
| GBR | 1 | 2 | 100 ms | $10^{-2}$ | Voice |
| | 2 | 4 | 150 ms | $10^{-3}$ | Live streaming video |
| | 3 | 3 | 50 ms | $10^{-3}$ | Real time gaming |
| | 4 | 5 | 300 ms | $10^{-6}$ | Buffered streaming video |
| | 5 | 1 | 100 ms | $10^{-6}$ | IMS signaling |
| | 6 | 6 | 300 ms | $10^{-6}$ | Buffered streaming video, progressive video, HTTP, e-mail, chat, FTP, P2P file sharing |
| | 7 | 7 | 100 ms | $10^{-3}$ | Voice, live streaming video, interactive gaming |
| | 8 (Privileged subscribers) 8 | 8 | 300 ms | $10^{-6}$ | Buffered streaming video, progressive video, HTTP, e-mail, chat, FTP, P2P file sharing |
| Non-GBR | 9 (Non-privileged subscribers) 9 | 9 | 300 ms | $10^{-6}$ | Buffered streaming video, progressive video, HTTP, e-mail, chat, FTP, P2P file sharing |

Error Loss Rate (PELR) is used for link layer protocol configurations. Both PDB and PELR have the same values in uplink and downlink for certain QCI.

## 2.5 Piggybacking of Control Messages

### 2.5.1 Piggybacking of NAS Messages

#### 2.5.1.1 Piggybacking on S1AP Messages

The NAS messages are always piggybacked on the S1AP messages for transfer over S1-MME interface. The S1AP messages can be categorized as follows based on their inclusion of NAS messages.

1. S1AP messages with piggybacked NAS messages including S1AP signaling information. These S1AP messages can include NAS-PDU IE, which contains the NAS message. The S1AP messages also include S1AP signaling information. This type includes the following messages:
   - E-RAB SETUP REQUEST message;
   - E-RAB MODIFY REQUEST message;
   - E-RAB RELEASE COMMAND message;
   - INITIAL CONTEXT SETUP REQUEST message;
   - NAS NON DELIVERY INDICATION message.

2. S1AP messages with piggybacked NAS messages but essentially with no S1AP signaling information. These S1AP messages can include NAS-PDU IE, which contains the NAS message. The S1AP messages do not include any significant S1AP signaling information and so these S1AP messages are used only to carry the NAS message. This type includes the following messages:
   - DOWNLINK NAS TRANSPORT message;
   - UPLINK NAS TRANSPORT message.

3. S1AP messages with no piggybacked NAS messages. This type includes all other S1AP messages except the messages shown for the above two types.

#### 2.5.1.2 Piggybacking on RRC Messages

The NAS messages are always piggybacked on the RRC messages for transfer over LTE-Uu interface. The RRC messages can be categorized as follows based on their inclusion of NAS messages.

1. RRC messages with both piggybacked NAS messages and RRC layer signaling information. These RRC messages can include DedicatedInfoNAS IE, which contains the NAS message. The RRC messages also include RRC layer signaling information. This type includes the following messages:
   - RRCConnectionReconfiguration message;
   - RRCConnectionSetupComplete message.

2. RRC messages with piggybacked NAS messages but with no RRC layer signaling information. These RRC messages can include DedicatedInfoNAS IE, which contains the NAS message. The RRC messages do not include any RRC layer signaling information and so these RRC messages are used only to carry the NAS message. This type includes the following messages:
   - DLInformationTransfer message;
   - ULInformationTransfer message.
3. RRC messages with no piggybacked NAS messages. This type includes all other RRC messages except the messages shown for the above two types.

#### 2.5.1.3 Piggybacking on NAS Messages

Certain EPS Mobility Management (EMM) messages can include ESM message container IE, which carries a single EPS Session Management (ESM) message. Thus, a NAS message gets piggybacked on another NAS message. The piggybacked ESM message does not include a NAS security header. The following EMM messages can contain ESM message piggybacked.

- ATTACH REQUEST;
- ATTACH ACCEPT;
- ATTACH COMPLETE;
- ATTACH REJECT.

### 2.5.2 Piggybacking of RRC Messages

The RRC message, MobilityFromEUTRACommand, includes the TargetRAT-MessageContainer IE, which contains one of the following RRC messages. However, these piggybacked RRC messages are generated in UTRAN or GERAN and not in E-UTRAN.

1. HANDOVER TO UTRAN COMMAND;
2. HANDOVER COMMAND;
3. PS HANDOVER COMMAND;
4. DTM HANDOVER COMMAND.

The X2AP runs on X2 interface between two eNodeBs. The following X2AP messages can have RRC messages piggybacked:

1. The HANDOVER REQUEST message has RRC message piggybacked in RRC Context IE. The piggybacked RRC message is the HandoverPreparationInformation message.
2. The HANDOVER REQUEST ACKNOWLEDGE message has RRC message piggybacked in Target eNodeB to Source eNodeB Transparent Container IE. The piggybacked RRC message is the HandoverCommand message.

## 2.6 NAS Layer States

### 2.6.1 EPS Mobility Management (EMM) States

The EPS Mobility Management (EMM) in both the UE and the MME can stay in different states. The major EMM states are as follows.

- *EMM-DEREGISTERED state:* The UE is not registered with the network in EMM-DEREGISTERED state, and, thus, there is no EMM context. The MME is not aware of the location of the UE and the MME cannot communicate with the UE. However, some UE context can still be stored in the UE and MME for quickening processes, for example, key set identifier (KSIASME) can be stored in order to avoid invoking the EPS authentication and key agreement (EPS AKA) procedure during every Attach procedure.
- *EMM-REGISTERED state:* The UE is registered with the network in EMM-REGISTERED state, and, thus, there is an EMM context established. Also, there is a default EPS bearer established. The MME is aware of the location of the UE and the MME can communicate with the UE.

The attach procedure is performed to switch the UE and the MME from EMM-DEREGISTERED state to EMM-REGISTERED state. In case of handover to an LTE cell from another RAT, the MME enters the EMM-REGISTERED state by a tracking area update procedure. The UE and the MME move from the EMM-REGISTERED state to the EMM-DEREGISTERED state in the following cases.

- Detach procedure;
- Attach reject;
- Tracking area update reject;
- Deactivation of all EPS bearers.

The EMM maintains connectivity between the UE and the EPC using EPS Connection Management (ECM) protocols. There are two ECM states as shown here.

- *ECM-IDLE state:* In the ECM-IDLE state of the UE and the MME, there is no NAS signaling connection between the UE and the MME. In this state, the MME is aware of the UE location with an accuracy of the TAI list of the UE. A TAI list provides a list of tracking areas within which the UE can move around without performing any tracking area updating procedure. The tracking areas in the TAI list correspond to the same MME area.

    The UE and the MME may enter ECM-IDLE state when there is a long period of inactivity in data transfer. This allows reduction of processing in the network and in the UE. The MME still keeps the UE context and information about the established bearers during the ECM-IDLE state. However, the UE and the network may have different sets of established EPS bearers.
- *ECM-CONNECTED state:* In the ECM-CONNECTED state of the UE and the MME, there is NAS signaling connection between the UE and the MME.

In this state, the MME is aware of the UE location with an accuracy of the serving eNodeB of the UE.

The UE and the MME enter the ECM-CONNECTED state when transmission of uplink or downlink signaling information or user data is required. In the ECM-CONNECTED state, the set of EPS bearers is synchronized between the UE and the network.

The UE and the MME move from the ECM-IDLE state to the ECM-CONNECTED state when the NAS signaling connection is established between the UE and the MME. The UE and the MME move from the ECM-CONNECTED state to the ECM-IDLE state when the NAS signaling connection is released. The NAS signaling connection establishment and release will be explained in Chapter 14.

Certain requirements cause dependency among different EMM and ECM states. For example, the attach procedure takes place during the transition from the EMM-DEREGISTERED state to the EMM-REGISTERED state. This attach procedure requires the establishment of NAS signaling connection; thus, it requires the transition from the ECM-IDLE state to the ECM-CONNECTED state. Also, the establishment of the RRC connection is a part of the establishment of the NAS signaling connection. Thus, the transition from the ECM-IDLE state to the ECM-CONNECTED state requires the transition from the RRC_IDLE state to the RRC_CONNECTED state.

### 2.6.2 EPS Session Management (ESM) States

The EPS Session Management (ESM) can stay in the following states.

- ESM-ACTIVE state: The EPS bearer context is active.
- ESM-INACTIVE state: The EPS bearer context is inactive.

The EPS bearer setup causes transition from the ESM-INACTIVE state to the ESM-ACTIVE state. The EPS bearer release causes transition from the ESM-ACTIVE state to the ESM-INACTIVE state.

IP connectivity is preserved in the ECM-IDLE state if there is an EMM-REGISTERED state. Thus, in the ECM-IDLE State, EPS bearer context also remains active.

# CHAPTER 3
# Physical Layer Properties

The physical layer performs a number of processes. The major processes are mentioned in Section 3.2. The physical layer uses orthogonal frequency division multiple access (OFDMA) for downlink transmission and single-carrier frequency division Mmultiple access (SC-FDMA) for uplink transmission.

## 3.1  OFDMA for Downlink

The orthogonal frequency division multiplex (OFDM) breaks the available bandwidth into many narrow parts with subcarriers at the center of each of these parts. Then the data symbols are modulated in parallel streams onto the subcarriers. The subcarriers are made orthogonal to each other by choosing a subcarrier spacing, $\Delta f = 1/T$ where $T$ is the OFDM symbol period. This allows increased spectral efficiency with no intercarrier interference (ICI) among adjacent subcarriers virtually as shown in Figure 3.1.

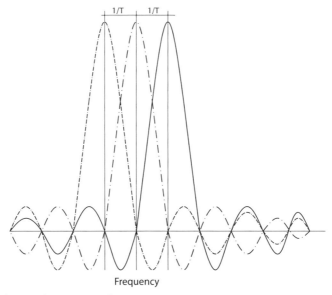

**Figure 3.1**  Orthogonality among subcarriers in OFDM.

The OFDMA employs OFDM for simultaneous access of multiple users. A time-frequency resource grid structure is created based on OFDM. Then multiple users are allocated different parts of the time-frequency resource grid structure.

### 3.1.1 Implications of OFDMA

#### 3.1.1.1 Time-Dispersive Radio Channel

In cellular applications, the radio channel is time-dispersive and a symbol received from different multipaths can have different delays. The delay spread can be several microseconds. Thus, a symbol received along a delayed path may overlap a subsequent symbol arriving at the receiver via a more direct path. This effect is referred to as intersymbol interference (ISI). As mentioned in Section 3.1.1.2, a cyclic prefix is added to overcome ISI.

The time dispersiveness of a radio channel is frequency dependent. Thus, a wide channel gets the different parts of its frequency range affected differently (i.e., it undergoes frequency-selective fading and gives rise to distortion). A coherence bandwidth is considered to limit within flat fading as opposed to frequency-selective fading. The coherence bandwidth depends on the level of time dispersiveness as characterized by the delay spread. OFDM uses a large number of subcarriers with narrow bandwidths instead of a wideband signal; this allows the signal on each subcarrier to be considered to undergo flat fading instead of frequency-selective fading. Thus, OFDM has the inherent property to obviate the need for an equalizer at the receiver and offer great improvement in the performance as well.

In OFDM, there can be intercarrier interference (ICI) between subcarriers in addition to ISI due to the time dispersiveness of the radio channel. The demodulator correlation interval for one multipath component can overlap the adjacent symbol period of a different multipath component. The integration interval can be chosen for the strong multipath component but this interval may not correspond to an integer number of periods of complex exponentials for a different multipath component. This will lead to the loss of orthogonality between subcarriers and will cause ICI between subcarriers. As mentioned in Section 3.1.1.2, the cyclic prefix helps overcome ICI.

#### 3.1.1.2 Cyclic Prefix (CP)

The transmission duration for a symbol is made little longer than the symbol period by adding a cyclic prefix (CP) to each OFDM symbol. For this purpose, the last part of the OFDM symbol is copied and inserted at the beginning of the OFDM symbol. The duration of the cyclic prefix is denoted as $T_{CP}$ and the total transmission duration for the symbol becomes $T + T_{CP}$. The cyclic prefix offers advantages and disadvantages.

*Advantages*

1. The cyclic prefix allows an extension from the actual symbol period. Thus, if the span of the time dispersion of the channel is shorter than the cyclic-prefix length, a symbol received along a delayed path will not bleed into a

subsequent symbol arriving at the receiver via a more direct path. Thus, the cyclic prefix helps overcome ISI from the time-dispersive channel.
2. The demodulator correlation interval is kept equal to the actual symbol period and it is not increased. Assuming that the span of the time dispersion of the channel is shorter than the cyclic-prefix length, the demodulator correlation interval for one multipath component will not overlap the adjacent symbol period of a different multipath component. This preserves the subcarrier orthogonality at the receiver side. Thus, the cyclic prefix helps overcome ICI between subcarriers.
3. The cyclic prefix allows more imperfection in the time alignment between the transmitter and the receiver. The summation of the time alignment error and the delay spread of the channel needs to be less than the cyclic prefix.

*Disadvantages*
The cyclic prefix increases the length of the OFDM symbol period and so, reduces the OFDM symbol rate. Also, a fraction $T/(T + T_{CP})$ of the received signal power is actually utilized by the OFDM demodulator.

### 3.1.1.3 Subcarrier Spacing

A constant subcarrier spacing is used regardless of the transmission bandwidth. Different sizes of transmission bandwidths are defined by assigning different number of OFDM subcarriers. The subcarrier spacing, $\Delta f$, is chosen as 15 kHz. For an MBMS dedicated cell, reduced subcarrier spacing, $\Delta f = 7.5$ kHz, can be used in downlink. The choice of subcarrier spacing, $\Delta f$, takes a few points into account. $\Delta f$ should be:

- Small enough to allow the signal on each subcarrier to undergo flat fading. This requirement depends on the delay spread in the multipath environment.
- Small enough to make T sufficiently larger than TCP to minimize inefficiency.
- Large enough to prevent the signal from significant distortion due to fast fading. Thus, the subcarrier spacing also depends on how fast the channel is varying. The Doppler spread characterizes the time varying nature of the frequency dispersiveness of the channel in the environment at the speed of the mobile user.
- Large enough so that frequency errors cannot cause intercarrier interference (ICI). The mismatch in the reference frequencies of local oscillators between the transmitter and the receiver is the major reason of frequency errors. Phase noise may also lead to frequency errors. The mismatch in the reference frequencies has to be typically limited within a few percent of the subcarrier spacing.

### 3.1.2 Transmission and Reception

#### 3.1.2.1 OFDMA Transmission

1. A particular number, $M$ modulation symbols from the data stream are treated as a block. During each OFDM symbol interval, these $M$ modulation symbols are transmitted in parallel. For this purpose, a serial-to-parallel (S/P) converter takes the data stream into $M$ parallel streams. The OFDMA transmission is explained next and it is schematically shown in Figure 3.2.
2. The $M$ parallel data streams are modulated onto $M$ number of OFDM subcarriers using an appropriate level of modulation (e.g., QPSK, 16-QAM, or 64-QAM). The level of modulation can vary among the streams.
3. The block of $M$ symbols, in each OFDM symbol period, passes through an inverse fast Fourier transform (IFFT) of size $N$, where $N$ is called the FFT size. $N$ is an integer power of 2 and can be 128, 256, 512, 1,024, 1,536, or 2,048 depending on the channel bandwidth as shown in Table 3.1. Here, $N > M$ (i.e., the FFT size is larger than the number of subcarriers). Zeros are used in addition to $M$ symbols in order to create $N$ inputs. This is like an over-sampled time-discrete OFDM signal with sampling rate, $fs = 1/Ts = N \times \Delta f$, where $\Delta f$ is the spacing between subcarriers. The sampling rate is always made a multiple of 3.84 MHz to enable smooth handover between LTE and UMTS. $N$ exceeds $M$ with a sufficient margin in an attempt to fulfill the sampling theorem. For example, as shown in Table 3.1, in the case of 10-MHz spectrum allocation, the number of subcarriers including the DC subcarrier is 601. This requires a bandwidth of 601 × 15 kHz = 9.015 MHz. The rest of the 10-MHz bandwidth is used as guard band. Here, the

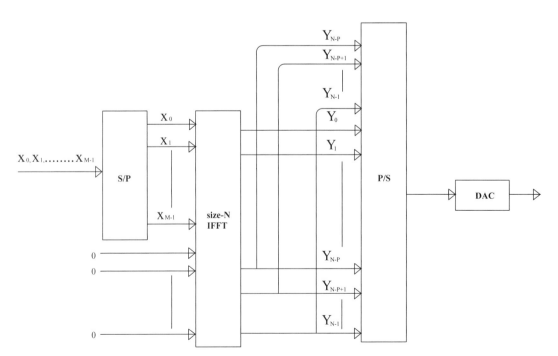

**Figure 3.2** OFDMA transmission.

## 3.1 OFDMA for Downlink

**Table 3.1** Different Parameters Used for Downlink Transmission

| Transmission Parameters | Value | | | | | |
|---|---|---|---|---|---|---|
| Channel BW | 1.4 MHz | 3 MHz | 5 MHz | 10 MHz | 15 MHz | 20 MHz |
| Number of subcarriers (without DC subcarrier) | 72 | 180 | 300 | 600 | 900 | 1,200 |
| FFT size, $N$ | 128 | 256 | 512 | 1,024 | 1,536 | 2,048 |
| Sampling rate, $f_s$ | 128 ×15 kHz = 1.92 MHz | 256 ×15 kHz = 3.84 MHz | 512 ×15 kHz = 7.68 MHz | 1,024 ×15 kHz = 15.36 MHz | 1,536 ×15 kHz = 23.04 MHz | 2,048 ×15 kHz = 30.72 MHz |
| Samples per slot | 960 | 1,920 | 3,840 | 7,680 | 11,520 | 15,360 |
| Number of resource blocks in frequency, $N_{RB}^{DL}$ | 6 | 15 | 25 | 50 | 75 | 100 |

FFT size, $N$, is 1,024 making the sampling rate $f_s = N \times \Delta f = 1{,}024 \times 15$ kHz = 15.36 MHz.

4. A DC subcarrier is assumed at the center of the bandwidth but the DC subcarrier is not actually used in transmission in order to avoid interference (e.g., due to local oscillator leakage).
5. Each OFDM symbol is preceded by a cyclic prefix. The cyclic prefix increases the length of the OFDM symbol period from $T$ to $T + T_{CP}$. In order to add the cyclic prefix, the last $P$ samples of the IFFT output block of length $N$ are copied and inserted at the beginning of the block. This increases the block length from $N$ to $N + P$.
6. The output of the IFFT is P/S converted.
7. The output is transmitted after D/A conversion.

### 3.1.2.2 OFDMA Reception

The OFDMA reception is explained next and it is schematically shown in Figure 3.3.

1. A/D conversion process samples the received signal.
2. An S/P converter takes the received data samples into $N$ parallel streams.
3. The P samples corresponding to the length of the cyclic prefix are discarded.
4. FFT processing with FFT size, $N$, is used for demodulation of OFDM signal. Only the $M$ number of outputs of the FFT is useful here and the rest of the outputs are left unused.
5. The parallel $M$ modulation symbols of the block are P/S converted.

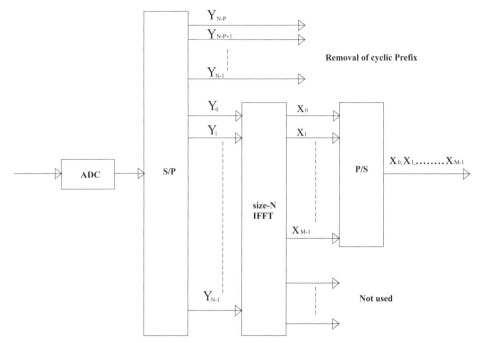

**Figure 3.3** OFDMA reception.

## 3.2 Downlink Time-Frequency Resource Grid

Using subcarriers in the frequency axis and symbols in the time axis, a time-frequency resource grid is considered. If multiple antenna elements are used at the eNodeB, then each antenna port is considered to have its own time-frequency resource grid. The antenna ports are numbered as $p = 0, 1, 2, …, 8$.

Each element in the time-frequency resource grid is called a *resource element*; thus, a resource element represents one subcarrier and one symbol resource. Each resource element is uniquely identified by the index pair $(k, l)$ where $k$ and $l$ are the indices for subcarriers in the frequency domain and symbols in the time domain, respectively.

To provide consistent and exact timing definitions, different time intervals within the LTE radio access specification are expressed as multiples of a basic time unit, $Ts$. $Ts$ is equal to the sampling time with the largest FFT size, $N = 2,048$ for subcarrier spacing, 15 kHz or the sampling time with FFT size, $N = 4,096$ for subcarrier spacing, 7.5 kHz.

$Ts = 1/ (2,048 \times \Delta f) = 1/(2,048 \times 15,000) = 1/30,720,000$ sec $\approx 32.552$ ns
$Ts = 1/ (4,096 \times \Delta f) = 1/(4,096 \times 7,500) = 1/30,720,000$ sec $\approx 32.552$ ns

### 3.2.1 Radio Frame Structure

The time domain structure includes radio frames appearing one after another in time. The following radio frame structures are supported.

## 3.2 Downlink Time-Frequency Resource Grid

1. Frame structure type 1 (FS1) applicable to FDD;
2. Frame structure type 2 (FS2) applicable to TDD.

TDD is beyond the scope of this book and only radio frame structure type 1 (FS1) will be discussed. The structure for FS1 is shown in Figure 3.4.

### 3.2.1.1 Radio Frame

A radio frame has length $T_{frame} = 307{,}200 \times Ts = 10$ ms. This provides an easy backward compatibility with UMTS, which also uses 10 ms as the length of radio frame. A radio frame consists of 10 equally sized subframes. The subframes in a radio frame are numbered from 0 through 9. Each subframe is further divided into two equally sized slots.

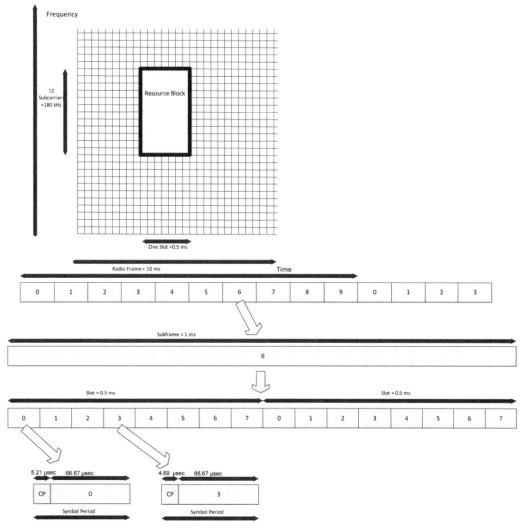

**Figure 3.4** Time-frequency resource grid with 15-kHz subcarrier spacing for normal cyclic prefix.

- A subframe has length $T_{subframe} = 30{,}720 \times Ts = 1$ ms;
- A slot has length $T_{slot} = 15{,}360 \times Ts = 0.5$ ms;
- A slot contains 3, 6, or 7 OFDM symbols.

The radio frames appearing one after another in time are numbered sequentially and these numbers are called the system frame numbers (SFN). Thus, SFN is incremented in every radio frame. SFN is 10 bits long and it is restarted after it has wrapped around. SFN helps in synchronization. A broadcast message allows determination of SFN as explained in Section 4.3.

#### 3.2.1.2 Length of Cyclic Prefix

A short cyclic prefix allows for less overhead and waste of bandwidth. However, the cyclic prefix should be long enough to account for the maximum time delay of the multipath channel. Besides, a very short cyclic prefix can degrade orthogonality between subcarriers. The length of cyclic prefix may also take frequency reuse or interference from other cells, transmission power, and so forth into account. The following particular lengths are defined for a cyclic prefix for 15-kHz and 7.5-kHz subcarrier spacings.

1. *15-kHz subcarrier spacing:* Two different types of length are used for cyclic prefix as shown here.
   - *Normal cyclic prefix:* The length of a normal cyclic prefix is optimized for a typical urban multipath environment. This normal cyclic prefix is expected to be used most commonly. In this case, 7 OFDM symbols are used per slot. The length of the cyclic prefix of the first OFDM symbol: $T_{CP} = 160 \times Ts = 5.21\ \mu s$. The length of the cyclic prefix of the other six OFDM symbols: $T_{CP} = 144 \times Ts = 4.69\ \mu s$.
   - *Extended cyclic prefix:* The length of extended cyclic prefix is optimized for a larger cell size and heavy urban multipath environment. In this case, 6 OFDM symbols are used per slot. The length of the cyclic prefix of each OFDM symbol: $T_{CP} = 512 \times Ts = 16.67\ \mu s$.
2. *7.5-kHz subcarrier spacing:* In case of MBMS, signals may be combined from eNodeBs located far away as explained in Section 24.1. For the provision of large multipath delay, MBMS can optionally be used with 7.5-kHz subcarrier spacing, which provides for a very long cyclic prefix. Only one type of length is used for cyclic prefix with 7.5-kHz subcarrier spacing. It uses three OFDM symbols per slot and a cyclic prefix of length: $T_{CP} = 1{,}024 \times Ts = 33.3\ \mu s$.

#### 3.2.1.3 Symbol Period

The symbol period is defined for as follows for 15-kHz and 7.5-kHz subcarrier spacing.

## 3.2 Downlink Time-Frequency Resource Grid

1. *15-kHz subcarrier spacing:* The symbol period without cyclic prefix is: T = 1/Δf = 1/15,000 = 2,048 × Ts = 66.67 μs. There are two types of configurations:
   - *Seven OFDM symbols used per slot (normal cyclic prefix):* The length of cyclic prefix for the first OFDM symbol: $T_{CP}$ = 160 × Ts = 5.21 μs. The length of cyclic prefix for the other six OFDM symbols: $T_{CP}$ = 144 × Ts ≈ 4.69 μs. Thus, the total slot has length: $T_{slot}$ = [(160 + 2,048) + 6 × (144 + 2,048)] × Ts = 15,360 × Ts = 0.5 ms.
   - *Six OFDM symbols used per slot (extended cyclic prefix):* The length of cyclic prefix for all OFDM symbols: $T_{CP}$ = 512 × Ts ≈ 16.67 μs. Thus, the total slot has length: $T_{slot}$ = [6 × (512 + 2,048)] × Ts = 15,360 × Ts = 0.5 ms.
2. *7.5-kHz subcarrier spacing:* The symbol period without cyclic prefix is: T = 1/Δf = 1/7,500 = 4,096 × Ts ≈ 0.1333 ms. Three OFDM symbols are used per slot. The length of cyclic prefix for all OFDM symbols is: $T_{CP}$ = 1,024 × Ts ≈ 33.3 μs. Thus, the total slot has length,

$$T_{slot} = [3 \times (1,024 + 4,096)] \times Ts = 15,360 \times Ts = 0.5 \text{ ms}.$$

### 3.2.2 Resource Block (RB)

A resource block (RB) is a basic unit for allocation of resources to the UE. A UE can be an allocated integer multiple of resource blocks. Two types of resource blocks are defined as follows.

#### 3.2.2.1 Physical Resource Block (PRB)

A physical resource block (PRB) defines a fixed amount of resource in both frequency and time. In time, the length of a PRB is always 1 slot, which is equal to 0.5 ms. In frequency, the length of a PRB is always 180 kHz. The number of subcarriers in a resource block is denoted as $N_{SC}^{RB}$.

$N_{SC}^{RB}$ = 12 for 15-kHz subcarrier spacing;

$N_{SC}^{RB}$ = 24 for 7.5-kHz subcarrier spacing.

The number of available physical resource blocks in the total frequency resource is denoted as $N_{RB}^{DL}$. The value of $N_{RB}^{DL}$ can be 6, 15, 25, 50, 75, or 100 depending on the channel bandwidth as shown in Table 3.1. Physical resource blocks are numbered in frequency domain as $n_{PRB}$. Thus, $n_{PRB}$ ranges as follows:

$n_{PRB}$ = 0, 1, 2, 3, ..., $N_{RB}^{DL}$ − 1

The number of subcarriers in total frequency resource is, $N_{RB}^{DL} N_{SC}^{RB}$. k is used as the index for subcarriers in the total frequency resource and so, k ranges as follows,

$k = 0, 1, 2, 3, ..., N_{RB}^{DL} N_{SC}^{RB} - 1$

Thus, $n_{PRB}$ of the resource block that includes subcarrier $k$ satisfies the following relation.

$n_{PRB} = \text{floor}(k/N_{SC}^{RB})$

The number of symbols in a resource block or in a slot is denoted as $N_{symb}^{DL}$. Thus,

$N_{symb}^{DL} = 7$ for 15-kHz subcarrier spacing with normal cyclic prefix;

$N_{symb}^{DL} = 6$ for 15-kHz subcarrier spacing with extended cyclic prefix;

$N_{symb}^{DL} = 3$ for 7.5-kHz subcarrier spacing.

$l$ is used as the index for symbols in a resource block or in a slot and so, l ranges as follows:

$l = 0, 1, 2, 3, ..., N_{symb}^{DL} - 1$

A physical resource block consists of $N_{SC}^{RB} \times N_{symb}^{DL}$ resource elements. The composition of resource elements in a resource block occur in one of three ways:

- $12 \times 7 = 84$ resource elements for 15-kHz subcarrier spacing with normal cyclic prefix;
- $12 \times 6 = 72$ resource elements for 15-kHz subcarrier spacing with extended cyclic prefix;
- $24 \times 3 = 72$ resource elements for 7.5-kHz subcarrier spacing.

In the time domain, the scheduling decision can be modified at every transmission time interval (TTI) for a UE, which is 1 ms, one subframe, or two slots long. Thus, a pair of resource blocks that are contiguous in time is used as the granularity of scheduling decision. $n_{PRB}$ is used to refer to this pair of resource blocks in frequency. The number of transport blocks transmitted to the UE in every TTI is one if there is no spatial multiplexing and it can be two if spatial multiplexing is applied.

### 3.2.2.2 Virtual Resource Block (VRB)

A virtual resource block (VRB) is of the same size as a physical resource block in both frequency and time. The virtual resource blocks are mapped onto different physical resource blocks; thus, they are used to indicate localized or distributed resource allocations in downlink. The virtual resource blocks are of two types as shown below. For both the following types, a pair of virtual resource blocks over two slots in the subframe is assigned together using a single virtual resource block number in frequency domain, $n_{VRB}$.

1. *Localized virtual resource block:* A localized virtual resource block is mapped exactly to a single physical resource block. Because of this direct mapping, $n_{VRB} = n_{PRB}$. The number of available localized virtual resource blocks in the total frequency resource is equal to the number of available physical resource blocks, (i.e., $N_{VRB}^{DL} = N_{RB}^{DL}$). $n_{VRB}$ ranges as: $n_{VRB} = 0, 1, 2, 3, \ldots N_{VRB}^{DL} - 1$.
2. *Distributed virtual resource block:* This distributed virtual resource blocks are mapped to the physical resource blocks, which may not be contiguous in frequency. In addition, the pair of disturbed virtual resource blocks in the two slots of a subframe is mapped to two physical resource blocks located at predefined distance (i.e., it provides for interslot hopping). The distributed virtual resource blocks help in frequency diversity. The virtual resource block $n_{VRB}$ is mapped to the physical resource block $n_{PRB}$ where $n_{PRB}$ is a function of $n_{VRB}$ and $n_S$, (i.e., $n_{PRB} = f(n_{VRB}, n_S)$. Here, $n_S$ is the slot number within the radio frame.

## 3.3 Downlink Physical Layer Process

There can be one or two transport blocks in a TTI. Each transport block in the TTI goes through the following general steps in physical layer processing.

1. *CRC insertion:* A 24-bit-long CRC is computed over the transport block and is added at the end of the transport block for error detection. The CRC is 16 bits long for PDCCH and PBCH.
2. *Coding:* A rate 1/3 turbo encoder is used with a quadratic permutation polynomial (QPP) interleaver. The code blocks size is 6,144 bits at the maximum as it is the maximum size allowed by the turbo coder. If the number of bits becomes more than 6,144 after the addition of CRC, then the transport block is segmented into code blocks (CBs). The CBs have a fixed size of 6,144 bits, and for this size requirement padding zero bits may be added at the beginning of the first CB. The CBs are coded separately. This segmentation allows parallel coding and decoding and speeds up the process. Each CB includes its own CRC. Thus, at the receiver, if decoding fails for one CB, the decoding of other CBs are not continued unnecessarily, and, in this case, the whole transport block should be retransmitted by HARQ. PBCH uses rate 1/3 tail biting convolutional coding (TBCC).
3. *Rate matching:* The rate matching allows matching the number of bits to be transmitted exactly with the number of available resource elements. For this purpose, the rate matching applies puncturing and repetition in the rate of 1/3 turbo encoder output. The rate matching attempts to maximize the number of new bits in the HARQ retransmissions. The rate matching is performed per code block using a circular buffer. The rate matching also generates redundancy versions (RVs), namely, RV 0, RV 1, RV 2, and RV

3 per code block for initial transmission, first HARQ retransmission, second retransmission, and third retransmission, respectively. The four redundancy versions are available at four different starting points in the circular buffer. The initial transmission includes most systematic bits.

4. *Code block concatenation:* The code blocks are concatenated to obtain a coded transport block.
5. *Scrambling:* The transport block is scrambled with a length-31 Gold sequence. There are $2^{31}$ sequences as such and they are not cyclic shifts of each other. The scrambling sequence generator is reinitialized for each transport block and the initialization value is a function of the PCI of the cell and the RNTI of the UE. This randomizes interference among cells and among UEs.
6. *Modulation:* The information bits are modulated with QPSK, 16-QAM, or 64-QAM. PDCCH and PBCH use QPSK.
7. *Mapping to transmission layers:* Multiple streams of data are generated for mapping to different transmit antenna ports identified by transmission layers.
8. *Precoding:* Each transmission layer is precoded with a certain matrix.
9. *Mapping to resource elements:* The symbols are mapped to resource elements in the resource blocks. It uses frequency-first mapping (i.e., the symbols are mapped by incrementing the subcarrier index first for all the allocated resource blocks in a symbol period and thereafter, the next symbol period is taken).
10. *OFDM modulation:* The symbols are fed to IFFT for OFDM modulation as shown in Section 3.1.2.

## 3.4 SC-FDMA for Uplink

OFDMA offers high peak to average power ratio (PAPR). The high PAPR requires linear power amplifiers with a large dynamic range. This makes the power amplifier less efficient, more expensive, and more power consuming. This is a more severe problem in the uplink compared to the downlink because it increases the cost of the UE and consumes the battery power faster. The high PAPR also degrades coverage and cell-edge performance. Therefore, in order to reduce PAPR, single-carrier FDMA (SC-FDMA) is used in the uplink instead of OFDMA. The time dispersive or, equivalently, the frequency-selective radio channel distorts the SC-FDMA signal. Thus, an equalizer is needed to compensate for the radio channel frequency selectivity. This is the disadvantage of the use of SC-FDMA in place of OFDMA. However, the equalization is not very difficult to perform at the base station because the base station can accommodate more power and cost.

SC-FDMA allows the use of a number of subcarriers and exactly the same kind of time-frequency resource grid in the uplink as used in downlink with OFDMA in downlink. This commonality between uplink and downlink helps in implementation and resource allocation.

## 3.4 SC-FDMA for Uplink

SC-FDMA signal generation is performed in the frequency domain using DFT-spread OFDM (DFTS-OFDM). DFTS-OFDM, as compared to time domain SC-FDMA generation, gives better bandwidth efficiency at the cost of slightly less PAPR reduction. In the case of OFDMA, in every block, each subcarrier contains information about just one transmit symbol from the block. But here in case of DFT-spread OFDM, each subcarrier has information of all transmit symbols in the block as the transmit symbols were first spread by the DFT transform over the available subcarriers. Thus, if there is a deep fade on one of the subcarriers, it contains only a part of the transmit symbol information on that subcarrier and the symbol may still be recovered after IDFT at the other end.

### 3.4.1 SC-FDMA Transmission

The SC-FDMA transmission is explained next and it is schematically shown in Figure 3.5.

1. A particular number, $M$ modulation symbols from the data stream, are treated as a block. During each OFDM symbol interval, these $M$ modulation symbols are transmitted in parallel. A serial to parallel (S/P) converter takes the data stream into $M$ parallel streams.
2. The $M$ parallel data streams are modulated using an appropriate level of modulation (e.g., QPSK, 16-QAM, or 64-QAM). The level of modulation can vary among the streams.
3. The symbols in the time domain from $M$ parallel streams are converted to the frequency domain using size-$M$ discrete Fourier transform (DFT).
4. The $M$ outputs of the DFT are mapped on subcarriers. The total number of subcarriers available is $N$ where $N > M$. Zeros are applied to the unused subcarriers. N is an integer power of 2 and it can be 128, 256, 512, 1,024, 1,536, or 2,048 depending on the channel bandwidth.
5. Data from $N$ subcarriers are applied to size-NIDFT. Since $N$ is an integer power of 2, implementation efficient inverse fast Fourier transform (IFFT) processing can be used. This use of DFTS-OFDM can be regarded as

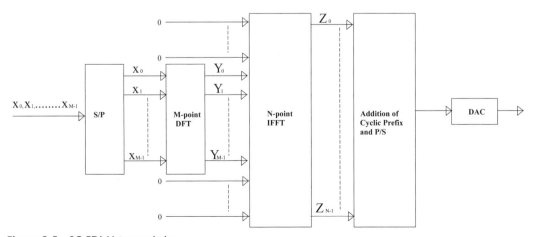

**Figure 3.5** SC-FDMA transmission.

OFDM modulation preceded by a DFT operation (i.e., precoded OFDM). If the DFT size $M$ was equal to the IDFT size $N$, the cascaded DFT and IDFT blocks would completely cancel out each other. Here $N > M$ and the remaining inputs to the IDFT are set to zeros. This gives the output of the IDFT single-carrier properties.
6. Similar to OFDM, a cyclic prefix is inserted for each block.
7. The output of the IFFT is P/S converted.
8. The output is transmitted after D/A conversion.

The nominal bandwidth of the transmitted signal will be, $M/N \times fs$ assuming a sampling rate, $fs$, at the output of the IDFT. Thus, by varying the block size, $M$, the instantaneous bandwidth of the transmitted signal can be varied; this allows flexible-bandwidth assignment. Besides, by shifting the mapping between IDFT inputs and DFT outputs, the transmitted signal can be shifted in the frequency domain.

### 3.4.2 SC-FDMA Reception

The SC-FDMA reception is explained next and it is schematically shown in Figure 3.6.

1. A/D conversion process samples the received signal.
2. An S/P converter takes the received data samples into $N$ parallel streams.
3. The cyclic prefix is removed.
4. FFT processing with FFT size $N$ is performed. Only the $M$ number of outputs of the FFT is useful here and the rest of the outputs are left unused.
5. The time dispersive or, equivalently, the frequency selective channel distorts the DFTS-OFDM signal and an equalizer is needed for the compensation of the frequency selectivity. A time domain equalizer would experience very long channel impulse response and require a large tap size for time domain filter because of broadband multipath channel. Therefore the equalization is rather performed in the frequency domain instead of time domain and this reduces the complexity. This is called frequency domain equalization (FDE). It requires the implementation of the frequency-domain filter after the FFT operation of M modulation symbols of the block.

    Since the cyclic prefix is allowed and the cyclic prefix is chosen longer than the channel response length, it causes the linear convolution of the time dispersive channel to appear as a circular convolution. When FFT is used, frequency domain multiplication is equivalent to time domain circular convolution. Thus, the frequency-domain filter taps can be easily determined. This low-complexity frequency-domain equalization is an additional advantage of the use of cyclic prefix in uplink compared to the advantages of cyclic prefix in downlink.
6. The parallel $M$ modulation symbols of the block carry out size-$M$ inverse DFT (IDFT) processing.
7. The parallel $M$ modulation symbols of the block are P/S converted.

## 3.5 Uplink Transmission Features

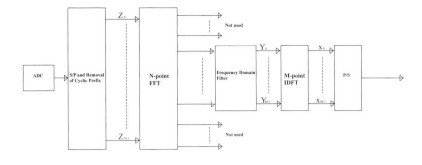

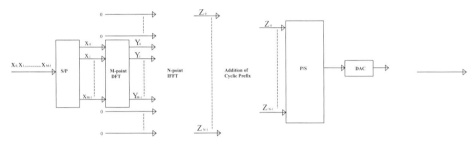

**Figure 3.6** SC-FDMA reception.

## 3.5 Uplink Transmission Features

The salient features of uplink transmission are shown here.

- The uplink uses the same type of time-frequency resource grid as used in downlink; this helps in implementation and resource allocation.
- The subcarrier spacing is always 15 kHz and there is no use of 7.5-kHz subcarrier spacing.
- The cyclic prefix can be normal or extended with exactly the same lengths as used in downlink for 15 kHz. The same length is used for cyclic prefix by all UEs in the cell. The radio frame has exactly the same configuration including its subframe and slot as in the downlink. The symbol period has the same configuration as used in the downlink for 15 kHz. A slot contains six or seven symbols. There is no slot configuration with three symbols.
- Resource elements are defined the same way as in downlink. Each resource element is uniquely identified by the index pair (k, l) where k and l are the indices for subcarriers in the frequency domain and symbols in the time domain, respectively.
- Resource blocks are defined the same way as in the downlink. Only the physical resource block (PRB) is defined and no virtual resource block (VRB) exists.
- The number of subcarriers in a resource block, $N_{SC}^{RB} = 12$. The number of available resource blocks in the total frequency resource is denoted as $N_{RB}^{UL}$.

The value of $N_{RB}^{UL}$ can be 6, 15, 25, 50, 75, or 100 depending on the channel bandwidth. The physical resource blocks are numbered in frequency domain as $nPRB$ and $nPRB$ ranges as follows:

$n_{PRB} = 0, 1, 2, 3, ..., N_{RB}^{UL} - 1$

- The number of subcarriers in total frequency resource is $N_{RB}^{UL} N_{SC}^{RB}$. $k$ is used as the index for subcarriers in the total frequency resource and so, $k$ ranges as follows:

$k = 0, 1, 2, 3, ..., N_{RB}^{UL} N_{SC}^{RB} - 1$

  - $n_{PRB}$ of the resource block that includes subcarrier $k$ satisfies the following relation.
  - $n_{PRB} = \text{floor}(k / N_{SC}^{RB})$

- The number of symbols in a resource block or in a slot is denoted as $Nsymb UL$. Thus,

$N_{symb}^{UL} = 7$ for normal cyclic prefix

$N_{symb}^{UL} = 6$ for extended cyclic prefix

  - $l$ is used as the index for symbols in a resource block or in a slot and so, $l$ ranges as follows,

$l = 0, 1, 2, 3, ... N_{symb}^{UL} - 1$

- A resource block consists of $N_{SC}^{RB} \times N_{symb}^{UL}$ resource elements. The composition of resource elements in a resource block occur in one of the following two ways:
  - $12 \times 7 = 84$ resource elements for normal cyclic prefix;
  - $12 \times 6 = 72$ resource elements for extended cyclic prefix.
- The scheduling decision can be modified at every subframe. A pair of resource blocks, which are contiguous in time, is used as the granularity of scheduling decision.
- There can be only one transport block per TTI as spatial multiplexing is not supported in uplink in Release 9.
- The transport block on PUSCH goes through similar processing as shown in Section 3.3 (i.e., the processing includes 24 bits long CRC insertion, segmentation into code blocks of 6,144 bits, rate 1/3 turbo coding, rate matching, code block concatenation, modulation, mapping to resource elements, and SC-FDM modulation). However, according to Release 9, there is no mapping to transmission layers as there is only one transport block per TTI. Also, there is no precoding for multiple input multiple output (MIMO).

# CHAPTER 4
# System Information

The RRC layer of the eNodeB is responsible for broadcasting system information in the cell in order to configure various necessary parameters for the UEs. The different parameters require different periodicity of transmission. This is facilitated by defining different types of system information where each type is a RRC message and contains certain parameters. The different types of system information are classified as follows.

1. Master Information Block (MIB) message.
2. System Information Block (SIB) type 1 message.
3. System Information (SI) messages. These SI messages include 12 different SIB messages, namely from SIB type 2 through SIB type 13. SIB types 12 and 13 are introduced in Release 9.

## 4.1 Mapping Among Layers

The mapping for MIB and SIB messages is shown here:

1. *MIB message:* The mapping for MIB messages among different layers is as follows.
   - The RRC layer at the eNodeB generates RRC PDU for MIB. The RRC PDU is directly mapped onto a RLC SDU without using the PDCP layer. The RLC layer applies the transparent mode (TM), so the RLC SDU is mapped straight onto a RLC PDU without any segmentation or concatenation and even without the addition of any header. The RLC layer delivers the RLC PDU to the BCCH logical channel.
   - The RLC PDU of MIB is treated as MAC SDU. The MAC layer uses transparent MAC. The MAC SDU is mapped straight to a MAC PDU without any segmentation and even without addition of any header. The MAC PDU is delivered to the BCH transport channel as a transport block. The MAC PDU includes data from only one logical channel and no multiplexing is applied. There is one to one mapping between BCCH and BCH. Thus, the whole layer 2 neither segments the RRC PDU of MIB nor adds any header to it.

- The physical layer receives the transport block of MIB, performs necessary processing, and transmits over the physical channel, PBCH. This is explained in the Section 5.1.1.
2. *SIB messages:* The mapping for SIB messages among different layers is as follows.
   - The RRC layer at the eNodeB generates RRC PDUs for SIB types from type 1 through type 13. The RRC PDUs are directly mapped onto RLC SDUs without using the PDCP layer. The RLC layer applies the transparent mode (TM), so the RLC SDUs are mapped straight to a RLC PDUs without any segmentation or concatenation and even without the addition of any header. The RLC layer delivers the RLC PDU to the BCCH logical channel.
   - RLC PDUs from different logical channels including SIB messages are multiplexed by the MAC entity into MAC PDUs. The MAC PDUs are delivered to the DL-SCH transport channel as transport blocks.
   - The physical layer receives the transport blocks, performs necessary processing, and transmits over the physical channel, PDSCH. The modulation scheme used is always QPSK. HARQ is not used.

## 4.2 Required System Information

The UE must read at least certain system information messages known as required system information in certain cases. The required system information is shown next for the RRC_IDLE state and the RRC_CONNECTED state.

- *RRC_IDLE state:* The required system information in the RRC_IDLE state includes:
  - MIB, SIB type 1 and SIB type 2.
  - Certain SIB types out of SIB types from 3 through 8 depending on the support of RAT. The requirement of SIB type 6, SIB type 7, and SIB type 8 depends on the support of UMTS, GSM, and CDMA2000, respectively.

  The UE must read the required system information if it receives notification about a change in system information or if the maximum validity duration of system information expires in the RRC_IDLE state. Also, the UE has to read the required system information before camping on the cell. The UE camps on a cell during selection of a cell after powering up or during reselection of a cell from another E-UTRA or other RAT cell in the RRC_IDLE state.

- *RRC_CONNECTED state:* The required system information in the RRC_CONNECTED state includes:
  - MIB, SIB type 1 and SIB type 2;
  - SIB type 8 if CDMA2000 is supported.

  The UE must read the required system information if it receives notification about a change in system information or if the maximum validity duration of system information expires in the RRC_CONNECTED state. Also, the UE has to read the required system information after handover from another E-UTRA or other RAT cell in RRC_CONNECTED state.

## 4.3 MIB Message

An MIB message form is contained in a BCH transport block and is transmitted as a PBCH instance. The MIB message on a PBCH is mapped on the central 72 subcarriers or six resource blocks of the system bandwidth around the DC subcarrier regardless of the size of the system bandwidth. It may be noted that the smallest system bandwidth consists of 72 subcarriers. A PBCH instance or a MIB message is repeated every 40 ms (i.e., one TTI of PBCH includes four radio frames). The PBCH instance has 14 information bits, 10 spare bits, and 16 CRC bits. After coding and rate matching, the BCH transport block is split into four parts such that the BCH transport block can be decoded from each of the four parts assuming sufficiently good channel conditions. This allows soft combining and it improves the demodulation performance. The four parts of the PBCH instance are then transmitted on the first subframe (subframe 0) of the four radio frames of every TTI. On each of these subframes, PBCH uses the first four symbols of the second slot. The CRC check allows determination of the boundary of 40-ms TTI.

The MIB message includes only the size of system bandwidth, System Frame Number (SFN), and PHICH configuration. This little information allows an MIB to be robust even at the cell edge. SFN is defined in Section 3.1.3.1. The MIB message indicates the value of SFN for radio frames. The MIB message includes the SystemFrameNumber field, which gives the eight most significant bits of the SFN. The two least significant bits of the SFN are not explicitly signaled but they are derived from the radio frame position of MIB message within its 40-ms TTI. The two least significant bits of SFN are 00 for the first radio frame, 01 for the second, 10 for the third, and 11 for the last radio frame within the 40 ms. The first radio frame of the four radio frames has an SFN multiple of 4.

## 4.4 SIB Type 1 Message

SIB type 1 is transmitted on PDSCH using a fixed time-domain schedule. SIB type 1 messages are sent one after another periodically in every eight radio frames or 80 ms. In this eight radio frames duration, the same SIB type 1 message is sent four times on subframe 5 of every other radio frame. An SIB type 1 message is first sent on the first radio frame of the eight radio frames, which has an SFN multiple of eight. Then, the same SIB type 1 message is repeated three times on alternate radio frame within the eight radio frames, which have SFN multiple of two.

The PDCCH indicates the presence of SIB type 1 and thus, provides the frequency-domain scheduling of SIB type 1. PDCCH uses DCI format 1A or DCI format 1C, to schedule one PDSCH codeword as shown in Section 5.1.3.2. The CRC appended to the PDCCH instance is scrambled by System Information RNTI (SI-RNTI). The SI-RNTI has a fixed value, FFFF in hexadecimals.

## 4.5 SI Messages

### 4.5.1 Scheduling

SIB type 1 can configure one or more SI messages. Each of the SI messages can carry one or more SIB types from SIB type 2 through SIB type 13. Each SI message is associated with a time domain SI window and only the particular SI message can be transmitted in its own SI window. The SI windows of different SI messages do not overlap. The SI windows for all SI messages have the same length. SIB type 1 includes the SI-WindowLength field, which gives the length of the SI window in milliseconds. The length can be 1, 2, 5, 10, 15, 20, or 40 ms.

SIB type 1 includes SchedulingInfoList IE, which gives the list of transmitted SI messages and schedules all these SI messages between SIB type 2 and SIB type 13. Thus, this IE gives the number of SI messages as well as the number of SI windows. One or more than one SIB types can be mapped to an SI message. Any SIB type must be included in a single SI message. The SchedulingInfoList IE includes SchedulingInfo IE for each of the SI messages. The first SI message in the list always includes SIB type 2. The SchedulingInfo IE includes the SIB-MappingInfo field for a particular SI message. The SIB-MappingInfo field gives the list of the SIB types mapped to the particular SI message. In case of the first SI message in the SchedulingInfoList IE, the SIB-MappingInfo field invariably includes SIB type 2 and so it does not explicitly mention SIB type 2. Thus, this particular SIB-MappingInfo field can include any SIB types from SIB type 3 through SIB type 13.

The SI windows are repeated at regular intervals for an SI message. The SchedulingInfo IE includes the SI periodicity field for a particular SI message and this field gives the interval of the periodic repetition in number of radio frames. The SI periodicity can be 8, 16, 32, 64, 128, 256 or 512 radio frames. Since a SIB type is included in a single SI message, only the SIB types that require the same periodicity can be mapped to the same SI message.

The SI-window for an SI message is determined as follows.

- $x$ is calculated as:

  $x$ = (order of the SI message in the list of SchedulingInfoList IE − 1) × SI-WindowLength

- The SI window starts at the radio frame with SFN satisfying:

  SFN mod (SI-Periodicity) = floor ($x$/10)

- The SI window starts at the subframe with subframe number, $x$ mod 10.
- The SI window ends after SI-WindowLength from its beginning. However, the UE can keep reading the SI message after the end of the SI window if the SI message does not finish.

The SI message can be repeated multiple times within its SI window. The SI messages can be transmitted in any subframe except MBSFN subframes and subframes used for SIB type 1. It may be noted that subframe 5 of radio frames with an SFN multiple of 2 carry SIB type 1. The SI messages are dynamically scheduled within their SI windows using PDCCH. To indicate the allocation for an SI message, the CRC appended to the PDCCH instance, is scrambled by SI-RNTI. The same SI-RNTI is used to address SIB type 1 and all SI messages. PDCCH uses DCI format 1A or DCI format 1C, which schedule one PDSCH codeword as shown in Section 5.1.3.2.

Once the UE receives any SIB type, it can apply the information of the SIB type right away. The UE does not need to wait for reception of all SI messages. While the UE is receiving any SIB type, if it detects that the SIB-MappingInfo has been updated and the SIB-MappingInfo no longer includes this SIB type, then the UE stops receiving the particular SIB type.

### 4.5.2 Modification

The modification of SIB type 10, 11, and 12 is explained in Section 26.3. In case of any change in other SIB types of SI messages, the updated SI messages begin to be transmitted in a new period, which is regarded as a new modification period. Therefore, modification periods of a fixed length are used, which appear one after another. The eNodeB first notifies the UEs about the change in the SI messages over a modification period. Thereafter, the eNodeB transmits the updated system information in the next modification period as shown in Figure 4.1. Such modification period and notification for modification allow the UE to wake up for a short period to check if there is any update in the SI messages and thus save battery power.

Once the UE receives the notification of change in the SI messages, it acquires the new SI messages from the beginning of the next modification period. The UE uses the old SI messages until it acquires the new SI messages. Regardless of any notification, the UE, after acquisition of SI messages, considers the SI messages to be valid for at most 3 hours; thereafter, the UE acquires the SI messages anew. There is an exception to this validity restriction as shown in Section 1.4.1.2.

#### 4.5.2.1 Modification Period

The SFN of the radio frames at the modification period boundaries is a multiple of the modification period. The modification period is calculated in number of radio frames as: (ModificationPeriodCoeff × DefaultPagingCycle).

The UE can identify the modification periods reading SIB type 2. SIB type 2 includes the RadioResourceConfigCommonSIB IE, as shown in Appendix A.1. This

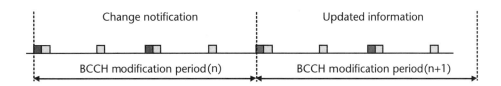

**Figure 4.1** Change of system information.

IE includes BCCH-Config IE, which contains the ModificationPeriodCoeff field. The value of ModificationPeriodCoeff can be 2, 4, 8, or 16.

The RadioResourceConfigCommonSIB IE also includes PCCH-Config IE, which contains the DefaultPagingCycle field. The DefaultPagingCycle can be 32, 64, 128, or 256 radio frames. It actually gives the length of the DRX cycle. The UE has one paging occasion (PO) in the DRX cycle.

#### 4.5.2.2 Notification of Modification

The UE receives the notification of change in the SI messages via paging message or SIB type 1 depending on the RRC state as shown here:

- RRC_IDLE state: A paging message indicates the change in the SI messages. The UE, in the RRC_IDLE state, monitors the paging message in order to detect the change in the SI messages.
- RRC_CONNECTED state: A paging message as well as SIB type 1 indicates the change in the SI messages. The UE, in the RRC_CONNECTED state, monitors the paging message and/or SIB type 1 in order to detect the change in the SI messages.

The notifications are sent as follows.

*Notification on Paging Message*

The eNodeB sends a paging message to indicate a change in the SI messages. If the SystemInfoModification field is present in the paging message, then it indicates a change in the SI messages at the next modification period.

The paging message does not inform which SIB type is going to change or any other details as such. In case of SIB types 10, 11, and 12, the paging message uses the ETWS-Indication field or the CMAS-Indication field instead of the SystemInfoModification field. This is explained in Section 26.3. Also, the eNodeB does not send any paging message for certain modifications, for example, change in the following fields of SIB type 8.

- SystemTimeInfo field, which gives CDMA2000 system time;
- LongCodeState1XRTT field, which gives the state of long code generation registers in CDMA2000.

*Notification on SIB Type 1*

SIB type 1 includes the SystemInfoValueTag field, which indicates if there is any change in the SI messages at the next modification period or not. The value of SystemInfoValueTag is incremented every time there is any change in the SI messages. The value of SystemInfoValueTag ranges from 0 to 31. The SystemInfoValueTag field does not inform which SIB type is going to change or any other details as such. The eNodeB does not update SystemInfoValueTag in SIB type 1 for changes in the following cases.

- SIB types 10, 11, and 12;
- SystemTimeInfo field in SIB type 8;
- LongCodeState1XRTT field in SIB type 8.

The UE may use SystemInfoValueTag to check if its stored SI messages are still valid after it returns from out of coverage. If the current value of SystemInfoValueTag is different from the value corresponding to the stored SI messages, then the UE acquires the SI messages anew.

# CHAPTER 5
# Physical Channels and Signals

## 5.1 Downlink Physical Channels and Signals

The different downlink physical channels and physical signals are multiplexed in time and frequency. The allocation of control channel instances involves the following terms.

- *Control region:* Each subframe uses the first few OFDM symbols for transmission of control information; this region is often referred to as the control region. The Physical Control Format Indicator Channel (PCFICH), Physical Downlink Control Channel (PDCCH), and Physical Hybrid ARQ Indicator Channel (PHICH) are transmitted only in the control region and no data is transmitted in the control region. The control region does not have a fixed size and it takes the first 1, 2, or 3 OFDM symbols in each subframe if $N_{RB}^{DL}$ > 10 (i.e., if the system bandwidth has more than 10 resource blocks in frequency). And the control region takes the first two, three, or four OFDM symbols in each subframe if the system bandwidth does not have more than 10 resource blocks in frequency (i.e., if $N_{RB}^{DL} = 6$). Thus, the size of the control region can vary from subframe to subframe. Such variation allows the use of resources for control purposes only as much as necessary. Since the beginning of the subframe is dedicated as the control region and the rest is dedicated for other purposes, the UE may sleep and save its power in the rest of the subframe provided the control region does not schedule any activities. This justifies concentrating the control signaling within a short period in time.
- *Resource Element Group (REG):* An REG consists of four physical resource elements to be used as control channel resources. The REG may cover a region of six resource elements in the time- frequency resource grid including two resource elements used for downlink reference signals. Nevertheless, the REG always includes four resource elements for control channel resources.
- *Control Channel Element (CCE):* A CCE consists of nine REGs and so, it consists of 4 × 9 = 36 resource elements. The control region consists of a set of CCEs, numbered from 0 to $N_{CCE,k} - 1$ where $N_{CCE,k}$ is the total number of CCEs in the control region of subframe $k$.

After allocation of cell-specific reference signals, PCFICH and PHICH, the rest of the control region can be used for PDCCH. Thus, the availability of resources for PDCCH depends on the number of symbols in the control region. However, typically, a part of the control region is left unused (i.e., some CCEs are not allocated). This allows the average interference to be less severe in the neighborhood. It may be noted that the control region applies intercell interference randomization among neighboring cells.

The downlink control physical channels and downlink reference signals are explained in the following sections. The synchronization signals are explained in Section 10.1.

### 5.1.1 Physical Broadcast Channel (PBCH)

The Physical Broadcast Channel (PBCH) carries only Master Information Block (MIB) messages and no other information. The modulation scheme is always QPSK for PBCH. The PBCH information bits are coded and rate matched. The PBCH information bits are then scrambled with a cell-specific scrambling sequence in order to avoid any confusion with PBCH transmissions from neighboring cells. The mapping of PBCH on subcarriers and the scheduling of PBCH instances in time are explained in Section 4.3. The UE attempts to decode PBCH blindly for all antenna diversities (i.e., one, two, or four antenna elements). Thus, the UE finds out the number of transmit antennas from PBCH transmission.

### 5.1.2 Physical Control Format Indicator Channel (PCFICH)

Only one PCFICH instance is transmitted in a subframe in a cell. PCFICH indicates the actual number of OFDM symbols for the control region in the subframe. Since the UE does not have prior information about the size of the control region, PCFICH is transmitted on the first symbol of every subframe. A PCFICH instance carries a Control Format Indicator (CFI) field. The CFI contains a 32-bit-long code word, which represents 1, 2, or 3 as shown in Table 5.1. CFI 4 is reserved for future use. The 32-bit-long code word for representation of only these few variations makes the CFI robust enough for por radio link quality.

Table 5.2 shows the possible number of OFDM symbols for control region in the beginning of every subframe and how the actual number is indicated by CFI. MBSFN subframes are defined in Section 24.3. The control region for MBSFN subframes is not more than two OFDM symbols to allow the UE to wake up for a very short period and save more power.

**Table 5.1** Code Words for CFI

| Control Format Indicator (CFI) | 32-Bit-Long CFI Code Word |
|---|---|
| 1 | 0,1,1,0,1,1,0,1,1,0,1,1,0,1,1,0,1,1,0,1,1,0,1,1,0,1,1,0,1,1,0,1 |
| 2 | 1,0,1,1,0,1,1,0,1,1,0,1,1,0,1,1,0,1,1,0,1,1,0,1,1,0,1,1,0,1,1,0 |
| 3 | 1,1,0,1,1,0,1,1,0,1,1,0,1,1,0,1,1,0,1,1,0,1,1,0,1,1,0,1,1,0,1,1 |

**Table 5.2** Possible Number of OFDM Symbols for Control Region in the Beginning of Every Subframe

| Case | The Possible Number of OFDM Symbols for Control Channel Information in the Beginning of Every Subframe | |
|---|---|---|
| | $N_{RB}^{DL} > 10$ | $N_{RB}^{DL} \leq 10$ |
| | The value of CFI gives the actual number of OFDM symbols. | The value of CFI + 1 gives the actual number of OFDM symbols. |
| MBSFN subframes are configured with one or two cell-specific antenna ports | 1 or 2 | 2 |
| MBSFN subframes are configured with four cell-specific antenna ports | 2 | 2 |
| Non-MBSFN subframes are configured with positioning reference signals | 1, 2, or 3 | 2 or 3 |
| Other cases | 1, 2, or 3 | 2, 3, or 4 |

PCFICH uses (32, 2) block code resulting in a 1/16 coding rate. PCFICH always uses QPSK as the modulation scheme. Since each PCFICH instance is 32 bits long, the PCFICH instance requires 16 symbols; it occupies 16 resource elements or four REGs. A PCFICH instance is divided into four parts and each part is mapped to one REG. The four REGs are allocated at subcarrier numbers, $k$, ($k + \frac{1}{4} \times N_{RB}^{DL} N_{SC}^{RB}$), ($k + \frac{1}{2} \times N_{RB}^{DL} N_{SC}^{RB}$), and ($k + \frac{3}{4} \times N_{RB}^{DL} N_{SC}^{RB}$) where $N_{RB}^{DL} N_{SC}^{RB}$ represents the total number of subcarriers in the system bandwidth. Thus, the PCFICH instance is spread within the whole bandwidth providing for maximum frequency diversity. The value of k depends on the physical cell identifier (PCI) of the cell. The PCI is explained in Section 10.1.2. Thus, a cell-specific frequency offset is applied to the position of the PCFICH resource elements and this provides for intercell interference randomization among neighboring cells. Also, the PCFICH information bits are scrambled and the initialization of the scrambling sequence depends on the PCI of the cell for intercell interference randomization.

### 5.1.3 Physical Downlink Control Channel (PDCCH)

The physical downlink control channel (PDCCH) is used to provide different types of control information. A PDCCH instance contains a message known as Downlink Control Information (DCI) and the DCI carries the control information for a UE or a group of UEs. The control information can be as follows.

1. Downlink resource scheduling;
2. Uplink resource grant;
3. Uplink power control instruction;
4. Indication for paging or system information;

5. Assignment of dedicated preamble signature for noncontention-based random access;
6. Notification of change in MCCH.

Several PDCCH instances can be transmitted in a subframe while one of them may be intended for a particular UE. The PDCCH information bits are coded using tail biting convolutional coding (TBCC). The rate-matching algorithm is used to generate different coding rates and the coding rate can depend on the radio link quality as indicated by CQI report. The PDCCH information bits are then scrambled and the initialization of the scrambling sequence depends on the PCI of the cell for intercell interference randomization. Then modulation is performed and PDCCH always uses QPSK as the modulation scheme. Thereafter, interleaving is applied to the symbol blocks to provide frequency diversity. Finally, a cell-specific cyclic shift is made to allow cell-specific mapping of PDCCH onto resource elements for intercell interference randomization.

#### 5.1.3.1 Identification of PDCCH Instance

In order to allow a UE to identify which PDCCH instances are addressed to it, each PDCCH instance is coded with a MAC ID. The MAC ID may address multiple UEs simultaneously. For example, a TPC command may address multiple users. The MAC ID is a Radio Network Temporary Identifier (RNTI). A 16-bit CRC is appended to each PDCCH instance for error detection. This CRC is scrambled by the RNTI. So, a UE checks on the CRC to identify its PDCCH instance. There are several types of RNTIs. A particular type of RNTI is selected depending on which logical channel type is indicated by the PDCCH instance and the specific purpose of the PDCCH instance. The different types of RNTIs and the indicated logical channel types are shown in Table 5.3. The values of different RNTIs are shown in Section 1.6.3.

#### 5.1.3.2 DCI Formats

Several DCI message format types are defined with different sizes. This allows the DCI length to be only as much as required for the particular scenario. The different

**Table 5.3** Different RNTI Types and the Indicated Logical Channel Types

| Logical Channel Type | Type of RNTI |
|---|---|
| DCCH and DTCH | C-RNTI, Temporary C-RNTI and SPS C-RNTI |
| CCCH during random access | Temporary C-RNTI |
| PCCH | P-RNTI |
| Random Access Response on DL-SCH | RA-RNTI |
| BCCH | SI-RNTI |
| MCCH | M-RNTI |

DCI formats are explained next. The downlink resource allocation method can be of type 0, type 1, or type 2 as explained in Section 19.2.1. When a DCI format is used for downlink resource allocation, it uses a particular type of resource allocation method.

*DCI Format 0*

The DCI format 0 is used to convey the resource grants on PUSCH for the uplink transmission of the UE. If the UE supports antenna selection for uplink transmissions, the requested antenna may be indicated using DCI format 0 by applying an antenna-specific mask to the CRC. This allows the use of a DCI message of the same size whether antenna selection is used or not. The DCI format 0 includes the following information.

- *Flag to differentiate between format 0 and format 1A (1 bit):* This flag takes on value 0 indicating DCI format 0. Value 1 of this flag indicates DCI format 1A.
- *Hopping flag (1 bit):* This flag is explained in Section 19.2.2.
- *Resource block A assignment and hopping resource allocation:* This field is explained in Section 19.2.2.
- *Modulation and coding scheme (5 bits):* It gives the modulation and coding scheme (MCS) index, $I_{MCS}$. The possible value of $I_{MCS}$ ranges between 0 and 28 for every new transmission. Each of these values corresponds to a modulation order (QPSK, 16-QAM or 64-QAM) and a Transport Block Size (TBS) index, $I_{TBS}$, for the data on PUSCH according to a specified mapping table. $I_{TBS}$ corresponds to the transport block size depending on the number of allocated resource blocks according to another specified mapping table. The code rate can be derived from the modulation order and the transport block size. The value of $I_{MCS}$ is set to 29, 30, or 31 to indicate the Redundancy Version (RV), RV1, RV2, or RV3 for HARQ retransmissions, respectively. In these cases, the modulation order and $I_{TBS}$ are not indicated as they are known.
- *New data indicator (NDI) (1 bit):* This field is explained in Section 20.3.3.
- TPC command for scheduled PUSCH (2 bits): It gives the TPC command field for calculation of transmit power on PUSCH as explained in Section 17.2.2.3.
- *Cyclic shift for demodulation reference signals (3 bits):* Twelve equally spaced cyclic time shifts are defined for demodulation RS (DM RS) allowing the multiplexing of 12 UEs. This field specifies a value, which is used in the calculation of the cyclic shift to be used by the particular UE.
- *CQI request (1 bit):* If it is set to 1 in subframe n, then the UE performs aperiodic CQI/PMI/RI reporting using PUSCH in subframe n + 4.

*DCI Format 1*

The DCI format 1 allocates single PDSCH code word using resource allocation type 0 or type 1. The DCI format 1 includes the following information.

- Resource Allocation Header (1 bit): This field is explained in Section 19.2.1.1.
- *Resource block assignment:* It gives resource allocation information as explained in Section 19.2.1.1.
- *Modulation and coding scheme (5 bits):* It gives the modulation and coding scheme (MCS) index, $I_{MCS}$. The value of $I_{MCS}$ ranges between 0 and 28. Each of these values corresponds to a modulation order (QPSK, 16-QAM, or 64-QAM) and a Transport Block Size (TBS) index, $I_{TBS}$, for the data on PDSCH according to a specified mapping table. $I_{TBS}$ corresponds to the transport block size depending on the number of allocated resource blocks according to another specified mapping table. The code rate can be derived from the modulation order and the transport block size.
- *HARQ process identifier (HARQ ID) (3 bits):* It gives the HARQ process identifier (HARQ ID), which uniquely identifies the associated HARQ process.
- *New data indicator (NDI) (1 bit):* This field is explained in Section 20.3.3.
- *Redundancy Version (RV) (2 bits):* This field is explained in Section 20.3.3.
- *TPC Command for PUCCH (2 bits):* It gives the TPC command field for calculation of transmit power on PUCCH as explained in Section 17.3.2.3.

*DCI Format 1A*

The DCI format 1A is used for two different purposes as shown here.

1. *Compact scheduling of one PDSCH code word:* The DCI format 1A can be used for compact scheduling of one PDSCH code word using resource allocation type 2. It assigns localized or distributed virtual resource blocks. The DCI format 1A carries the following information.
   - *Flag to differentiate between format 0 and format 1A (1 bit):* This flag takes on value 1 indicating DCI format 1A. Value 0 of this flag indicates DCI format 0.
   - *Resource block assignment:* It carries resource allocation information as explained in Section 19.2.1.2.
   - *Localized/distributed VRB assignment flag (1 bit):* It indicates whether localized virtual resource blocks or distributed virtual resource blocks are assigned for resource allocation type 2. Value 0 indicates localized and value 1 indicates distributed VRB assignment.
   - *Modulation and coding scheme, HARQ process identifier (HARQ ID), new data indicator (NDI), redundancy version (RV), and TPC command for PUCCH*: These fields are explained above for DCI format 1.
2. *Assignment of a dedicated preamble signature:* The DCI format 1A can be used as a PDCCH order in case of noncontention-based random access for initiation of downlink data transfer. The PDCCH order assigns a dedicated

random access preamble to the UE. The CRC is scrambled with C-RNTI. The following fields in DCI format 1A have settings as shown below.
- *Preamble index (6 bits):* It is explained in Section 11.3.1.1.
- *PRACH mask index (4 bits):* It is explained in Section 11.3.1.1.
- *Resource block assignment:* This field does not indicate any resource assignment. All bits of this field are set to 1.
- *Localized/distributed VRB assignment flag (1 bit):* This field does not correspond to any resource assignment. It is set to 0.

3. All the remaining bits of DCI format 1A are set to zeroes here. Those bits are only used when DCI format 1A is used for compact scheduling of one PDSCH code word.

*DCI Format 1B*

The DCI format 1B is used for compact scheduling of one PDSCH code word using closed loop precoding with rank 1 transmission, which is regarded as transmission mode 6. It uses resource allocation type 2 and it assigns localized or distributed virtual resource blocks. The DCI format 1B carries the following information.

- *Resource block assignment:* It carries resource allocation information as explained in Section 19.2.1.2.
- *Localized/distributed VRB assignment flag (1 bit):* This field is explained above for DCI format 1A.
- *Modulation and coding scheme, HARQ process identifier (HARQ ID), new data indicator (NDI), redundancy version (RV), and TPC command for PUCCH:* These fields are explained above for DCI format 1.
- *TPMI information for precoding:* (The number of bits is 2 or 4 for two or four antenna ports at eNodeB, respectively.) This field is explained in Section 18.2.1.1.
- *PMI confirmation for precoding (1 bit):* This field is explained in Section 18.2.1.1.

*DCI Format 1C*

The DCI format 1C is very compact and the smallest among the DCI formats. It is used for two different purposes as shown here:

1. *Scheduling of one PDSCH code word:* The DCI format 1C allocates one PDSCH code word. It may be used to indicate random access responses, paging messages and system information messages. The indicated PDSCH transmission always uses QPSK modulation qand it does not use HARQ. It uses resource allocation type 2 and it assigns only distributed virtual resource blocks and not localized ones. The DCI format 1C carries the following information:
    - *Resource block assignment:* It carries resource allocation information as explained in Section 19.2.1.2 using only distributed virtual resource blocks.

- *Modulation and coding scheme (5 bits):* It indicates modulation order and transport block as explained above for DCI format 1. The indicated modulation scheme is always QPSK.

2. *Notification of change in MCCH:* The DCI format 1C can be used to indicate a change in MCCH and this use is introduced in Release 9. DCI format 1C includes 8-bit information. The 8-bit information works as a bitmap and each of these bits indicate if there is any change in MCCH for an MBSFN area. Value 1 of the bit indicates change in MCCH and value 0 indicates no change. This is explained in Section 24.5.3.

*DCI Format 1D*

The DCI format 1D is used for compact scheduling of one PDSCH code word using Multiple Users MIMO (MU-MIMO), which is regarded as transmission mode 5. It uses resource allocation type 2 and it assigns localized or distributed virtual resource blocks. The DCI format 1D carries the following information.

- *Resource block assignment:* It carries resource allocation information as explained in Section 19.2.1.2.
- *Localized/distributed VRB assignment flag (1 bit):* This field is explained above for DCI format 1A.
- *Modulation and coding scheme, HARQ process identifier (HARQ ID), new data indicator (NDI), redundancy version (RV), and TPC command for PUCCH:* These fields are explained above for DCI format 1.
- *TPMI information for precoding:* (The number of bits is 2 or 4 for two or four antenna ports at eNodeB, respectively.) This field indicates which codebook index is used from the two-antenna codebook or four-antenna codebook as explained in Section 18.2.2.
- *Downlink power offset (1 bit):* This indicates whether a power offset is applied to the data symbols as explained in Section 18.2.2.

*DCI Format 2*

The DCI format 2 is used for scheduling PDSCH to UEs configured in closed-loop spatial multiplexing mode, which is regarded as transmission mode 4. It uses resource allocation type 0 or type 1. The DCI format 2 carries the following information.

- *Resource allocation header (1 bit):* This field is explained in Section 19.2.1.1.
- *Resource block assignment:* It gives resource allocation information as explained in Section 19.2.1.1.
- *Modulation and coding scheme, HARQ process identifier (HARQ ID), new data indicator (NDI), redundancy version (RV), and TPC command for PUCCH:* These fields are explained above for DCI format 1. Separate $I_{MCS}$, NDI, and RV fields are included for each of the transport blocks if two transport blocks are scheduled.

- *Transport block to code word swap flag (1 bit):* This field is explained in Section 18.2.1.
- *Precoding information:* (The number of bits is 3 or 6 for two or four antenna ports at eNodeB, respectively.) This field is explained in Section 18.2.1.1.

*DCI Format 2A*

The DCI format 2A is used for scheduling PDSCH to UEs configured in open-loop spatial multiplexing mode, which is regarded as transmission mode 3. It uses resource allocation type 0 or type 1. The DCI format 2A carries the following information.

- *Resource allocation header (1 bit):* This field is explained in Section 19.2.1.1.
- *Resource block assignment:* It gives resource allocation information as explained in Section 19.2.1.1.
- *Modulation and coding scheme, HARQ process identifier (HARQ ID), new data indicator (NDI), redundancy version (RV), and TPC command for PUCCH:* These fields are explained above for DCI format 1. Separate $I_{MCS}$, NDI, and RV fields are included for each of the transport blocks if two transport blocks scheduled.
- *Transport block to code word swap flag (1 bit):* This field is explained in Section 18.2.1.
- *Precoding information:* (The number of bits is 0 or 2 for two or four antenna ports at eNodeB, respectively.) This field is explained in Section 18.2.1.2.

*DCI Format 2B*

DCI format 2B is introduced in Release 9. It is used for dual layer transmission using transmission mode 8.

*DCI Format 2C*

DCI format 2C is introduced in Release 10. It is used for transmission using transmission mode 9.

*DCI Format 3*

The DCI format 3 is used for uplink power control. It contains TPC commands for PUCCH and PUSCH indicating 2-bit power adjustments for multiple users as explained in Chapter 17.

*DCI Format 3A*

The DCI format 3A is used for uplink power control. It contains TPC commands for PUCCH and PUSCH indicating single bit power adjustments for multiple users as explained in Chapter 17.

Please add the following:

*DCI Format 4*

The DCI format 4 is used to convey the resource grants on PUSCH for the uplink transmission of the UE using closed-loop spatial multiplexing. It is introduced in Release 10.

### 5.1.3.3  Allocation of PDCCH

Each PDCCH instance is transmitted using one or more control channel elements (CCEs). Since the modulation scheme is QPSK for PDCCH, each REG contains four symbols or 8 bits and each CCE contains 72 bits. For mapping onto the REG, the symbols are divided into groups consisting of 4 symbols. Each group is called symbol quadruplet and it is mapped on one REG. The aggregation level of a PDCCH instance specifies if it is constructed from the concatenation of 1, 2, 4, or 8 CCEs. Thus, the possible PDCCH positions can be every CCE position, every second CCE position, every fourth CCE position, and every eighth CCE position for aggregation level, 1, 2, 4, and 8, respectively. The four aggregation levels give rise to four types of PDCCH formats.

The number of CCEs for a PDCCH instance is determined by the eNodeB based on the channel condition. If the radio link is poor, possibly because the user is at the cell edge, then a higher aggregation level (i.e., a larger number of CCEs) can be used to achieve stronger coding while sending a PDCCH instance. Conversely, the same PDCCH instance can be sent with a lower aggregation level if the radio link is good. Moreover, the power level may be adjusted for PDCCH based on the channel condition.

The UE searches for PDCCH candidates in every non-DRX subframe. The UE needs to check for a possible PDCCH instance with CRC scrambled by the particular RNTI. The PDCCH instance has a particular combination of PDCCH format and DCI format but it is not known to the UE. This requires blind decoding for the UE. The number of blind decoding attempts has been restricted to 44 for the sake of simplicity of the UE. In order to help reduce the blind decoding, a limited set of locations are defined where a PDCCH instance may be placed. These locations are called search spaces. The search spaces allow fewer blind decoding attempts and ensure that the number of combinations of PDCCH format and DCI format is not more than 44. There are two kinds of search spaces as shown below.

1. *Common search space:* A common search space can be configured for any UE in the cell and all UEs are informed about the common search space. The common search space is typically used for signaling common to the UEs; it can also be used to schedule resources dedicated for the UE if the UE-specific search space is blocked by other UEs. The common search space is used by only DCI format 0, 1A, 1C, 3, and 3A and they use only the aggregation levels, 4 and 8. The UE monitors one common search space at each of the aggregation levels, 4 and 8. The UE always monitors the common search space and, thus, the UE always decodes DCI format 0, 1A, 1C, 3, and 3A. The common search space does not vary with time.
2. *UE-specific search space:* A UE-specific search space is configured for a particular UE and the UE performs a search in this search space in addition to the common search space. The UE-specific search space is used to schedule

resources dedicated for the UE. Thus, the UE-specific search space is similar with HS-SCCH in HSDPA. Only DCI format 1, 1B, 1D, 2, and 2A use a UE-specific search space. The current transmission mode of PDSCH depicts which DCI format needs to be decoded from 1, 1B, 1D, 2, and 2A. The transmission modes are semistatically configured as shown in Chapter 18. The UE monitors one UE-specific search space at each of the aggregation levels, 1, 2, 4, and 8 (i.e., there is one search space at aggregation level 1, one at aggregation level 2, one at aggregation level 4, and one at aggregation level 8 for the UE). The starting offset of PDCCH allocations depends on the C-RNTI of the UE. The UE-specific search space keeps varying over time and among UEs. The UE-specific search space may overlap for a particular UE with the common search space.

DCI format 1C has the smallest payload size. The DCI format 0, 1A, 3, and 3A have equal size and it is next larger than DCI format 1C. DCI formats 0 and 1A may require different numbers of bits for indication of resource assignments, but the smaller format is extended by appending zero bits to make its size the same as the larger format. The equal size requires a single attempt and, thus, helps in blind decoding of the UE. Thereafter, in general, DCI formats 1B, 1D, 1, 2A, and 2 have sequentially increasing payload size. The number of bits required for resource assignment depends on the system bandwidth and therefore the sizes of DCI formats also vary with the system bandwidth. The UE needs to look for particular DCI formats in search spaces according to Table 5.4.

Each PDCCH format has a particular size and scope of search as shown in Table 5.5.

PCFICH and PHICH groups are first mapped onto REGs. Then PDCCH instances are mapped by mapping onto REGs first in time and then in frequency (i.e., REGs are mapped on the available symbol periods one after another and then the subcarrier index is incremented). This avoids imbalance in power requirement among symbol periods.

### 5.1.4 Physical Hybrid ARQ Indicator Channel (PHICH)

The PHICH carries the HARQ ACK/NACK indicating if a transport block on PUSCH has been correctly received or not. The HARQ indicator is 1 bit and it is set to 0 for ACK and 1 for NACK. A PHICH group is formed by including multiple PHICH instances that are mapped onto the same set of resource elements. The PHICH instances in a PHICH group are intended for separate UEs. The different PHICH instances in a PHICH group are code multiplexed and separated through different orthogonal sequences shown in Table 5.6 (i.e., a particular PHICH instance is identified by its group number and its associated orthogonal sequence). The PHICH instances in the PHICH group are multiplied with different complex orthogonal spreading sequences and a cell-specific scrambling sequence. The sequence length is four for normal cyclic prefix and two for extended cyclic prefix. Therefore, the maximum number of PHICH instances in a PHICH group is eight for normal cyclic prefix and four for extended cyclic prefix. The code multiplexing allows the use of large dynamic range for power adjustment on PHICH. The PHICH power level can be increased enough to support cell edge users without

**Table 5.4** Search Spaces for Various Purposes

| Purpose | RNTI to Scramble CRC | DCI Format | Search Space |
|---|---|---|---|
| Indication for system information | SI-RNTI | DCI format 1A or 1C | Common |
| Indication for paging | P-RNTI | DCI format 1A or 1C | Common |
| Indication for random access response | RA-RNTI | DCI format 1A or 1C | Common |
| | | DCI format 1A | Common or UE specific |
| | C-RNTI | DCI format 1, 1B, 1D, 2 or 2A | UE specific |
| Indication for dynamic downlink allocation | | DCI format 1A | Common or UE specific |
| | Temporary C-RNTI | DCI format 1 | UE specific |
| Indication for semipersistent downlink allocation | | DCI format 1A | Common or UE specific |
| | SPS C-RNTI | DCI format 1, 2, or 2A | UE specific |
| | C-RNTI | DCI format 0 | Common or UE specific |
| Indication for dynamic uplink allocation | Temporary C-RNTI | DCI format 0 | Common |
| Indication for semipersistent uplink allocation | SPS C-RNTI | DCI format 0 | Common or UE specific |
| TPC command | TPC-PUSCH-RNTI or TPC-PUCCH-RNTI | DCI format 3 or 3A | Common |

**Table 5.5** PDCCH Format Types and Their Search Spaces

| PDCCH Format | Number of CCEs | Number of REGs | Number of bits | Common Search Space | | UE-Specific Search Space | |
|---|---|---|---|---|---|---|---|
| | | | | Size in Number of CCEs | Number of PDCCH Candidates | Size in Number of CCEs | Number of PDCCH Candidates |
| 0 | 1 | 9 | 72 | — | — | 6 | 6/1 = 6 |
| 1 | 2 | 18 | 144 | — | — | 12 | 12/2 = 6 |
| 2 | 4 | 36 | 288 | 16 | 16/4 = 4 | 8 | 8/4 = 2 |
| 3 | 8 | 72 | 576 | 16 | 16/8 = 2 | 16 | 16/8 = 2 |

causing much power variation among subcarriers. The initialization of the scrambling sequence depends on the PCI allowing intercell interference randomization.

The HARQ indicator uses repetition coding with factor 3 before spreading with orthogonal spreading sequences. So, the HARQ indicator takes 3 × 4 = 12 bits after spreading for a normal cyclic prefix. These bits are BPSK modulated and mapped on 3 REGs, that is, a PHICH group requires 3 REGs or 12 resource elements. The HARQ indicators from at most 7 other UEs are multiplexed on the same PHICH group. The position of the first REG in frequency domain depends on

**Table 5.6** Orthogonal Sequence for PHICH

| Sequence Index | Orthogonal Sequence for Normal Cyclic Prefix | Orthogonal Sequence for Extended Cyclic Prefix |
|---|---|---|
| 0 | [+1 +1 +1 +1] | [+1 +1] |
| 1 | [+1 −1 +1 −1] | [+1 −1] |
| 2 | [+1 +1 −1 −1] | [+j +j] |
| 3 | [+1 −1 −1 +1] | [+j −j] |
| 4 | [+j +j +j +j] | — |
| 5 | [+j −j +j −j] | — |
| 6 | [+j +j −j −j] | — |
| 7 | [+j −j −j +j] | — |

PCI allowing intercell interference randomization. The other two REGs are located with equal spacing and this spacing is large enough to achieve frequency diversity.

The eNodeB configures the transmission of PHICH using PHICH-Config IE. The PHICH-Config IE is included in the broadcast message, MIB. Also, the source eNodeB sends the PHICH-Config IE to the UE for use at the target eNodeB when handover takes place and Appendix A.3 shows how the PHICH-Config IE is transmitted.

PHICH is transmitted within the control region in the subframe. It is typically transmitted only on the first symbol of the subframe. However, if the radio link is very poor then PHICH is extended to a number of symbols for robustness. This extension uses first three symbols for non-MBSFN subframes and of course, in this case, the control region must also be three OFDM symbols long in the subframe. The extension uses first two symbols for MBSFN subframes. The PHICH-Config IE contains the PHICH-Duration field, which indicates whether the extension has been applied or not for PHICH. So, the value of this field can be normal or extended. If extended is indicated, the three OFDM symbols are used and the indication on PCFICH is ignored.

The number of PHICH groups is constant in all subframes. The PHICH-Config IE includes the PHICH-Resource field, which gives the value of $N_g$. The number of PHICH groups is calculated using $N_g$ as follows.

CEIL [$1/8 \times N_g \times N_{RB}^{DL}$] for the normal cyclic prefix

$2 \times$ CEIL [$1/8 \times N_g \times N_{RB}^{DL}$] for the extended cyclic prefix

$N_{RB}^{DL}$ is the total number of resource blocks.

The PHICH-Resource field, $N_g$, can have values, 1/6, 1/2, 1, or 2.

The PHICH allocation can be derived from the resource block allocation for which it is carrying ACK/NACK. As shown in Section 20.3.6, the uplink HARQ has a fixed gap of 4 TTI between uplink transmission and ACK/NACK on PHICH (i.e., if the uplink transmission occurs on subframe n, its PHICH instance is sent on subframe n + 4). The UE derives the PHICH group number and the orthogonal sequence using the following information.

1. The lowest resource block number of the corresponding uplink resource allocation;
2. The spreading factor size;
3. The cyclic shift used by the UE in transmission of DM RS;
4. The number of PHICH groups configured.

### 5.1.5 Reference Signals

A number of reference signals are defined which transmit known symbols, primarily to allow the UE to estimate the downlink propagation channel and use coherent detector. The antenna port refers to the transmission of different reference signals from antennas as received by the UE. An antenna port may not always represent a physical antenna element and it is rather associated with the transmission of a reference signal. The designated antenna ports different reference signals are shown below. In fact, the antenna ports for cell-specific reference signals represent actual physical antenna elements and the antenna ports for other reference signals include logical ports. However, all antenna ports used for a particular type of reference signal represent separate physical antenna elements.

- *Cell-specific reference signals:* The eNodeB uses one, two, or four layer transmission for cell-specific reference signals using the same number of antenna elements. If the eNodeB transmits using single antenna, then antenna port, p = 0 is considered to be used. If the eNodeB uses two layer transmission, then antenna ports, p = 0 and 1 are considered to be used. If there is four layer transmission, possibly using a 4 × 4 antenna configuration, then antenna ports, p = 0, 1, 2, and 3 are considered to be used.
- *MBSFN reference signal:* The MBSFN reference signal uses antenna port, p = 4 for transmission.
- *UE-specific reference signal:* The UE-specific reference signal uses antenna port, p = 5, for transmission in Release 8. It also uses port, p = 7 and 8 in Release 9. It additionally uses port, p = 9, 10, 11, ..., 14 in Release 10.
- *Positioning reference signal:* The positioning reference signal uses antenna port, p = 6 for transmission.
- *Channel-state information reference signal:* The channel-state information reference signal uses port p = 15 if it is transmitted on one antenna port. It uses port p = 15 and 16 if it is transmitted on two antenna ports. It uses port p = 15, 16, 17, and 18 if it is transmitted on four antenna ports. It uses port p = 15, 16, ..., 22 if it is transmitted on eight antenna ports.

### 5.1.5.1 Cell-Specific Reference Signal

The cell-specific reference signal is formed by inserting known symbols into the downlink time-frequency resource grid. The functionality of cell-specific reference signal is similar to the Common Pilot Channel (CPICH) in UMTS. The cell-specific reference signal serves the following purposes.

1. The UE estimates the propagation channel based on the cell-specific reference signals and the estimate allows the use of a coherent detector. The coherent detector provides similar performance with much simpler implementation as compared to a noncoherent detector.
2. The UE measures downlink channel quality based on the cell-specific reference signal and determines CQI/PMI/RI to be reported based on this measurement.
3. The UE calculates RSRP and RSRQ based on the cell-specific reference signal. RSRP and RSRQ are used as metrics to rank different LTE candidate cells.

When handover takes place, the source eNodeB informs the UE about the number of antenna ports for a cell specific reference signal at the target eNodeB by sending AntennaInfoCommon IE. Appendix A.3 shows how the source eNodeB sends the AntennaInfoCommon IE to the UE. The AntennaInfoCommon IE includes the AntennaPortsCount field, which specifies the number of antenna ports and its value can be 1, 2, or 4.

The cell-specific reference signal sequence uses a complex pseudo-random sequence generated as length 31 gold sequence. The second m sequence used to generate the gold sequence, uses a seed that is a function of the symbol number, slot number, cyclic prefix length, and physical cell identifier (PCI) of the cell. The PCI is explained in Section 10.1.2. Thus, the cell-specific reference signal sequence is cell specific and it is unique in a neighborhood.

The mapping of modulation symbols onto the resource elements in a resource block depends on the number of antenna ports. The mapping is shown in Figure 5.1 for a normal cyclic prefix and in Figure 5.2 for extended cyclic prefix. The notation RP is used to denote a resource element used for reference signal transmission on antenna port, p. For transmission of cell-specific reference signal on port, p = 0, each slot has reference symbols located on symbol 0 and symbol 4 for normal cyclic prefix and on symbol 0 and symbol 3 for an extended cyclic prefix. For each symbol time, reference symbols are available once in every six subcarriers. Between the two symbol periods in a slot, the reference symbols are staggered every three subcarriers. The availability of a symbol in every 45 kHz allows good channel estimation under maximum expected delay spread. In time domain, two symbols are available in every 0.5 ms, which allows good channel estimation under the maximum expected Doppler spread. The equidistant position of the symbols in the grid allows minimum mean squared error channel estimation. When two or four antenna ports are used, the resource elements used for transmission of the reference signal on one antenna port is left unused on other antenna ports. This avoids interantenna interference between reference signals.

The particular subcarriers carrying the reference symbols are shifted among cells in the neighborhood. Since symbols are available once in every six subcarriers,

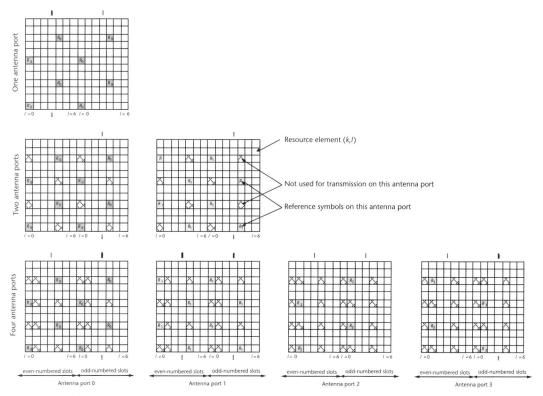

**Figure 5.1** Mapping of cell-specific reference signals for normal CP.

every six adjacent cells use distinct frequency shift. The frequency shift depends on the PCI of the cell (i.e., the frequency shift can be derived from the PCI). Thus, the frequencies of reference signal are not repeated in neighboring cells. This allows the use of relatively higher power level for reference signals and provides for better channel estimation. Also, when the UE measures signal quality of a neighboring cell, the signal is not interfered by another cell.

When the cell-specific reference signal uses transmission on port $p = 1$, this transmission is just 3 subcarriers shifted from the transmission on port $p = 0$. The four-layer transmission with a 4 × 4 antenna can typically be used when there is low Doppler spread. A less frequent transmission of reference symbols can be sufficient with low Doppler spread, so transmission on port p = 2 and 3 uses one reference symbol per slot instead of two. When more than one antenna port is used, the reference symbols are repeated every three subcarriers; then, there are three effective frequency shifts available instead of six.

The downlink cell-specific reference signals can apply frequency hopping. The frequency-hopping pattern has a period of one radio frame. Each frequency-hopping pattern corresponds to one cell identity group.

### 5.1.5.2 UE-Specific Reference Signal

The UE-specific reference signal is transmitted only within resource blocks allocated to the particular UE (i.e., the UE-specific reference symbols are embedded in the resource blocks within allocations for PDSCH). The UE-specific reference symbols

## 5.1 Downlink Physical Channels and Signals

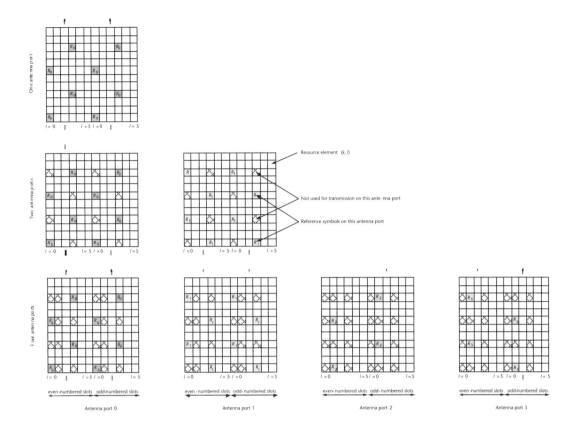

**Figure 5.2** Mapping of cell-specific reference signals for extended CP.

are allocated in the resource elements where cell-specific reference symbols are not transmitted (i.e., the cell-specific reference signal and its associated features are not affected). For each symbol time, the UE-specific reference symbols are available once in every four subcarriers, as shown in Figure 5.3.. Thus, it is more frequent than cell-specific reference symbols in frequency. This allows the UE to have a better estimate of the channel response, which is required for the particular application.

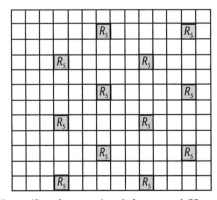

**Figure 5.3** Mapping of UE-specific reference signals for normal CP.

The ratio between PDSCH EPRE and UE-specific RS EPRE is maintained at a constant value in each OFDM symbol of a resource block. The UE-specific reference symbols are located in a similar way to the cell-specific reference symbols so that the UE can apply similar channel estimation algorithms. The UE finds the transmission of a specific reference signal from a distinct antenna port in addition to a cell-specific reference signal. In Release 9, the UE-specific reference signal is extended for dual-layer transmission and it can use two additional antenna ports.

When the UE-specific reference signal is transmitted, the UE uses it to estimate the downlink propagation channel for coherent demodulation of PDSCH transmission in the associated resource blocks. The UE-specific reference signal is typically used when PDSCH transmission applies beamforming in the direction of a particular UE. Then the PDSCH transmission may use a correlated array of physical antenna elements for beamforming towards the particular UE. This PDSCH transmission has a distinct channel response, which the cell-specific reference signal cannot represent. Therefore, the eNodeB beamforms and transmits the UE-specific reference signal in the same way as the PDSCH transmission occurs. This allows the UE to coherently demodulate PDSCH.

#### 5.1.5.3 MBSFN Reference Signal

As shown in Chapter 24, in case of Multimedia Broadcast Multicast Service (MBMS), the same data content is simultaneously transmitted on Physical Multicast Channel (PMCH) from all eNodeBs in the MBSFN area and the UE combines them. The eNodeBs also transmit time-synchronized MBSFN reference signals to allow the UE to estimate propagation channel and use coherent detector for PMCH transmissions. Thus, the eNodeBs transmit the MBSFN reference signal only when MBMS is functional (i.e., there is transmission of PMCH).

The MBSFN reference signal is defined only for the extended cyclic prefix when subcarrier spacing is 15 kHz. In addition, the MBSFN reference signal is defined for a 7.5-kHz subcarrier spacing.

The MBSFN reference signals consist of known reference symbols and fixed mapping. For 15-kHz subcarrier spacing, the symbols of MBSFN reference signal are mapped onto every other subcarrier in the third, seventh, and eleventh OFDM symbol of the subframe as shown in Figure 5.4(a). For the reduced subcarrier spacing, 7.5 kHz, the symbols of the MBSFN reference signal are mapped one in every four subcarriers in the second, fourth, and sixth OFDM symbol of the subframe as shown in Figure 5.4(b). The MBSFN reference signal uses a pseudo-random sequence mapped on the symbols. The initialization of the pseudo-random sequence generator is a function of the ID of the respective MBSFN area. This ID is broadcast using MBSFN-AreaId field on SIB type 13 as shown in Section 24.4.

#### 5.1.5.4 Positioning Reference Signal

The Positioning Reference Signal (PRS) is introduced in Release 9. The eNodeB transmits the PRS to assist UE positioning using OTDOA. The UE measures the timing of the positioning reference signal received from multiple eNodeBs and observes their time differences. Then the UE position is computed based on these time differences and the geographical coordinates of the respective eNodeBs. The

## 5.1 Downlink Physical Channels and Signals

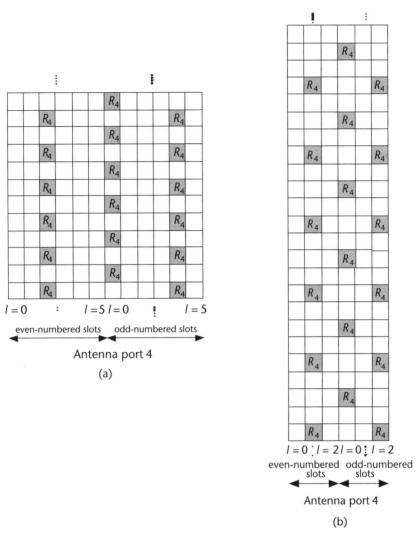

**Figure 5.4** (a) Mapping of MBSFN reference signals for 15-kHz subcarrier spacing. (b) Mapping of MBSFN reference signals for 7.5-kHz subcarrier spacing.

positioning reference signal can be transmitted only with subcarrier spacing, 15 kHz. The resource symbol mapping for PRS is shown in Figure 5.5(a) for the normal cyclic prefix and in Figure 5.5(b) for the extended cyclic prefix.

The positioning reference signal is transmitted over an $N_{PRS}$ number of consecutive subframes. This transmission over $N_{PRS}$ subframes is repeated in every $T_{PRS}$ number of subframes. Also, $\Delta_{PRS}$ number of subframes can be assigned as offset to the period of every $T_{PRS}$ subframes. The first subframe of the $N_{PRS}$ subframes satisfies the following relation.

$$(10 \times \text{SFN} + \text{floor}(Ns/2) - \Delta_{PRS}) \bmod T_{PRS} = 0$$

Here $Ns$ is the slot number in the radio frame.

A location server is defined in the network for the purpose of UE positioning as shown in Section 1.5.1. The location server sends ProvideAssistanceData message

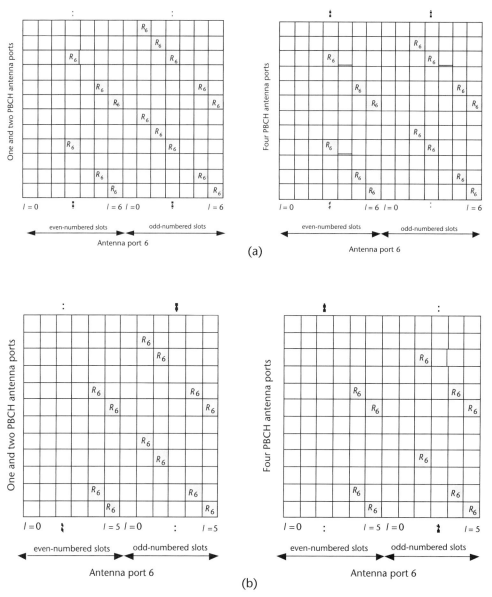

**Figure 5.5** (a) Mapping of positioning reference signals for the normal cyclic prefix. (b) Mapping of positioning reference signals for the extended cyclic prefix.

to the UE, which includes PRS-Info IE and, thus, informs the configuration of the positioning reference signal in the cell. The PRS-Info IE includes the following fields.

- *NumDL-Frames:* The NumDL-Frames field specifies the value of $N_{PRS}$. It can define 1, 2, 4, or 6 consecutive subframes.
- *PRS-ConfigurationIndex:* The PRS-ConfigurationIndex field gives the value of PRS configuration index $I_{PRS}$. The values of $T_{PRS}$ and $\Delta_{PRS}$ are determined from $I_{PRS}$ as shown in Table 5.7.

**Table 5.7** Evaluating $T_{PRS}$ and $D_{PRS}$

| $I_{PRS}$ | $T_{PRS}$ | $\Delta_{PRS}$ |
|---|---|---|
| 0–159 | 160 | $I_{PRS}$ |
| 160–479 | 320 | $I_{PRS} - 160$ |
| 480–1,119 | 640 | $I_{PRS} - 480$ |
| 1,120–2,399 | 1,280 | $I_{PRS} - 1,120$ |

#### 5.1.5.5 Channel-State Information (CSI) Reference Signals

This channel-state information (CSI) reference signal is introduced in Release 10 in order to support transmission mode 9. The CSI reference signal essentially provides an enhancement to the functionality of cell-specific reference signals.

## 5.2 Uplink Physical Channels and Signals

The different uplink physical channels and physical signals are time and frequency multiplexed. The UE sends its control signaling to the eNodeB using a message piece called the Uplink Control Information (UCI). The UCI is sent on physical uplink control channel (PUCCH) if there is no uplink data transfer on the physical uplink shared channel (PUSCH) in the subframe. Conversely, if there is an uplink data transfer and appropriate resources are available on PUSCH for UCI in the subframe, then the UCI is also sent on PUSCH to maintain the single carrier properties. Thus, the UE uses only the subframes for PUCCH in which the UE has not been allocated any resource blocks for PUSCH transmission (i.e., the UE does not transmit PUCCH and PUSCH simultaneously). The UCI carries the following information.

- *Downlink data transfer related:* This includes Channel Quality Indicator (CQI), Precoding Matrix Index (PMI), Rank Indicator (RI), and HARQ ACK/NACK. A CQI instance includes CQI, PMI, and RI reports.
- *Uplink data transfer related:* This includes scheduling request (SR). Besides, there are certain information (e.g., BSR and PHR) that can be carried by only PUSCH and not PUCCH.

### 5.2.1 Uplink Reference Signals

There are two types of uplink physical signals as follows.

1. Demodulation reference signal (DM RS);
2. Sounding reference signal (SRS).

The reference signal sequence for both DM RS and SRS is constructed from a frequency-domain Zadoff-Chu (ZC) sequence. The ZC sequences have the following desirable properties.

- The ZC sequences have ideal cyclic autocorrelation. The cyclic shifted versions of the same sequence are orthogonal to each other provided that the propagation delay plus the delay spread is smaller than the time domain equivalent of the cyclic shift.
- The ZC sequence has constant amplitude. This allows for equal excitation of all the allocated subcarriers for unbiased channel estimates. A small variation in frequency allows simple channel estimation. The constant amplitude property of ZC sequence allows low cubic metric (CM) in the time domain. CM is the metric of the reduction in power capability or power derating of the power amplifier.

The ZC sequence is generated by a cyclic shift, $\alpha$, of a base sequence. The base sequence is given by,

$$r^{\alpha}(n) = e^{j\alpha n} x_q(n\, Mod\, N) \qquad n = 0,1,2,\cdots, M_{SC}^{RS}$$

Here, $M_{SC}^{RS}$ is the length of the reference signal sequence and $M_{SC}^{RS}$ is equal to the number of assigned subcarriers. Thus, if m number of resource blocks are allocated, then the length of the reference signal sequence, $M_{SC}^{RS} = mN_{SC}^{RB}$. The maximum value of $M$ can be $N_{RB}^{UL} N_{SC}^{RB}$.

$x_q$ is the $q$th root Zadoff-Chu (ZC) sequence of length $N$ and given by

$$x_q(m) = e^{-j\pi qm(m+1)/N} \qquad m = 0,1,2,\cdots, N-1$$

$N$ is made the largest prime number such that $N < M_{SC}^{RS}$. For $N$ equal to a prime number, there is maximum number of basic ZC sequences, which enhances flexibility in design. The ZC sequence is cyclically extended to the number of assigned subcarriers, $M_{SC}^{RS}$. The cyclic extension preserves the constant amplitude in the frequency domain as well as the zero autocorrelation. The cyclic extension of the ZC sequence provides better CM compared to truncation.

A number of reference signal sequences are generated from the same base sequence using different values of cyclic shift, $\alpha$. This phase rotation in the frequency domain is equivalent to cyclic shift in time domain. The ZC sequences have ideal cyclic autocorrelation meaning that the correlation of a ZC sequence with any cyclic shift of the same sequence is zero. The reference signal sequences used in the cell for both DM RS and SRS are derived from different values of cyclic shift of the base sequences.

### 5.2.1.1 Demodulation Reference Signal (DM RS)

DeModulation Reference Signals (DM RS) uses the transmission of known symbols and are primarily used for enabling the eNodeB to estimate the propagation channel, and, thus, to aid in coherent demodulation. DM RS is used with transmission of PUSCH and PUCCH. One or two symbols in every uplink slot carry DM RS and the rest of the symbols in the slot carry either data from PUSCH or control signaling from PUCCH. For PUSCH transmission, the DM RS uses the fourth symbol in the slot for normal cyclic prefix and the third symbol in the slot for the extended cyclic prefix as shown in Figure 5.6. For PUCCH transmission, the position and number of DM RS symbols in a slot depends on the PUCCH format.

The DM RS spans the same bandwidth as the allocated uplink data. For PUSCH, resource blocks are always allocated as multiples of two, three, or five resource blocks. Thus, the DM RS sequence length, $M_{SC}^{RS}$, also takes on values multiples of 24, 36, and 60.

The base sequences are divided into 30 groups known as sequence-shift patterns. For each of the 30 sequence-shift patterns, base sequences are defined for each possible RB allocation size. If the number of resource blocks is less than or equal to 5, there is one base sequence defined for any number of resource blocks. Thus, there is one sequence of length one resource block, another one sequence of length two resource blocks, and so on. If the number of resource blocks is more than 5, then more ZC sequences are available because of longer sequence and there are two base sequences individually defined for any number of resource blocks. Thus, there are two sequences of length 6 resource blocks, two sequences of length 8 resource blocks, and so on. All UEs in the same cell use base sequences from the same group. This provides for orthogonality within the cell. The cells in an area can be divided into clusters of 30 cells so that all cells in one cluster use different groups. When DM RS is used with PUCCH, the group number is derived from the PCI of the cell. When DM RS is used with PUSCH, the eNodeB configures the grouping using UL-ReferenceSignalsPUSCH IE. This IE is included in PUSCH-ConfigCommon IE. Appendix A.1 shows how the eNodeB broadcasts PUSCH-ConfigCommon IE on SIB type 2. Appendix A.3 shows how the source eNodeB sends PUSCH-Config-Common IE to the UE for use at the target eNodeBwhen handover takes place. The

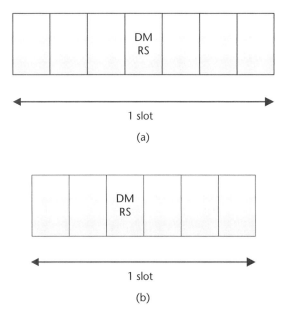

**Figure 5.6** (a) DM RS position for normal cyclic prefix. (b) DM RS position for extended cyclic prefix.

UL-ReferenceSignalsPUSCH IE includes GroupAssignmentPUSCH field, which has a value between 0 and 29 indicating the group. The sequence-shift pattern is determined from both the value of GroupAssignmentPUSCH field and PCI of the cell.

It is also possible to configure hopping among the groups for intercell interference randomization. Then the UE hops to different groups from slot to slot in a pseudo-random manner. The hopping pattern has a length equal to a radio frame with different sequence groups in each slot. The UL-ReferenceSignalsPUSCH IE includes the GroupHoppingEnabled field, which indicates if hopping is activated or not. There are 17 different group hopping patterns because there are 504 PCIs and 30 base sequences. The group hopping pattern is the same for PUSCH and PUCCH but the sequence-shift patterns may be different between PUSCH and PUCCH.

Multiple UEs may transmit on the same resource blocks using reference signal sequences derived through different values of cyclic shift from the base sequence. Thus, intracell orthogonality among the rotated sequences is attained, which provides for multiple users MIMO (MU-MIMO). Also, between neighboring cells, the same group may be used with different cyclic shifts providing for intercell orthogonality. 12 equally spaced cyclic time shifts of base sequences are defined for DM RS. So, the cyclic shift for a particular UE is $\alpha = 2\pi n_{CS}/12$. Since, the symbol period without cyclic prefix is, 66.67 $\mu$s, each UE takes up 66.67/12 = 5.55 $\mu$s for transmission of its DM RS in the slot. Thus, it allows delay spread up to 5.55 $\mu$s. The UL-ReferenceSignalsPUSCH IE includes CyclicShift field, which is used to calculate $n_{CS}$. $n_{CS}$ also depends on the slot number allowing different cyclic shifts in different slots for a particular UE.

### 5.2.1.2 Sounding Reference Signal (SRS)

The Sounding Reference Signal (SRS) is used for various purposes:

- Propagation channel estimation by the eNodeB in order to aid frequency-selective scheduling in the uplink. This is the primary purpose of SRS.
- Uplink power control.
- Timing estimation.
- Direction-of-arrival estimation to support downlink beamforming.
- Initial MCS selection.

The UE sends SRS in different parts of the bandwidths where no uplink data transmission is available. Within its bandwidth, the SRS is transmitted on every alternate subcarrier with a comb-like structure. It uses either even or odd comb offset. The SRS is transmitted on the last symbol of the subframe. The SRS uses the same sequences in the cell as the DM RS does.

The eNodeB sends SoundingRS-UL-ConfigIE in order to specify the SRS configuration and it includes SoundingRS-UL-ConfigCommonIE and SoundingRS-UL-ConfigDedicatedIE. The SoundingRS-UL-ConfigCommonIE part of the SoundingRsUl-ConfigIE is used to specify the common SRS configuration. Appendix A.1 shows how the eNodeB broadcasts SoundingRS-UL-ConfigCommonIE on SIB type 2. Appendix A.3 shows how the source eNodeB sends SoundingRS-UL-ConfigCommonIE to the UE for use at the target eNodeB when handover takes place. On the other hand, the SoundingRS-UL-ConfigDedicatedIE part of the SoundingRsUl-ConfigIE is used to specify UE specific SRS configuration. Appendix A.2 shows how the eNodeB sends SoundingRS-UL-ConfigDedicatedIE to the UE while setting up or modifying radio bearers.

The SoundingRS-UL-ConfigCommonIE includes Simultaneous-AN-and-SRS IE, which depicts if the UE might transmit ACK/NACK on PUCCH and SRS in one subframe. The SoundingRS-UL-ConfigCommonIE and SoundingRS-UL-ConfigDedicatedIE provide bandwidth of the SRS to be used. A wide bandwidth for SRS causes lower power availability per resource block. On the other hand, limiting SRS to a small bandwidth makes it difficult for the eNodeB to make good use of SRS and find an optimal scheduling solution.

Multiple UEs may transmit SRS using the same resource blocks and same comb offset but using different cyclic shifts providing for orthogonality. Up to eight UEs can be multiplexed as such. The cyclic shift, $\alpha$, is given as, $\alpha = 2\pi \cdot n_{SRS}/8$. The SoundingRS-UL-ConfigDedicatedIE includes the CyclicShift field, which assigns the value of $n_{SRS}$ to be used by the UE. The value can be 0, 1, 2, 3, 4, 5, 6, or 7.

The SoundingRS-UL-ConfigDedicatedIE also includes a 1-bit duration field to configure the UE for an individual SRS transmission or for periodic transmission of SRS. When periodic SRS transmissions are configured, the SRS-ConfigIndex field included SoundingRS-UL-ConfigDedicatedIE indicates the periodicity. The periodicity can be 2, 5, 10, 20, 40, 80, 160, or 320 ms. A more frequent transmission of SRS requires more resources, but it allows better adjustments of resource scheduling

### 5.2.2 Physical Uplink Control Channel (PUCCH)

The Physical Uplink Control channel (PUCCH) contains UCI to provide for various control signaling. There are different PUCCH formats defined so that PUCCH uses only as much resources as required for the particular scenario. The PUCCH formats are shown in Table 5.8. Release 10 adds a new format, PUCCH format 3 supporting carrier aggregation.

A UE transmits in a PUCCH region, which consists of a resource block but with its two slots located at two different sets of subcarriers. These two sets of subcarriers are located at or near the opposite edges of the system bandwidth providing maximum frequency diversity. The PUCCH regions take up more and more resource blocks at the two edges of the bandwidth as the required number of PUCCH instances increase. It leaves the whole middle part of the bandwidth for the allocation of PUSCH. The exact position and number of DM RS symbols in a slot depends on the PUCCH format.

A PUCCH region (Figure 5.7) is only 12 subcarriers wide and so the UE may have out of band (OOB) emissions in a narrow frequency region. Since, the PUCCH regions are located at the edges of the system bandwidth, it ensures small OOB emissions beyond the system bandwidth.

When there is no downlink resource allocated on PDCCH, it can be recognized as discontinuous transmission (DTX). Then the UE sends its feedback the same way as it does for NACK. This allows the DTX indication to trigger a downlink retransmission, if required, just like a NACK does.

**Table 5.8** PUCCH Format Types

| PUCCH Format | Uplink Control Information (UCI) | Modulation Scheme | Bits per Subframe | Slot Configuration |
|---|---|---|---|---|
| Format 1 | SR | N/A | N/A | No specific configuration |
| Format 1a | 1 bit HARQ ACK/NACK with/without SR | BPSK | 1 | Normal CP with three DM RS symbols in a slot or extended CP with two DM RS symbols in a slot |
| Format 1b | 2-bit HARQ ACK/NACK with/without SR | QPSK | 2 | Normal CP with three DM RS symbols in a slot or extended CP with two DM RS symbols in a slot |
| Format 2 | CQI/PMI or RI | QPSK | 20 | Normal CP with two DM RS symbols in a slot |
| Format 2 | CQI/PMI or RI and 1- or 2-bit HARQ ACK/NACK | QPSK | 20 | Extended CP with one DM RS symbol in a slot |
| Format 2a | CQI/PMI or RI and 1-bit HARQ ACK/NACK | QPSK + BPSK | 21 | Normal CP with two DM RS symbols in a slot |
| Format 2b | CQI/PMI or RI and 2-bit HARQ ACK/NACK | QPSK + QPSK | 22 | Normal CP with two DM RS symbols in a slot |

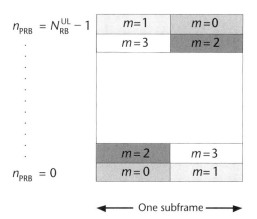

**Figure 5.7** PUCCH region.

The PUCCH-ConfigCommonIE is used to specify the common PUCCH configuration. Appendix A.1 shows how the eNodeB broadcasts PUCCH-ConfigCommonIE on SIB type 2. Appendix A.3 shows how the source eNodeB sends PUCCH-ConfigCommonIE to the UE for use at the target eNodeB when handover takes place. The PUCCH-ConfigDedicatedIE is used to specify UE specific PUCCH configuration. Appendix A.2 shows how the eNodeB sends PUCCH-ConfigDedicatedIE to the UE while setting up or modifying radio bearers.

### 5.2.2.1 PUCCH Formats 1, 1a, and 1b

In a pair of resource blocks for PUCCH, multiple UEs are code multiplexed to save resources. For this purpose, a particular base sequence based on Zadoff-Chu sequence is used in a cell. In a given SC-FDMA symbol, different cyclic time shifts of the base sequence are modulated with the signaling information symbol prior to SC-FDM modulation. As the resource block spans 12 subcarriers, it uses a base sequence of length 12 and supports up to 12 equally spaced cyclic shifts. The cyclic time shifts allow 12 different UEs to be multiplexed on the same resource block. The cyclic shift depends on the PCI of the cell allowing intercell interference randomization. Since, the users in the same cell use the same base sequence, they are orthogonal to each other. However, the users in the neighboring cells are nonorthogonal as they use different sequences. Thus, the PUCCH capacity is typically limited by intercell interference. In order to further increase the capacity on PUCCH, the symbols with PUCCH information are multiplied by the Walsh sequence of length 4 known as orthogonal covers. This allows multiple UEs to use the same phase rotation by using different orthogonal cover sequences. This may result in 12 × 3 = 36 combinations for the resource block and a resource element uses one the combinations. The number of the combinations can be reduced by time-dispersion of the channel or intercell interference. The particular combination assigned to a resource element, is identified by a resource index that refers to one of the 12 phase rotations in the frequency domain and one of the cover sequences in the time domain. A cell specific mapping of the resource indices on resource elements is used for intercell interference randomization. Also, the mapping is varied between the two slots in the subframe in order to increase the randomization.

A UE sends the scheduling request (SR) using PUCCH format 1 in order to seek uplink resources allocation. PUCCH format 1 does not contain any explicit bit and the presence or absence of its transmission indicates if uplink resources are sought or not. The SchedulingRequestConfigIE provides the resource index, $n^{(1)}_{PUCCH,SRI}$, and SR configuration index, $I_{SR}$, which indicate the PUCCH region for SR. Appendix A.2 shows how the eNodeB sends SchedulingRequestConfigIE to the UE while setting up or modifying radio bearers.

Formats 1a and 1b transmit only HARQ ACK/NACK. They use either normal CP with three DM RS symbols in the middle of a slot or extended CP with two DM RS symbols in the middle of a slot. In both cases, they have four SC-FDMA symbols in a slot for ACK/NACK transmission. Since there is no CQI/PMI/RI information, more DM RS symbols are used in a slot to improve the channel estimation accuracy for coherent demodulation.

Format 1a sends HARQ ACK/NACK for single code word downlink transmission. It transmits one ACK/NACK bit using BPSK modulation on each of the four SC-FDMA symbols in a slot.

Format 1b sends HARQ ACK/NACK for two code words downlink transmission, which may be the case for spatial multiplexing. It transmits two ACK/NACK bits using QPSK modulation on each of the four SC-FDMA symbols in a slot.

The eNodeB provides delta-PUCCH-shift, $\Delta^{PUCCH}_{Shift}$, and the resource index, $n^{(1)}_{PUCCH}$ via PUCCH-ConfigCommon IE. The resource index, $n(1)_{PUCCH}$, and delta-PUCCH-shift, $\Delta^{PUCCH}_{Shift}$, specify both the PUCCH region and the cyclic time shift to be used.

### 5.2.2.3 PUCCH Formats 2, 2a, and 2b

The PUCCH formats 2, 2a, and 2b send a CQI instance for a UE. The CQI instance is allocated 10 CQI information bits to include CQI/PMI/RI reports. The 10 bits are coded with rate 1/2 giving 20 coded bits. These 20 bits are then scrambled with a length 31 Gold sequence. Thereafter, they are QPSK modulated for transmission on symbol 10 SC-FDMA symbols. As the resource block spans 12 subcarriers against one subframe, it can include CQI/PMI/RI information bits from 12 UEs.

Format 2 transmits 20 bits CQI/PMI/RI information for a normal CP. It transmits HARQ ACK/NACK bits along with CQI/PMI/RI only for an extended CP. The 1- or 2-bit HARQ ACK/NACK is jointly encoded with the CQI/PMI/RI using Reed-Muller-based block code resulting in a 20-bit instance. The 20 coded bits are QPSK modulated for transmission on 10 SC-FDMA symbols. Format 2 uses normal CP with first and fifth symbols in the slot for DM RS transmission. It also uses extended CP with only a third symbol in the slot for DM RS transmission. In both cases, a slot has five symbols except the DM RS slot. Thus two slots or one subframe on one subcarrier accommodate all information bits from the UE.

Format 2a sends one ACK/NACK bit and a CQI instance for single code word downlink transmission. Format 2b sends ACK/NACK bit and CQI instance for two code words downlink transmission, which may be the case for spatial multiplexing. It transmits one ACK/NACK bit and one CQI instance for each of the two code words.

Formats 2a and 2b use only normal CP with two DM RS symbols in a slot. They transmit HARQ ACK/NACK bits along with CQI/PMI/RI. The 20 bits for CQI/PMI/RI are QPSK modulated for transmission on 10 SC-FDMA symbols. A slot has five symbols except the DM RS slots. Thus, two slots or one subframe on one subcarrier accommodate all CQI/PMI/RI information bits from the UE. For format 2a, the second DM RS symbol in the slot is BPSK modulated to indicate HARQ ACK/NACK. For format 2b, the second DM RS symbol in the slot is QPSK modulated to indicate HARQ ACK/NACK for both code words.

The eNodeB provides simultaneous-AN-and-CQI, the resource index, $n^{(2)}_{PUCCH}$, via CQI-ReportConfig IE. The resource index, $n^{(2)}_{PUCCH}$, specifies both the PUCCH region and the cyclic time shift to be used. The Simultaneous-AN-and-CQI field determines if a UE can transmit a combination of CQI and HARQ-ACK on PUCCH in the same subframe.

### 5.2.3 Physical Random Access Channel (PRACH)

The UE sends random access preambles on the physical random access channel (PRACH) as mentioned in Chapter 11. This is the only nonsynchronized transmission of the UE in LTE. The UE has already synchronized to the downlink signal before transmitting on PRACH. The downlink and uplink propagation delays leave timing uncertainty for transmission on PRACH. PRACH does not use frequency hopping. PRACH transmissions may use time Switched Transmit diversity (TSTD). TSTD allows transmit antenna switching in the time domain.

While PRACH carries both RACH preamble and RACH message in UMTS, the PRACH instance includes preamble but no user data or signaling in LTE. The PRACH instance is composed of one or two preamble sequences, a cyclic prefix, and a guard period. The length of a preamble sequence, $T_{SEQ}$, is 800 μs. The preamble sequence may be repeated with a view to enable the eNodeB to decode the preamble when the radio link is poor. In this case, since the preamble sequence is sent twice, $T_{SEQ}$ is 2 × 800 μs = 1,600 μs. Nothing is transmitted during the guard period. The guard period, $T_{GT}$, needs to be enough long since the PRACH transmission is not synchronized. The PRACH instance from a nearby user will reach the eNodeB earlier within the subframe than a user located far away from the eNodeB. Thus, the guard period is not explicitly specified, but the location of PRACH transmission in the subframe provides the guard period. The length of cyclic prefix, $T_{CP}$, attempts to accommodate timing uncertainty, especially for cell edge users. A longer cyclic prefix can support the tolerance for a larger cell size.

Four different preamble formats are defined to adapt with various scenarios as shown in Table 5.9. They are, namely, preamble formats 0, 1, 2, and 3 as shown in Figure 5.8. Preamble format 0 is expected to be used most commonly. Preamble formats 1, 2, and 3 are expected to support a large cell size and poor radio links. Also, they can help UEs perform handover at a low SNR condition. The velocity of radio wave, $c \approx 3 \times 10^5$ km/sec. For TCP = 103.13 μs, the two-way propagation delay covers, $(103.13 \times 10^{-6} \times 3 \times 10^5)/2 = 103.13/6.7 \approx 15$ km cell radius. The DELTA_PREAMBLE is used to determine the power for initial transmission of the random access preamble as shown in Section 11.2.1.

The PRACH transmission is restricted to specific time and frequency resources. The eNodeB configures PRACH using PRACH-ConfigInfo IE. Appendix A.1 shows how the eNodeB broadcasts PRACH-ConfigInfo IE on SIB type 2. Appendix A.3 shows how the source eNodeB sends PRACH-ConfigInfo IE to the UE for use at the target eNodeB when handover takes place.

A subframe cannot contain more than one random access instance. Also, the eNodeB restricts the transmission of random access preamble within certain

**Table 5.9** Preamble Formats

| Preamble Format | $T_{CP}$ | $T_{SEQ}$ | Transmission Time | DELTA_PREAMBLE Value | Scenario |
|---|---|---|---|---|---|
| 0 | 3,168 Ts = 103.13 μs | 24,576 Ts = 800 μs | 1 ms | 0 dB | The cyclic prefix is sufficient for about 103.13/6.7 ≈ 15 km cell radius. |
| 1 | 21,024 Ts = 684.38 μs | 24,576 Ts = 800 μs | 2 ms | 0 dB | The cyclic prefix is quite long to support about 684.38/6.7 ≈ 100 km cell radius. |
| 2 | 6,240 Ts = 203.13 μs | 49,152 Ts = 2 × 800 μs | 2 ms | −3 dB | The long cyclic prefix supports about 203.13/6.7 ≈ 30 km cell radius. The preamble sequence is repeated to compensate for poor radio link. |
| 3 | 21,024 Ts = 684.38 μs | 49,152 Ts = 2 × 800 μs | 3 ms | −3 dB | The cyclic prefix is quite long to support about 684.38/6.7 ≈ 100 km cell radius. The preamble sequence is repeated to compensate for poor radio link. |

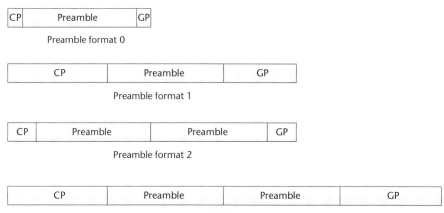

**Figure 5.8** Preamble formats.

subframes. It may be restricted to radio frames with even or odd SFN. The PRACH-ConfigInfo IE includes the PRACH-ConfigIndex field, which gives the value of PRACH-ConfigurationIndex. Its value can range between 0 and 63. The PRACH-ConfigurationIndex indicates the preamble format and subframe numbers in which the random access preamble is allowed to be transmitted as shown in Table 5.10. For example, if the PRACH-Configuration index is 12, then the preamble format 0 can be transmitted in any subframe out of the five subframes with subframe number 0, 2, 4, 6, and 8.

PRACH-ConfigIndex 0-15 indicates preamble format 0, PRACH-ConfigIndex 16-31 indicates preamble format 1, while PRACH-ConfigIndex 30 is not used. PRACH-ConfigIndex 32-47 indicates preamble format 2 while PRACH-ConfigIndex 46 is not used. PRACH-ConfigIndex 48-63 indicates preamble format 3 while PRACH-ConfigIndex 60, 61, 62 are not used. A PRACH-ConfigIndex providing more subframe chances for random access may be chosen when the channel bandwidth is larger and the cell is lighter loaded.

A PRACH instance takes up 6 resource blocks or (6 ×180 kHz) = 1.08 MHz in the frequency domain. PRACH resources are semistatically allocated within the PUSCH region in uplink. The PRACH-ConfigInfo IE includes RACH-FreqOffset field, which gives the resource block number of the first physical resource block allocated to PRACH opportunity. The value of PRACH-FrequencyOffset can vary between 0 and $(N_{RB}^{UL} - 6)$.

The subcarrier spacing for PRACH transmission is different from 15 kHz as the preamble duration is different. The subcarrier spacing is calculated from the length of the preamble sequence as follows.

$T_{SEQ}$ = 24,576 Ts = 24,576/30,720,000 sec = 800 $\mu$s

The subcarrier spacing is: $\Delta f = 1/T_{SEQ}$ = 1/ 800 $\mu$s = 1.25 kHz

Thus, 1.08 MHz/1.25 kHz = 864 subcarriers are available for PRACH transmission.

The random access preambles are generated from a frequency-domain ZC sequence similar with reference signal sequence explained in Section 5.2.1. The

## 5.2 Uplink Physical Channels and Signals

**Table 5.10** Subframe Number Based on PRACH Configuration Index

| PRACH Configuration Index | Preamble Format | SFN | Subframe Number |
|---|---|---|---|
| 0 | 0 | Even | 1 |
| 1 | 0 | Even | 4 |
| …… | …… | …… | …… |
| 12 | 0 | Any | 0, 2, 4, 6, 8 |
| 13 | 0 | Any | 1, 3, 5, 7, 9 |
| …… | …… | …… | …… |
| 17 | 1 | Even | 4 |
| 18 | 1 | Even | 7 |
| 19 | 1 | Any | 1 |
| …… | …… | …… | …… |
| 27 | 1 | Any | 3, 6, 9 |
| 28 | 1 | Any | 0, 2, 4, 6, 8 |
| 29 | 1 | Any | 1, 3, 5, 7, 9 |
| …… | …… | …… | …… |
| 43 | 2 | Any | 3, 6, 9 |
| 44 | 2 | Any | 0, 2, 4, 6, 8 |
| 45 | 2 | Any | 1, 3, 5, 7, 9 |
| 56 | 3 | Any | 3, 8 |
| 57 | 3 | Any | 1, 4, 7 |
| …… | …… | …… | …… |
| 63 | 3 | Even | 9 |

random access preambles are generated first by different cyclic shifts of the same root ZC sequence and then by different root ZC sequences (i.e., to generate the 64 preamble sequences in a cell, all possible cyclic shifts of the same root ZC sequence are used first). Then root ZC sequences with consecutive logical indexes are used to obtain the rest of the 64 preamble sequences. The cyclic shifts of the same ZC sequence have zero cross-correlation provided; the cyclic shift is less than the maximum round trip time.

The $q$th root ZC sequence is given by

$$x_q(m) = e^{-j\pi qm(m+1)/839} \qquad m = 0,1,2,\cdots,838$$

From the $q$th root ZC sequence, random access preambles are generated by cyclic shifts as follows.

$$x_{q,v}(m) = x_q[(n + C_v) \bmod N_{zc}]$$

The cyclic shift, $C_v$, is calcualted as follows. The eNodeB can restrict the use of certain cyclic shifts to avoid false correlation peaks due to frequency errors. The PRACH-ConfigInfo IE includes the HighSpeedFlag field, which indicates if unrestricted set or restricted set is to be used.

$$C_v = \begin{cases} vN_{cs} & v = 0,1,\cdots,\lfloor N_{zc}/N_{cs} \rfloor - 1, N_{cs} \neq 0 & \text{for unrestricted sets} \\ 0 & N_{cs} = 0 & \text{for unrestricted sets} \\ d_{start}\lfloor v/n_{shift}^{RA} \rfloor + (v \bmod n_{shift}^{RA})N_{cs} & v = 0,1,\cdots,n_{shift}^{RA}n_{group}^{RA} + \overline{n}_{shift}^{RA} - 1 & \text{for restricted sets} \end{cases}$$

The PRACH-ConfigInfo IE includes ZeroCorrelationZoneConfig field, which indicates the value of $N_{CS}$ as shown in Table 5.11.

The eNodeB notifies the first logical index using RACH_ROOT_SEQUENCE field. Appendix A.1 shows how the eNodeB broadcasts RACH_ROOT_SEQUENCE on SIB type 2. Appendix A.3 shows how the source eNodeB sends RACH_ROOT_SEQUENCE to the UE for use at the target eNodeB when handover takes place. The value of RACH_ROOT_SEQUENCE can be between 0 and 837. The logical indexes are cyclic (i.e., the logical index 0 appears after 837). The number of root sequences used to generate preamble sequence depends on the cell radius. If the cell is too small, then all 64 preamble sequences are generated using cyclic shifts of the same root ZC sequence.

The length of the ZC sequence, $N_{ZC}$, is made a large prime number but less than the number of available subcarriers. The preamble sequence uses the length $N_{ZC} = 839$, which is a prime number fitting in the available resource properly. The preamble is positioned in the middle of the 864 available subcarriers with 12.5 null subcarriers on each side.

The generation of random access preambles can be easier in the frequency domain compared to the time domain. The generation has steps as follows in frequency domain.

**Table 5.11** PDCCH Format Types and Their Search Spaces

| ZeroCorrelationZoneConfig | Value of $N_{CS}$ for Unrestricted Set | Value of $N_{CS}$ for Restricted Set |
|---|---|---|
| 0 | 0 | 15 |
| 1 | 13 | 18 |
| 2 | 15 | 22 |
| 3 | 18 | 26 |
| 4 | 22 | 32 |
| 5 | 26 | 38 |
| 6 | 32 | 46 |
| 7 | 38 | 55 |
| 8 | 46 | 68 |
| 9 | 59 | 82 |
| 10 | 76 | 100 |
| 11 | 93 | 128 |
| 12 | 119 | 158 |
| 13 | 167 | 202 |
| 14 | 279 | 237 |
| 15 | 419 | |

1. The ZC sequence is generated in the time domain. Then it is fed to 839-point DFT for time to frequency domain conversion.
2. The output of DFT is mapped to the central 839 subcarriers of the 864 available subcarriers.
3. The 864-point IDFT is used for frequency to time domain conversion.
4. In the case of preamble format 2 and 3, the output of IDFT is repeated to generate two preamble sequences.
5. The cyclic prefix is added.

# CHAPTER 6

# Medium Access Control

The Medium Access Control (MAC) is the bottom sublayer within layer 2. The various functions of the MAC layer and the format of MAC PDUs are explained in the following sections.

The RRC layer at eNodeB configures the MAC layer of the UE for signaling and data radio bearers by sending MAC-MainConfig IE. Appendix A.2 shows how the eNodeB sends MAC-MainConfig IE to the UE while setting up or modifying radio bearers.

## 6.1 MAC Layer Functions

The MAC layer performs the following major functions.

1. *Multiplexing, demultiplexing and mapping between logical channels and transport channels:* This is explained in Section 2.3.
2. *Scheduling of radio resources among UEs:* The scheduler in the MAC layer of the eNodeB allocates the available radio resources among different UEs for both uplink and downlink transmission in a cell. This is explained in Chapter 19.
3. *Random access procedure:* This is explained in Chapter 11.
4. *Uplink time alignment:* This is explained in Section 10.2.
5. *HARQ operation:* This is explained in Section 20.3.
6. *Discontinuous reception (DRX):* This is explained in Chapter 16.
7. *Transport format selection:* The MAC layer of the eNodeB is responsible for transport format selection in both uplink and downlink. The transport format includes the modulation and coding scheme (MCS) and transport block size (TBS) to be used for the transport block. Section 5.1.3.2 shows how the eNodeB indicates the transport format for both uplink and down transport blocks using PDCCH.
8. *Sending scheduling request:* This is explained in Section 6.2.1.2 and Section 19.2.2.1..
9. *Buffer status reporting:* This is explained in Section 6.2.1.2.
10. *Power headroom reporting:* The MAC layer of the UE performs power headroom reporting as explained in Section 6.2.1.2.
11. *Apportionment of resources among logical channels:* This is explained in Section 19.3.

## 6.2 MAC Protocol Data Unit (PDU)

A MAC PDU is fitted into a transport block on DL-SCH, UL-SCH, or MCH and the transport block contains exactly one MAC PDU. The MAC layer generates each transport block and protects it with a CRC.

The MAC PDU on MCH is called MCH MAC PDU. Both MCCH and MTCH logical channels are mapped on the MCH transport channel. Therefore, the MAC SDU in MCH MAC PDU can belong to either MCCH or MTCH. The eNodeB cannot transmit more than one MCH MAC PDU in a subframe. However, more than one transport block can be transmitted in a subframe on DL-SCH for a UE as shown in Section 18.2.1.

The MAC PDUs have the format shown in Section 6.2.1 with the exception that MAC PDUs for random access response and transparent MAC have special formats shown in Section 6.2.2.

### 6.2.1 Basic MAC PDU Format

A MAC PDU consists of the following parts as shown in Figure 6.1.

1. *MAC header:* There must be one and only one MAC header. The MAC header is placed in the beginning of the MAC PDU.
2. *MAC control element:* The MAC control element may or may not be present in a MAC PDU. If it is present, then there can be one or more MAC control elements in the MAC PDU. MAC control elements are placed right after the MAC header and before MAC SDUs.
3. *MAC Service Data Units (MAC SDU):* An RLC PDU is treated as a MAC SDU by the MAC layer. A MAC PDU may or may not include any MAC SDU. If included, then there can be one or more MAC SDUs in the MAC PDU. MAC SDUs are placed right after the MAC control elements and before padding.

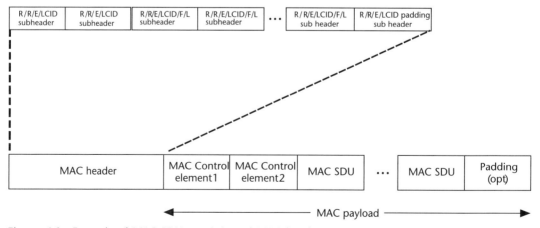

**Figure 6.1** Example of MAC PDU consisting of MAC header, MAC control elements, MAC SDUs, and padding.

## 6.2 MAC Protocol Data Unit (PDU)

4. *Padding*: Padding bits are added at the end of the MAC PDU in order to complete the transport block size. Thus, padding may or may not be present in a MAC PDU.

#### 6.2.1.1 MAC Header

The MAC header is of variable size. It consists of separate MAC subheaders corresponding to each of the MAC control elements, each of the MAC SDUs, and padding. The MAC header and subheaders are byte aligned. A MAC PDU subheader can be of the following three types, as shown in Figure 6.2.

1. *Subheader with R/R/E/LCID/F/L fields in which L field has 7 bits*: This subheader is used for MAC SDUs. The L (length) field indicates the length of the MAC SDU in bytes. The L field is 7 bits long if the size of the MAC SDU is less than 128 bytes and, thus, can be indicated in 7 bits.
2. *Subheader with R/R/E/LCID/F/L fields in which L field has 15 bits*: This subheader is used for MAC SDUs. The L (length) field indicates the length of the MAC SDU in bytes. The L field is 15 bits long if the size of the MAC SDU is not less than 128 bytes and, thus, cannot be indicated in 7 bits.
3. *Subheader with R/R/E/LCID fields*: This subheader is used for MAC control elements and padding.

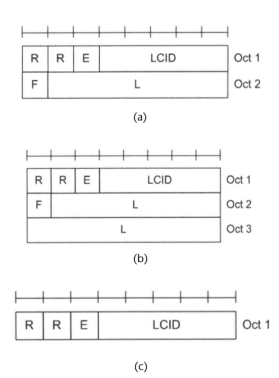

**Figure 6.2** MAC PDU subheader with (a) R/R/E/LCID/F/L fields in which L field has 7 bits, (b) R/R/E/LCID/F/L fields in which L field has 15 bits, and (c) R/R/E/LCID fields.

The MAC PDU subheader includes the following fields:

- *Logical Channel ID (LCID):* The LCID field is 5 bits long. It indicates the identity of the logical channel instance of the MAC SDU, and the type of the MAC control element or padding according to Tables 6.1–6.3. If one or two bytes of padding is required in the MAC PDU, then one or two dummy LCID fields can be included in the MAC PDU instead of padding at the end of the MAC PDU.
- *Length (L):* The L field indicates the length of the MAC SDU in bytes. The L field can be 7 bits long or 15 bits long depending on the size of the MAC SDU.
- *Format (F):* The 1-bit F field gives the size of the L field. The F field is set to 0, if the L field has 7 bits and the F field is set to 0, if the L field has 15 bits.

**Table 6.1** LCID Values for Downlink MAC PDUs on DL-SCH

| LCID Values | Indication |
| --- | --- |
| 00000 | MAC SDU for CCCH. |
| 00001–01010 | MAC SDU: TheLCID value gives the identity of the logical channel instance for the MAC SDU. |
| 11100 | UE contention resolution identity MAC control element. |
| 11101 | Timing advance command MAC control element. |
| 11110 | DRX command MAC control element. |
| 11111 | Padding. |

**Table 6.2** LCID Values for Downlink MAC PDUs on MCH

| LCID Values | Indication |
| --- | --- |
| 00000 | MAC SDU for MCCH. However, if there is no MCCH used then it can be used to indicate MAC SDU for MTCH. |
| 0000–11100 | MAC SDU for MTCH. |
| 11110 | MCH scheduling information MAC control element. |
| 11111 | Padding. |

**Table 6.3** LCID Values for Uplink MAC PDUs

| LCID Values | Indication |
| --- | --- |
| 00000 | MAC SDU for CCCH. |
| 0000–01010 | MAC SDU: TheLCID value gives the identity of the logical channel instance for the MAC SDU. |
| 11010 | Power headroom report MAC control element. |
| 11011 | C-RNTI MAC control element. |
| 11100 | Buffer status report (BSR) MAC control element with truncated BSR. |
| 11101 | Buffer status report (BSR) MAC control element with short BSR. |
| 11110 | Buffer status report (BSR) MAC control element with long BSR. |
| 11111 | Padding. |

- *Extension (E):* The 1-bit E field indicates whether this is the last subheader in the MAC header or not. The E field is set to 1 if there are more subheaders. The E field is set to 0 if there are no more subheaders and a MAC SDU or a MAC control element starts at the next byte.
- *Reserved (R):* The R bits are set to 0.

#### 6.2.1.2 MAC Control Element

The MAC control elements are used for signaling between the MAC layers of the eNodeB and the UE. There are seven types of MAC control elements, which can be categorized as follows.

1. MAC control elements on DL-SCH:
   - DRX command MAC control element;
   - UE contention resolution identity MAC control element;
   - Timing advance command MAC control element.
2. MAC control element on MCH:
   - MCH scheduling information MAC control element.
3. MAC control elements on UL-SCH:
   - Buffer status report (BSR) MAC control element;
   - C-RNTI MAC control element;
   - Power headroom MAC control element.

*MAC Control Elements on DL-SCH*

*DRX Command MAC Control Element*
As explained in Section 16.2.2, the eNodeB may use the DRX command MAC control element in the MAC PDU in order to initiate the DRX cycle in RRC_CONNECTED state for saving battery power in the UE. The DRX command MAC control element has a length of zero bits. Thus, it has only a corresponding MAC subheader in the MAC header and there is actually nothing to add as a control element after the MAC header.

*UE Contention Resolution Identity MAC Control Element*
The UE contention resolution identity MAC control element is 6 bytes long as shown in Figure 6.3. It includes the UE contention resolution identity field. This field contains the uplink CCCH SDU.

*Timing Advance Command MAC Control Element*
The timing advance command MAC control element is one byte long as shown in Figure 6.4. It includes the timing advance command field, which is 6 bits long. The timing advance command has the value $T_A$, which ranges between 0 and 63. It is used to maintain timing alignment of the uplink transmission of the UE as explained in Section 10.2.2.

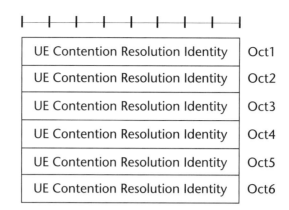

**Figure 6.3** UE Contention resolution identity MAC control element.

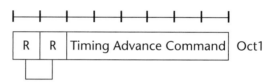

**Figure 6.4** Timing advance command MAC control element.

*MAC Control Element on MCH: MCH Scheduling Information MAC Control Element*
The MCH scheduling information MAC control element is used to indicate the position of each MTCH on the MCH and unused subframes in the MCH Subframe Allocation (MSA) as explained in Section 24.7. The MCH scheduling information MAC control element includes the following fields for each of the MTCHs as shown in Figure 6.5. Therefore, it has a variable size depending on the number of MTCH.

- LCID: It is a 5-bit-long field. This field gives the logical channel ID of the MTCH, which is indicated by the corresponding stop MTCH field.
- Stop MTCH: It is an 11-bits-long field. This field gives the ordinal number of the subframe within the MCH scheduling period where the particular MTCH stops. This field can have a special value, 2,047, which indicates that the MTCH is not scheduled.

*MAC Control Elements on UL-SCH*
*Buffer Status Report (BSR) MAC Control Element*
The UE may use the Buffer Status Report (BSR) MAC control element in the MAC PDU in order to report pending data status in its uplink buffers while the UE has PUSCH resources allocated for uplink transmission.

The UE performs BSR reporting based on logical channels. The logical channels are grouped for BSR reporting depending on the delay and rate constraints of

## 6.2 MAC Protocol Data Unit (PDU)

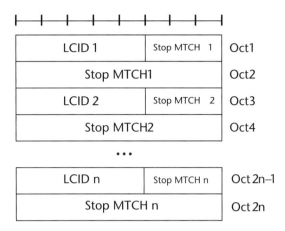

**Figure 6.5** MCH Scheduling information MAC control element.

different logical channels. Four Logical Channel Groups (LCGs) can be used at the maximum. A LCG is identified by a 2-bit LCG ID, which ranges from 0 through 3. The BSR reporting is performed individually for each of the LCGs notifying the data queue of the respective LCG.

The eNodeB assigns LCG ID for the logical channel while setting up or modifying radio bearers by sending LogicalChannelConfig IE. Appendix A.2 shows how the eNodeB sends this IE. The LogicalChannelConfig IE includes theLogicalChannelGroup field. The LogicalChannelGroup field gives the value of LCG ID, which can be between 0 and 3.

The BSR MAC control element has two different formats as shown below.

- Short BSR and truncated BSR: The short BSR and the truncated BSR have the same format and they both report the amount of data that is buffered in for one LCG. They contain a 2-bit field for the LCG ID and a buffer size field for the LCG as shown in Figure 6.6(a).
- Long BSR: The Long BSR reports the amount of data that is buffered in for four LCGs. It contains four buffer size fields, which correspond to the four LCGs with LCG ID from 0 through 3 LCG as shown in Figure 6.6(b).

The buffer size field is 6 bytes long. It carries an index, which indicates the amount of data in the buffer in number of bytes as partially shown in Table 6.4. The transmission of BSR MAC control element is explained in Section 19.2.2.1.

When a MAC PDU includes a MAC BSR control element, it includes only one MAC BSR control element. The BSR is classified as follows depending on the triggering of BSRs.

1. Regular BSR;
2. Periodic BSR;
3. Padding BSR.

The transmissions of regular BSR and periodic BSR have precedence over the transmission of padding BSR.

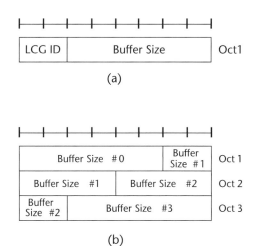

**Figure 6.6** BSR MAC control element for (a) short BSR and truncated BSR and (b) long BSR.

**Table 6.4** Buffer Size Levels for BSR

| Index | Buffer Size in Bytes |
|---|---|
| 0 | Buffer size = 0 |
| 1 | 0 < buffer size <= 10 |
| 2 | 10 < buffer size <= 12 |
| 3 | 12 < buffer size <= 14 |
| 4 | 14 < buffer size <= 17 |
| 5 | 17 < buffer size <= 19 |
| ... | ...... |
| 31 | 967 < buffer size <= 1,132 |
| 32 | 1,132 < buffer size <= 1,326 |
| 33 | 1,326 < buffer size <= 1,552 |
| 34 | 1,552 < buffer size <= 1,817 |
| ... | ...... |
| 58 | 68,201 < buffer size <= 79,846 |
| 59 | 79,846 < buffer size <= 93,479 |
| 60 | 93,479 < buffer size <= 109,439 |
| 61 | 109,439 < buffer size <= 128,125 |
| 62 | 128,125 < buffer size <= 150,000 |
| 63 | buffer size > 150,000 |

*Regular BSR*

The UE may send a BSR in the following cases and in these cases, the BSR is called regular BSR.

1. Data arrives for a logical channel, which has a higher priority than the logical channels that already had data available for transmission.
2. Data arrives for a logical channel and other logical channels do not have data available for transmission.

3. Data is available for transmission for a logical channel and a certain time has elapsed after the last allocation of uplink resources. The elapsed time is indicated by the expiry of a timer called RetxBSR-Timer. The RetxBSR-Timer is started when the uplink resource grant is received. The eNodeB configures the RetxBSR-Timer. The MAC-MainConfig IE includes the UL-SCH-Config IE as shown in Appendix A.2. The UL-SCH-Config IE contains the RetxBSR-Timer field. The value of RetxBSR-Timer can be 320, 640, 1,280, 2,560, 5,120, or 10,240 subframes.

The BSR format is selected as follows for regular BSR.

- *Long BSR:* The UE sends long BSR if more than one LCG have data available for transmission in the TTI in which the BSR is sent.
- *Short BSR:* The UE sends short BSR if only one LCG has data available for transmission in the TTI in which the BSR is sent.

*Periodic BSR*

The UE may send a BSR if data is available for transmission for a logical channel and a certain time has elapsed after the last transmission of BSR. In this case, the BSR is called periodic BSR. The elapsed time is indicated by the expiry of a timer called PeriodicBSR-Timer. The PeriodicBSR-Timer is started when a BSR is transmitted. However, the PeriodicBSR-Timer is not started in case of transmission of truncated BSR, which is used for padding BSR. The eNodeB configures the PeriodicBSR-Timer. The UL-SCH-Config IE contains the PeriodicBSR-Timer field. The value of PeriodicBSR-Timer can be 5, 10, 16, 20, 32, 40, 64, 80, 128, 160, 320, 640, 1,280, or 2,560 subframes. It can also be infinity, which means no repetition of BSR. The BSR format is selected for periodic BSR the same way as done for regular BSR.

*Padding BSR*

The UE may send a BSR when padding is required to fill in the uplink resources allocated. In this case, the BSR is called padding BSR. This utilizes the resources, which would be left unused otherwise. Here, the size of the required padding has to be equal to or larger than the size of the BSR MAC control element plus its subheader. The BSR format is selected as follows for padding BSR.

- *Long BSR:* The UE sends long BSR if the required padding is equal to or larger than the size of the long BSR plus its subheader.
- *Short BSR:* The UE sends short BSR if the required padding is equal to or larger than the size of the short BSR plus its subheader but smaller than the size of the long BSR plus its subheader. Also, only one LCG has data available for transmission in the TTI in which the BSR is sent.
- *Truncated BSR:* The UE sends truncated BSR to report one LCG when the available resources are not enough to accommodate a long BSR and report all the LCGs, which have data available for transmission. Thus, the truncated BSR is used if the required padding is equal to or larger than the size of the

short BSR plus its subheader but smaller than the size of the long BSR plus its subheader. Also, in contrast to the case above, more than one LCG has data available for transmission in the TTI in which the BSR is sent. As shown in Section 6.2.1.2, the truncated BSR has the same format as the short BSR and it reports the amount of data for only one LCG. The truncated BSR reports the LCG with the logical channel of the highest priority. As mentioned above, PeriodicBSR-Timer is not started when a truncated BSR is sent.

*C-RNTI MAC Control Element*
The C-RNTI MAC control element includes only a 16 bits long C-RNTI field as shown in Figure 6.7. The C-RNTI field contains the C-RNTI of the UE. This MAC control element is used in case of contention-based random access as explained in Section 11.2.

*Power Headroom MAC Control Element*
The UE may use the power headroom MAC control element in the MAC PDU in order to send a power headroom report (PHR). The power headroom report notifies the eNodeB how much more or less power the UE is capable of transmitting. The power headroom reports provide the eNodeB with the difference between the UE's maximum transmit power capability and the estimated power for transmission on the received uplink resource grant with allocated HARQ and RV configuration. The UE can send its power headroom report only in subframes in which it has an uplink transmission grant. The power headroom report in a subframe corresponds to the same subframe in which it is sent. Thus, the power headroom report is a prediction rather than a direct measurement. The power headroom reporting assists the eNodeB to allocate uplink resources among different UEs appropriately as shown in Section 19.1.2. The power headroom reporting also assists in uplink power control as shown in Section 17.1.

The one byte long power headroom MAC control element has a Power Headroom (PH) field, which is 6 bits long as shown in Figure 6.8. The value of PH ranges from 0 to 63 and it indicates the power headroom level from −23 dB to 40 dB with steps of 1 dB. The negative values are to indicate that the UE has less transmit power capability compared to the received uplink resource grant.

The UE can send power headroom reports either periodically or when the path loss has changed by a specific value since the last power headroom report was sent. The eNodeB configures PH reporting using PHR-Config IE. The MAC-MainConfig IE contains the PHR-Config IE. The PHR-Config IE includes the PeriodicPHR-Timer field for periodic power headroom reporting. The PeriodicPHR-Timer gives

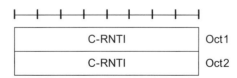

**Figure 6.7** C-RNTI MAC control element.

## 6.2 MAC Protocol Data Unit (PDU)

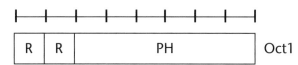

**Figure 6.8** Power headroom MAC control element.

the period between successive PH reporting in number of subframes. Its value can be 10, 20, 50, 100, 200, 500, or 1,000 subframes. It can also be infinity, which means no repetition of power headroom reporting.

The PHR-Config IE also includes the DL-PathLossChange field, which gives the specific value in dB by which the path loss has to change in order to trigger power headroom reporting. The value of DL-PathLossChange field can be 1 dB, 3 dB, 6 dB, or infinity. The UE triggers this power headroom reporting only if ProhibitPHR-Timer has expired. The PHR-Config IE includes ProhibitPHR-Timer and its value can be 0, 10, 20, 50, 100, 200, 500, or 1,000 subframes.

### 6.2.2 Special Format for MAC PDU

There are two special types of format for MAC PDU to be used for specific purposes. They are MAC PDU for transparent MAC and MAC PDU for random access response.

#### 6.2.2.1 MAC PDU for Transparent MAC

The MAC SDU and the MAC PDU have the same format as shown in Figure 6.9 for transparent MAC. The transparent MAC is used to transport data from BCCH and PCCH logical channels as shown in Section 4.1 and 13.1. In this case, the MAC PDU includes data from only one logical channel and no multiplexing is applied. Therefore, the LCID field of the MAC header is unnecessary. Also, the MAC SDU is directly mapped onto the MAC PDU. So, the size of the MAC SDU can be determined from the transport block size and the L field of the MAC header is also unnecessary. Thus, the whole MAC header is unnecessary and the MAC header is omitted. This makes the MAC SDU and the MAC PDU identical.

#### 6.2.2.2 MAC PDU for Random Access Response

The eNodeB uses this special MAC PDU to transport random access response (RAR) messages during the random access procedure, which is explained in Chapter 11. The MAC PDU consists of a MAC header, one or more MAC random access

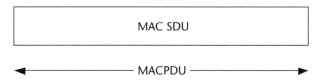

**Figure 6.9** MAC PDU for transparent MAC.

responses (MAC RARs), and optional padding as shown in Figure 6.10. The MAC PDU can contain multiple MAC RARs to address multiple UEs together. The MAC header and the MAC RAR are explained below.

*MAC Header*
The MAC PDU header consists of separate MAC PDU subheaders corresponding to each MAC RAR of the MAC PDU. Additionally, The MAC PDU header can optionally include a back-off indicator subheader. If the back-off indicator subheader exists, it is only included once and then it must be the first subheader in the MAC PDU header. The back-off indicator subheader is used only for contention-based random access and not for noncontention-based random access.

*Back-Off Indicator*
The back-off indicator subheader contains the back-off indicator (BI) field as shown in Figure 6.11. The back-off indicator field is 4 bits long and it carries an index, which corresponds to a back-off parameter value as shown in Table 6.5. The UE selects a random back-off time from a uniform distribution between 0 ms and the back-off parameter value. The UE delays retransmission of the random access preamble by the back-off time in case of unsuccessful random access response reception and unsuccessful contention resolution as explained in Section 11.2. A high back-off parameter value implies overload condition in the cell. If the back-off indicator subheader does not exist, the back-off time is set 0 ms.

*Random Access Preamble Identifier*
A MAC PDU subheader in the MAC header that corresponds to a MAC RAR contains the random access preamble identifier (RAPID) field as shown in Figure 6.12. The RAPID field is 6 bits long and it is an identifier of the random access preamble that the UE transmitted earlier.

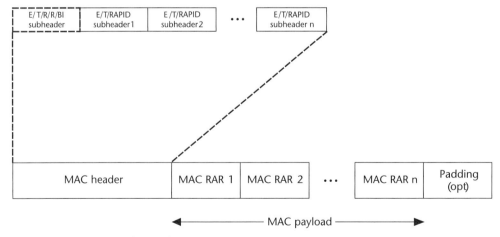

**Figure 6.10** MAC PDU for random access response.

## 6.2 MAC Protocol Data Unit (PDU)

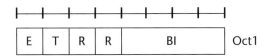

**Figure 6.11** Back-off indicator subheader.

**Table 6.5** Back-Off Parameter Values

| Back-Off Indicator (BI) | Back-Off Parameter Value (ms) |
|---|---|
| 0 | 0 |
| 1 | 10 |
| 2 | 20 |
| 3 | 30 |
| 4 | 40 |
| 5 | 60 |
| 6 | 80 |
| 7 | 120 |
| 8 | 160 |
| 9 | 240 |
| 10 | 320 |
| 11 | 480 |
| 12 | 960 |

```
| E | T |    RAPID    | Oct 1
```

**Figure 6.12** MAC subheader with Random Access Preamble Identifier.

*MAC RAR*

Each MAC RAR in the MAC payload has the format shown in Figure 6.13. Section 11.2.2.1 explains the MAC RAR allocates resources on PUSCH for transmission of message 3. The various fields in the MAC RAR are explained here:

1. *Timing Advance Command:* The Timing Advance Command field is 11 bits long. The timing advance command has the value $T_A$, which ranges between 0 and 1,282. It is used to establish timing alignment of the uplink transmission of the UE as explained in Section 10.2.1. The UE applies this timing alignment for transmission of message 3.
2. *UL grant:* The Uplink Grant field allocates resources on PUSCH for the UE and it is also called the random access response grant. This field is 20 bits long and it contains the following information.
   - *Resource Block Assignment (10 bits):* This field contains information of resources allocated on PUSCH for the UE. It allocates at least 80 bits.
   - *Hopping Flag (1 bit):* The UE performs PUSCH frequency hopping if this flag is set to 1. The UE does not perform PUSCH frequency hopping if

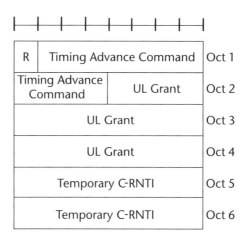

**Figure 6.13** MAC RAR.

this flag is set to 0. If the UE performs frequency hopping, then the UE interprets the 10 bits long resource block assignment field as follows.

- If $N_{RB}^{UL} \leq 44$, then the UE considers only the ceil $[(\log^2 (N_{RB}^{UL}(N_{RB}^{UL} + 1)/2)]$ number of least significant bits from the 10 bits long resource block assignment field and uses them for frequency hopping information in the same way as shown in Section 19.2.2.3.
- If $N_{RB}^{UL} > 44$, then $N_{UL\_hop}$ is determined as shown in Section 19.2.2.3. The UE inserts $[\text{ceil}[(\log^2 (N_{RB}^{UL}(N_{RB}^{UL} + 1)/2)] - 10]$ number of bits of 0 value into the 10 bits long resource block assignment field after its $N_{UL\_hop}$ number of MSB bits. Thereafter, this expanded set of bits is used for frequency hopping information in the same way as shown in Section 19.2.2.3.

- *Truncated modulation and coding scheme (4 bits):* This field provides the truncated modulation and coding scheme index, $I_{MCS}$, in 4 bits instead of 5 bits. So, the MCS indices corresponding to $I_{MCS}$ range from 0 through 15 only. These MCS indices are taken on the same table, which is used for DCI format 0 as explained in Section 5.1.3.2. The $I_{MCS}$ corresponds to a modulation order and a transport block size the same way as done for DCI format 0.
- *TPC command for scheduled PUSCH (3 bits):* After the UE transmits random access preamble without any good estimate of required uplink power level, this TPC command field instructs for necessary adjustment in subsequent uplink transmission on PUSCH. This is explained in Section 17.2.2.3.
- *UL delay (1 bit):* When the UL delay field is set to zero, the UE sends message 3 in the first available subframe on or after subframe $n + 6$, if the random access response is received in subframe $n$. The UE postpones the transmission to the next available subframe if the UL delay field is set to 1.
- *CQI request (1 bit):* If this field is set to 1, the UE includes aperiodic CQI/PMI/RI reporting in message 3 on PUSCH for noncontention based random access procedure. If it is set to 0, the UE does not include CQI/PMI/RI

reporting. The CQI request field is not used for contention-based random access procedure.

3. *Temporary C-RNTI:* This field assigns a temporary C-RNTI to the UE for use as a C-RNTI temporarily. Section 11.2.2.1 shows that this field may include previously allocated C-RNTI instead.

# CHAPTER 7
# Radio Link Control (RLC)

The Radio Link Control (RLC) is the middle sublayer within layer 2. The transmitting and receiving RLC entities work as peer entities for data transfer in both uplink and downlink. The RLC entity is configured in any of the following three modes.

1. Transparent mode (TM);
2. Unacknowledged mode (UM);
3. Acknowledged mode (AM).

All these three RLC modes transfer SDUs from the upper layer as shown in Section 2.3. The MAC layer notifies when there is a transmission opportunity. It also notifies the possible size and number for RLC PDUs to be generated based on the availability of transport blocks. The transport block size depends on the resources available and the Modulation and Coding Scheme (MCS) to be used on the basis of physical channel conditions.

The TM transfers SDUs without any processing. The UM performs in-sequence delivery, concatenation, segmentation, reassembly, and the discarding of duplicates during transfer of SDUs. In addition to these services, the AM performs error correction using Automatic Repeat Request (ARQ) and resegmentation of PDUs, if required in retransmission.

Another RLC function supported by all three RLC modes is protocol error detection and recovery. The RRC layer can reset the RLC layer in case of a protocol error. Then the transmitting and receiving RLC buffers discard all incomplete SDUs and they may even discard all SDUs waiting in the buffers. Besides, the PDCP layer can instruct the RLC layer to discard any SDUs that are not properly mapped.

The RRC layer at eNodeB configures the RLC layer of the UE for signaling and data radio bearers by sending RLC-Config IE. Appendix A.2 shows how the eNodeB sends RLC-Config IE to the UE while setting up or modifying radio bearers.

## 7.1 Transparent Mode (TM)

The TM operation does not guarantee the delivery of information. The transparent mode is used in the following cases.

- BCCH logical channel;

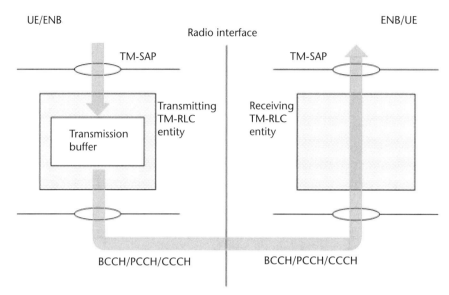

**Figure 7.1** Transparent mode peer entities.

- PCCH logical channel;
- SRB0 mapped on CCCH logical channel.

While LTE does not use the TM for transferring any user data, UMTS uses the TM for transferring circuit-switched (CS) user data (e.g., for voice transfer).

### 7.1.1 TM Functions

The TM provides a unidirectional data transfer. Two RLC entities are used for a bearer, a transmitting TM RLC entity, and a receiving TM RLC entity.

#### 7.1.1.1 Transmitting TM RLC Entity

The transmitting TM RLC entity performs the following functions as shown in Figure 7.1.

1. The transmitting TM RLC entity receives RLC SDUs from the RRC layer and puts them in the transmission buffer until there is any transmission opportunity. The RRC PDUs are directly mapped on RLC SDUs without using PDCP layer.
2. No segmentation of RLC SDUs is performed.
3. No concatenation of RLC SDUs is performed.
4. No header is added to RLC SDUs.
5. An RLC SDU is directly mapped onto an RLC PDU, which is called TMD PDU.
6. The TM RLC entity passes TMD PDUs to MAC layer for transmission over the air.

## 7.1.1.2 Receiving TM RLC Entity

The receiving TM RLC entity receives TMD PDUs from MAC layer and passes them on to RRC layer as RLC SDUs.

### 7.1.2 TM PDU Format

The RLC PDU is known as Transparent Mode Data PDU (TMD PDU). The TMD PDU contains only the data received from RRC layer and it does not include any header as shown in Figure 7.2.

## 7.2 Unacknowledged Mode (UM)

The UM does not guarantee the delivery of information and thus, packet losses may occur. It is suitable for delay sensitive and error tolerant real time applications. Thus, VoIP and other streaming traffic use the UM. The UM is used for DRBs mapped on DTCH logical channels when the RLC-Config IE specifies the UM instead of the AM. The unacknowledged mode is also used for MBMS information transfer over PMCH. Unlike UMTS, the unacknowledged mode is not used for signaling. Also, unlike UMTS, the RLC layer does not perform ciphering for the unacknowledged mode.

### 7.2.1 UM Functions

The unacknowledged mode provides a unidirectional data transfer. Two RLC entities are used for a unidirectional bearer; a transmitting UM RLC entity, and a receiving UM RLC entity. A bidirectional bearer requires two sets of such RLC entities.

#### 7.2.1.1 Transmitting UM RLC Entity

The transmitting UM RLC entity performs the following functions as shown in Figure 7.3.

1. The transmitting UM RLC entity receives RLC SDUs from PDCP layer and puts them in transmission buffer. The PDCP PDUs are mapped on RLC SDUs.
2. It performs segmentation and/or concatenation of RLC SDUs depending on the transport block information provided by the MAC layer. The segmentation and concatenation are explained in Section 7.2.1.3.

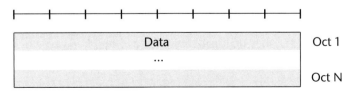

**Figure 7.2** TMD PDU.

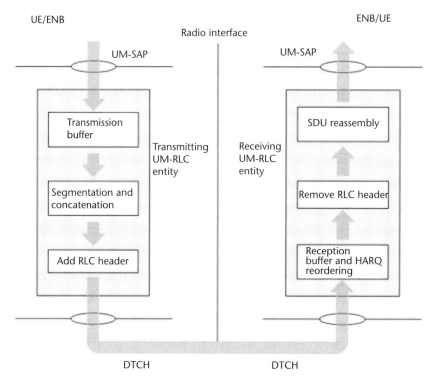

**Figure 7.3** Unacknowledged mode peer entities.

3. It adds the RLC header and generates RLC PDU.
4. It passes RLC PDUs to MAC layer for transmission over the air.

#### 7.2.1.2  Receiving UM RLC Entity

The receiving UM RLC entity performs the following functions.

1. The receiving UM RLC entity receives RLC PDUs from the MAC layer and puts them in the reception buffer.
2. It reorders out of sequence RLC PDUs. The reordering is explained in Section 7.2.1.4.
3. It detects lost RLC PDUs.
4. It discards duplicate RLC PDUs. The discard of duplicate RLC PDUs is explained in Section 7.2.1.5.
5. It removes the RLC header and reassembles complete RLC SDUs. The reassembly is the reverse process of the segmentation and concatenation. If any segment of a RLC SDU is missing, the whole RLC SDU is discarded. It also discards any RLC PDUs that have been received but cannot be reassembled into RLC SDUs.
6. It passes complete RLC SDUs on to PDCP layer.

### 7.2.1.3 Segmentation and Concatenation

The transmitting UM RLC entity performs segmentation and/or concatenation of RLC SDUs and produces RLC PDUs depending on the size of the RLC SDUs available and the size of the RLC PDUs to be generated. The RLC SDU is usually not expected to be segmented if the whole RLC SDU can be fitted into RLC PDU. If the RLC SDU is large compared to RLC PDU, then the RLC SDU is segmented to fit in the size of RLC PDU. On the other hand, if the RLC PDU is large compared to the RLC SDU or the RLC SDU segment, then concatenation of RLC SDUs and/or RLC SDU segments is performed to fit in the size of RLC PDU. Thus, a RLC PDU can contain zero or more full RLC SDUs. Also, the RLC PDU can contain zero or one RLC SDU segment in the beginning and at the end of the RLC PDU. The RLC SDUs are included in the order of their arrival. Section 7.2.2 shows that FI, E, and LI fields in the RLC PDU header depict how RLC SDU segments have been included and this allows the receiving UM RLC entity to reassemble RLC SDUs. Apart from fitting in RLC PDU size, segmentation may often be performed to improve the coverage because the smaller transport blocks have better coding and they are more likely to be decoded correctly.

The RLC SDUs can have various sizes because the size of IP packets varies depending on the application and data path. Video streaming produces large IP packets and VoIP produces small IP packets. On the other hand, the RLC PDU size is also allowed to vary dynamically in order to cope with the varying data rate. A small size of RLC PDU would cost a large amount of overhead from the RLC PDU headers. Conversely, a large size of RLC PDU might not match the transport block size as the data rate becomes lower. Therefore, the RLC PDU size is allowed to vary according to the transport block size available at the MAC layer. The MAC layer notifies the possible size and number for RLC PDUs to be generated. This is in contrast to HSPA where the RLC PDU size is semistatically configured by RRC layer. However, in Release 7, HSPA+ supports variable RLC PDU size.

There may be small padding in the RLC PDU header but no padding is necessary in the payload of the RLC PDU to satisfy a certain size level for the RLC PDU. The RLC PDU size can rather vary without requiring the addition of padding. It is the responsibility of the MAC layer to add necessary padding and complete the transport block size as shown in Section 6.2.1.

### 7.2.1.4 Reordering

The receiving RLC entity may receive out of sequence RLC PDUs. Since multiple HARQ processes run in the MAC layer, there will be extra delays for transport blocks, which are retransmitted in the HARQ protocol. This is the key reason for the possibility of receiving out of sequence RLC PDUs. Also, different propagation delays can make the PDUs out of sequence.

The RLC PDU includes the Sequence Number (SN) field in its header, which indicates the sequence number of the RLC PDU as shown in Section 7.2.2.1. The transmitting UM RLC entity increases the sequence number by one in every successive PDU. The receiving RLC entity reorders the RLC PDUs based on their sequence numbers. If the receiving RLC entity finds a gap in the sequence of the received RLC PDUs, it assumes that the missing RLC PDU has been delayed due to HARQ retransmission of the corresponding transport block; it then starts a reordering timer. It stores out of sequence RLC PDUs, does not deliver them to the higher layer, and waits for the missing RLC PDU until the timer expires. If the timer expires without arrival of the missing RLC PDU, the receiving RLC entity infers that HARQ has failed to recover the RLC PDU and the missing RLC PDU has been lost. It then reassembles only the RLC PDUs, which can generate complete RLC SDUs. This precludes the receiving RLC entity from waiting for the missing RLC PDUs indefinitely.

The eNodeB configures the reordering timer of the UE for downlink data transfer. The RLC-Config IE includes DL-UM-RLC IE. The DL-UM-RLC IE includes the T-Reordering field, which sets the value of the reordering timer. Its value can be any duration between 0 ms and 100 ms with 5-ms gaps or any duration between 100 ms and 200 ms with 10-ms gaps.

#### 7.2.1.5 Discarding Duplicate PDUs

If there is any duplicate RLC PDUs, the receiving RLC entity detects them based on the sequence number and discards them. It may happen that the feedback is decoded incorrectly in the HARQ process. If an ACK is misinterpreted as NACK, then the correctly received transport block can be retransmitted; this can create a duplicate RLC PDU.

### 7.2.2 UM PDU Format

The RLC PDU is known as Unacknowledged Mode Data PDU (UMD PDU). It consists of one or more than one data fields and an UMD PDU header as shown in Figure 7.4. Each data field contains a full RLC SDU or a segmented RLC SDU. The UMD PDU header consists of a fixed part and an extension part. Both the fixed part and the extension part are byte aligned.

#### 7.2.2.1 Fixed Part of Header

The fields of the fixed part are always present in every UMD PDU. The fields are explained here.

- *Sequence Number (SN):* The SN field is 5 or 10 bits long. It indicates the sequence number of the PDU, which is incremented by one in every successive PDU. The eNodeB configures the length of the SN in both uplink and downlink. The RLC-Config IE includes UL-UM-RLC IE and DL-UM-RLC IE for uplink and downlink, respectively. Both UL-UM-RLC IE and DL-UM-RLC IE include the SN-FieldLength IE, which specifies if the length of sequence number is 5 bits or 10 bits.
- *Framing Info (FI):* The FI field is 2 bits long. The 2 bits indicate if the first and the last data fields are full RLC SDUs or segmented RLC SDUs. If the first bit of the FI field is set to 1, then there is segmented RLC SDU in the beginning. Similarly, if the last bit of FI field is set to 1, then there is segmented RLC SDU at the end. Thus, the 0 value in the FI field is used to indicate no

## 7.2 Unacknowledged Mode (UM)

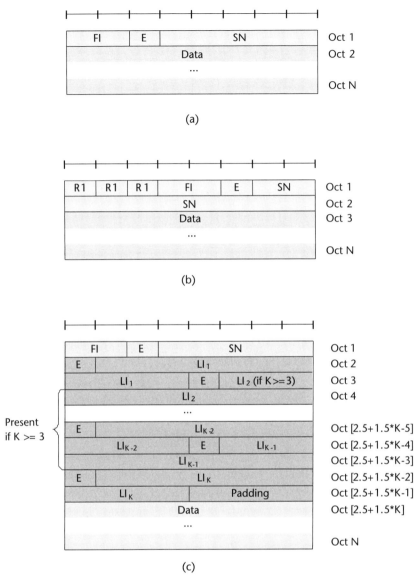

**Figure 7.4** (a) UMD PDU with 5-bit SN and only one data field element. (b) UMD PDU with 10 bit SN and only one data field element. (c) UMD PDU with 5-bit SN, more than one data field element, and odd number of LIs. (d) UMD PDU with 5-bit SN, more than one data field element, and even number of LIs. (e) UMD PDU with 10-bit SN, more than one data field element, and odd number of LIs. (f) UMD PDU with 10-bit SN, more than one data field element, and even number of LIs.

segmentation in the beginning or at the end. In other words, the 0 value for first or last bit in FI field indicates that the first or last byte of the data in RLC PDU corresponds to first or last byte of a RLC SDU respectively. When there is segmented RLC SDU, the receiving RLC entity needs to assemble the RLC SDU from multiple RLC PDUs.

- *Extension (E):* The E field has 1 bit. It indicates whether the data field or the E and LI fields exist right after the fixed part of the header.
- *Reserved 1 (R1):* Each of the R1 fields has 1 bit. This field is reserved.

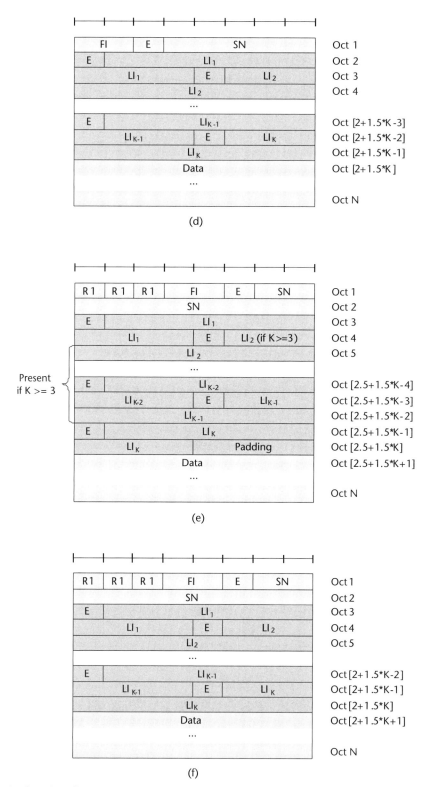

**Figure 7.4** (continued)

#### 7.2.2.2 Extension Part of Header

The extension part is needed and present only when more than one data field elements are present in the UMD PDU. For each of the data field elements except the last one, an E field and a LI field are present.

- *Length indicator (LI):* The LI field is 11 bits long. It indicates the length of the corresponding data field in number of bytes. The order of the LI fields and the order of the data fields are the same. There is no LI field for the last data field because the UMD PDU size gives the length of the last data field.
- *Extension (E):* The extension field has 1 bit. It indicates whether the data field or the E and LI fields exist right after the LI field following this E field.
- *Padding:* If there is odd number of LI fields, then padding is added after the last LI. The padding is 4 bits long.

## 7.3 Acknowledged Mode (AM)

The Acknowledged Mode (AM) provides reliable in-sequence delivery of data using the automatic repeat request (ARQ). The ARQ is an error control method for data transmission to achieve reliable data transmission. The receiver sends an indication if it has correctly received the PDU or not. If the packet is not received, the transmitter retransmits the PDU. Because of the use of ARQ, the acknowledged mode is suitable for error-sensitive and delay-tolerant applications. AM is the most commonly used RLC mode and it is especially used for interactive and background type applications (e.g., HTTP and FTP transfer). It is particularly suitable for TCP-based traffic. Streaming applications may also use AM RLC if they are not too delay intolerant. The AM RLC is also used for RRC signaling over SRB1 and SRB2. Thus, the AM RLC is used in the following cases.

- DRBs mapped on DTCH logical channels when the RLC-Config IE specifies AM instead of UM;
- SRB1 mapped on the DCCH logical channel;
- SRB2 mapped on the DCCH logical channel.

Unlike HSPA, the RLC layer does not perform ciphering in acknowledged mode.

### 7.3.1 AM Functions

The acknowledged mode provides a bidirectional data transfer. A single AM RLC entity is used for a bearer. It has two sides, a transmitting side and a receiving side as shown in Figure 7.5. The transmitting side and the receiving side use the sliding window for ARQ protocol. The sliding window in the transmitting side covers the RLC PDUs, which are not sent, as well as the RLC PDUs, which are sent but not yet acknowledged. The sliding window in the receiving side covers the expected RLC PDUs. All RLC PDUs may not be received with sequentially ordered numbers and

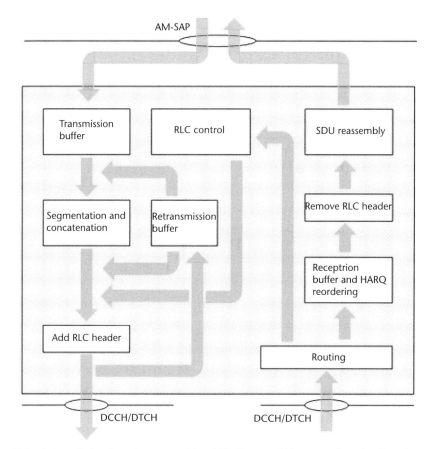

**Figure 7.5** Acknowledged mode peer entities. {AU: Please add in-text callout for figure}

they may be rather intermittent (i.e., some RLC PDUs may be missing while RLC PDUs with higher sequence numbers are received. In this case the sliding window is moved ahead of the sequentially received RLC PDUs and it covers intermittent RLC PDUs.

The same sequence number and reception buffer are used for both ARQ operation in the RLC layer and HARQ operation in the MAC layer. Thus, the HARQ reordering does not need an additional sequence number and reception buffer.

#### 7.3.1.1 Initial Transmission

*Transmitting Side*
The transmitting side of the AM RLC entity performs the following functions.

1. The transmitting side of the AM RLC entity receives RLC SDUs from the RRC layer for SRB1 and SRB2 and from the PDCP layer for DRBs. The RRC PDUs and PDCP PDUs are mapped onto RLC SDUs. It then puts the RLC SDUs in the transmission buffer.
2. It performs segmentation and/or concatenation of RLC SDUs as explained in Section 7.2.1.3.

3. It makes a copy of the transmit buffer for possible retransmissions using ARQ.
4. It adds the RLC header and generates RLC PDU.
5. It passes the RLC PDUs on to the MAC layer for transmission over the air.

*Receiving Side*
The receiving side of the AM RLC entity performs the following functions.

1. The receiving side of the AM RLC entity receives the RLC PDUs from the MAC layer.
2. It reorders out of sequence RLC PDUs as explained in Section 7.2.1.4. The T-Reordering field, which sets the value of the reordering timer, is included in the DL-AM-RLC IE. The RLC-Config IE includes the DL-AM-RLC IE. Like the UM mode, if the receiving side finds a gap in the sequence of the received RLC PDUs, it assumes that the missing RLC PDU has been delayed due to HARQ retransmission of the corresponding transport block. So, it then starts the reordering timer. It stores out of sequence RLC PDUs, does not deliver them to the higher layer, and waits for the missing RLC PDU until the timer expires. If the timer expires without the arrival of the missing RLC PDU, the receiving RLC entity infers that HARQ has failed to recover the RLC PDU and the missing RLC PDU has been lost. At this juncture, unlike the UM mode, it sends a status report as shown in Section 7.3.1.3.
3. It detects lost RLC PDUs.
4. It discards duplicate RLC PDUs as explained in Section 7.2.1.5.
5. If a RLC PDU is received correctly, it is marked accordingly.
6. It moves the sliding window ahead of all sequentially received RLC PDUs with no missing PDUs. The sliding window covers intermittently received RLC PDUs.
7. For sequentially received RLC PDUs, it removes the RLC header and reassembles complete RLC SDUs. The reassembly is the reverse process of the segmentation and concatenation.
8. It passes complete RLC SDUs onto the RRC layer or PDCP layer.
9. It sends a status report to the transmitting side of the AM RLC entity when needed. The status report includes positive and negative acknowledgements of reception. This is explained in Section 7.3.1.3.

### 7.3.1.2 Actions After Acknowledgment

*Transmitting Side*
The transmitting side of the AM RLC entity performs the following functions after reception of positive acknowledgement over the status report from the peer receiving side of the AM RLC entity.

1. The transmitting side of the AM RLC entity moves its sliding window ahead of the acknowledged RLC PDUs. The sliding window covers the RLC PDUs, which are not acknowledged whether they are sent or not.
2. It removes the acknowledged data from the retransmission buffer.

The transmitting side of the AM RLC entity performs the following functions after reception of negative acknowledgement over the status report from the peer receiving side of the AM RLC entity.

1. The transmitting side of the AM RLC entity retrieves the negatively acknowledged data from the retransmission buffer.
2. It retransmits the negatively acknowledged data. If the available resource for retransmission is smaller than the original RLC PDU size, then the RLC PDU is resegmented for retransmission. This is explained in Section 7.3.2.1.

*Receiving Side*

The receiving side of the AM RLC entity performs the following functions after reception of the retransmitted data from the peer transmitting side of the AM RLC entity.

1. The receiving side of the AM RLC entity updates its buffer with retransmitted RLC PDUs. It moves the sliding window ahead of the sequentially received RLC PDUs with no missing PDUs. The sliding window covers intermittently received RLC PDUs, if there are still any.
2. For sequentially received RLC PDUs, it removes the RLC header and reassembles complete RLC SDUs.
3. It passes complete RLC SDUs on to the RRC or PDCP layer.

### 7.3.1.3 Status Report

The receiving side of the AM RLC entity sends a STATUS PDU as the status report. As shown in Section 7.3.2.2, the status report includes positive acknowledgement for correctly received RLC PDUs and negative acknowledgement for RLC PDUs, which are fully and partially missing. More frequent transmission of STATUS PDUs reduces overall transmission delay but requires more radio resources. Thus, there is a trade-off in the frequency of transmission of STATUS PDUs. The STATUS PDUs are sent in the following cases.

1. *Polling:* The transmitting side of the AM RLC entity can request a status report from the receiving side of AM RLC entity when it feels it necessary. It may do so, for example, after transmission of its last RLC PDU to ensure that all data have been duly received by the receiver. Besides, it may request the status report periodically (i.e., after transmission of a particular number of RLC PDUs in order to advance its window). It makes the request by setting the Polling (P) field in the header to 1 as shown in Section

7.3.2.1. Then the receiving side of the AM RLC entity sends a STATUS PDU at the earliest transmission opportunity.
2. *Missing RLC PDUs:* As shown in Section 7.3.1.1, if the reordering timer expires without arrival of the missing RLC PDU, the receiving RLC entity infers that the missing RLC PDU has been lost. Then, unlike the UM mode, it sends a STATUS PDU.

The MAC layer treats the STATUS PDU as any other data and thus, it applies HARQ operation to the STATUS PDU. Thus, if there is any error or loss of the STATUS PDU, it can be detected and recovered by the HARQ process.

*Status Prohibit*
While retransmissions triggered by a STATUS PDU are on the way, transmission of another STATUS PDU may cause some unnecessary retransmissions wasting radio resources. This can result in duplicate RLC PDUs at the receiving side of the AM RLC entity. Therefore, the status prohibit function can be activated to achieve proper frequency of status reporting. In this case, the receiving side of the AM RLC entity starts a timer called T-StatusProhibit when it sends a status report. Then the receiving side of the AM RLC entity is prohibited from sending further STATUS PDUs until the T-StatusProhibit timer expires.

The eNodeB configures the T-StatusProhibit timer of the UE for downlink data transfer. The RLC-Config IE includes DL-AM-RLC IE. The DL-AM-RLC IE includes the T-StatusProhibit field, which gives the duration of the T-StatusProhibit timer. Its value can be any duration between 0 ms and 250 ms with 5 ms gaps, 300, 350, 400, 450, or 500 ms. If its value is 0 ms, then no prohibition is activated.

### 7.3.1.4 Retransmission of Data

The transmitting side of the AM RLC entity retransmits missing RLC PDUs or their missing portions. The transmitting side of the AM RLC entity prioritizes retransmission of RLC PDUs over transmission of new RLC PDUs. The retransmission occurs based on the following indications.

1. *STATUS PDU:* After reception of the STATUS PDU, according to its indication, the transmitting side of the AM RLC entity retransmits completely missing AMD PDUs and all the missing portions of AMD PDUs if they are partly missing.
2. *HARQ/ARQ interaction:* If the HARQ process on the transmitting side reaches its maximum retransmission limit, it can notify the transmitting side of the AM RLC entity and the corresponding AMD PDU can be retransmitted. This allows a quicker detection of the requirement of retransmission compared to the indication using STATUS PDU.

In case of retransmission, if the available resource for retransmission is smaller than the original PDU size, then the transmitting side of the AM RLC entity re-segments the original RLC data PDUs. This is explained in Section 7.3.2.1. The segments of the original RLC data PDU can be transmitted in different transport blocks. The resegmentation for retransmission is not allowed in HSPA.

The retransmission of an RLC PDU or a portion of an RLC PDU recurs if needed. However, the transmitting side of the AM RLC stops retransmission when it reaches a maximum limit. This maximum limit is specified by the MaxRetxThreshold. For uplink data transfer, the eNodeB provides the UE with the value of the MaxRetxThreshold field. The RLC-Config IE includes UL-AM-RLC IE, which contains the MaxRetxThreshold field. The value of MaxRetxThreshold can be 1, 2, 3, 4, 6, 8, 16, or 32. When the maximum number of retransmissions is reached, the RLC layer reports it to RRC layer and the RRC layer determines that a radio link failure has occurred.

### 7.3.2 AM PDU Format

The RLC PDU in the acknowledged mode can be of two types.

1. *RLC Data PDU:* The RLC Data PDU can be of two types as follows:
   - Acknowledged Mode Data PDU (AMD PDU);
   - Acknowledged Mode Data PDU segment (AMD PDU segment).
2. *RLC Control PDU:* The RLC Control PDU has only one type called the STATUS PDU.

#### 7.3.2.1 RLC Data PDU

*Acknowledged Mode Data PDU (AMD PDU)*

The AMD PDU consists of a data field and an AMD PDU header as shown in Figure 7.6. The AMD PDU header consists of a fixed part and an extension part. Both the fixed part and the extension part are byte aligned.

*Fixed Part of Header*

The fields of the fixed part are always present in every AMD PDU. The fields are explained here.

- *Data/Control (D/C):* The D/C field has 1 bit. If it is set to 1, then the RLC PDU is an RLC data PDU. If it is set to 0, then the RLC PDU is an RLC control PDU. Thus, the D/C field is set to 1 here.
- *Resegmentation Flag (RF):* The RF field has 1 bit. It is set to 0 in this case indicating that it is an AMD PDU and not an AMD PDU segment.
- *Polling (P):* The P field has 1 bit. It is set to 1 to indicate that a status report is requested from the receiving side of RLC entity. It is set to 0 otherwise.
- *Sequence Number (SN):* The SN field is always 10 bits long. It indicates the sequence number of the corresponding AMD PDU. The sequence number is incremented by one for every successive AMD PDU. The receiving RLC entity identifies an AMD PDU using the sequence number and performs reordering and duplicate detection.

## 7.3 Acknowledged Mode (AM)

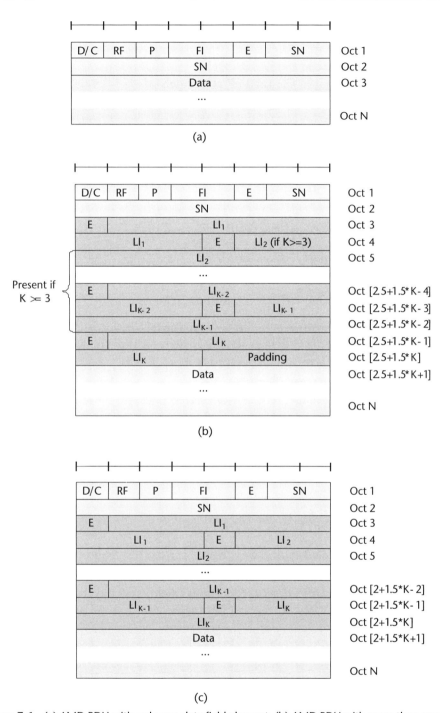

**Figure 7.6** (a) AMD PDU with only one data field element. (b) AMD PDU with more than one data field element and odd number of LIs. (c) AMD PDU with more than one data field element and an even number of LIs.

- Framing info (FI), extension (E), and reserved 1 (R1): These fields are explained in Section 7.2.2.1.

*Extension Part of Header*
The extension part is needed and present only when more than one data field elements are present in the AMD PDU. The extension part of the AMD PDU header is identical to that of the UMD PDU header, which is explained in Section 7.2.2.2.

*Acknowledged Mode Data PDU Segment (AMD PDU Segment)*
The AMD PDU segment is used to retransmit a PDU after its resegmentation. This is required when the available resource for retransmission is smaller than the original PDU size. The AMD PDU segment consists of a data field and an AMD PDU segment header. The AMD PDU segment header consists of a fixed part and an extension part as shown in Figure 7.7. Both the fixed part and the extension part are byte aligned.

*Fixed Part of Header*
The fields of the fixed part are always present in every AMD PDU segment. The fields are explained here.

- *Data/Control (D/C):* The D/C field has 1 bit. If it is set to 1, then the RLC PDU is an RLC data PDU. If it is set to 0, then the RLC PDU is an RLC control PDU. Thus, the D/C field is set to 1 here.
- *Resegmentation Flag (RF):* The RF field has 1 bit. It is set to 1 in this case indicating that it is an AMD PDU segment and not an AMD PDU.
- *Polling (P):* The P field has 1 bit. It is set to 1 to indicate that a status report is requested from the receiving side of RLC entity. It is set to 0 otherwise.
- *Sequence Number (SN):* The SN field is always 10 bits long. It indicates the sequence number of the original AMD PDU from which the AMD PDU segment has been created.
- *Framing Info (FI), extension (E), and reserved 1 (R1):* These fields are explained in Section 7.2.2.1.
- *Last Segment Flag (LSF):* The LSF has 1 bit. It indicates whether the AMD PDU segment is the last segment of the AMD PDU or not.
- *Segmentation Offset (SO):* The SO field is 15 bits long. It indicates the starting position of the AMD PDU segment within the original AMD PDU.

*Extension Part of Header*
The extension part is needed and present only when more than one data field elements are present in the AMD PDU segment. The extension part of the AMD PDU segment header is identical to that of the UMD PDU header, which is explained in Section 7.2.2.2.

### 7.3.2.2 RLC Control PDU: STATUS PDU

The only type of RLC control PDU is STATUS PDU, which has the format shown in Figure 7.8. The STATUS PDU indicates completely missing AMD PDUs and also,

## 7.3 Acknowledged Mode (AM)

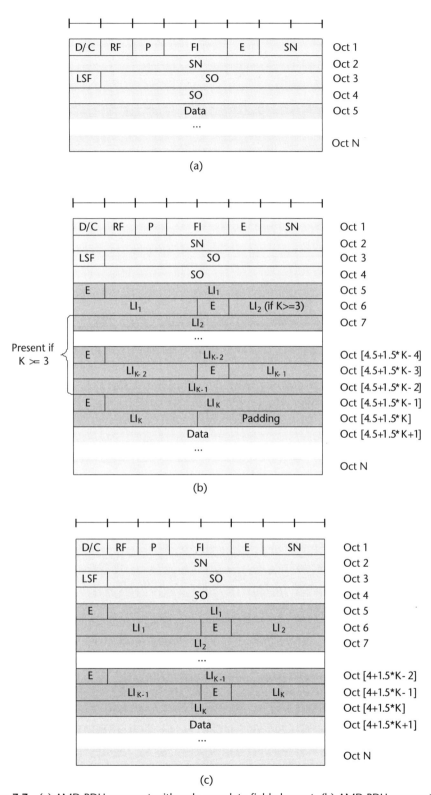

**Figure 7.7** (a) AMD PDU segment with only one data field element. (b) AMD PDU segment with more than one data field element and an odd number of LIs. (c) AMD PDU segment with more than one data field element and an even number of LIs.

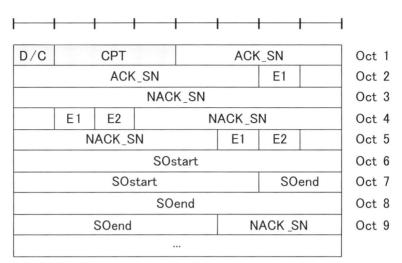

**Figure 7.8** STATUS PDU.

all the missing portions of AMD PDUs if there are partially missing PDUs. The transmitting side of the AM RLC entity prioritizes transmission of RLC control PDUs over RLC data PDUs. The fields of STATUS PDU are explained here.

- *Data/Control (D/C):* The D/C field has 1 bit. If it is set to 1, then the RLC PDU is an RLC data PDU. If it is set to 0, then the RLC PDU is an RLC control PDU. Thus, the D/C field is set to 0 here.
- *Control PDU type (CPT):* The CPT field is 3 bits long. It is for indication of the type of the RLC control PDU. In the current release, its only valid value is 0, which indicates STATUS PDU.
- *Acknowledgment SN (ACK_SN):* The ACK_SN field is 10 bits long. It gives a sequence number. All AMD PDUs up to but not including this sequence number have been correctly received by the receiver with the exception that the AMD PDUs or portions of AMD PDUs indicated in the NACK_SN fields have been missing.
- *Negative Acknowledgment SN (NACK_SN):* The STATUS PDU contains zero or more number of NACK_SN fields. The NACK_SN field is 10 bits long. A NACK_SN field indicates the sequence number of an AMD PDU that has been detected as lost at the receiving AM RLC entity. The NACK_SN field also indicates the sequence number of the AMD PDU whose portions has been detected as lost at the receiving AM RLC entity. An E1 field and an E2 field exist after each NACK_SN field and these three fields together can be considered as a set.
- *Extension 1 (E1):* An E1 field exists right after ACK_SN field. Also, every NACK_SN field is associated with an E1 field and an E2 field. The E1 field has 1 bit. It indicates whether a set of NACK_SN, E1, and E2 fields exist after this E1 field.
- *Extension 2 (E2):* Every NACK_SN field is associated with an E1 field and an E2 field. A SOstart field and a SOend field may or may not exist for each

## 7.3 Acknowledged Mode (AM)

NACK_SN field. The E2 field has 1 bit. It indicates whether a set of SOstart and SOend fields exist for the NACK_SN field associated with this E2 field.

- *SO start (SOstart):* The SOstart field is 15 bits long. It indicates the beginning of the portion of the AMD PDU, which has been detected as lost at the receiving AM RLC entity. It gives the position of the first byte of the lost portion of the AMD PDU in number of bytes considering that the first byte in the data field has the position 0.

- *SO end (SOend):* The SOend field is 15 bits long. It indicates the end of the portion of the AMD PDU, which has been detected as lost at the receiving AM RLC entity. It gives the position of the last byte of the lost portion of the AMD PDU in number of bytes considering that the first byte in the data field has the position, 0. A special value for SOend is 111111111111111. This is used to indicate that the missing portion of the AMD PDU includes all bytes up to the last byte of the AMD PDU.

# CHAPTER 8

# Packet Data Convergence Protocol (PDCP)

The Packet Data Convergence Protocol (PDCP) is the top sublayer within layer 2. The various functions of PDCP layer and the format of PDCP PDUs are explained in the following sections.

The RRC layer at eNodeB configures the PDCP layer of the UE for signaling and data radio bearers by sending PDCP-Config IE. Appendix A.2 shows how the eNodeB sends PDCP-Config IE to the UE while setting up or modifying radio bearers.

## 8.1 PDCP Layer Functions

The PDCP layer performs the following major functions as shown in Figure 8.1:

1. Processing of the user plane and control plane data packet;
2. Security function;
3. Header compression and decompression for user plane data;
4. Discard of data packets in queue;
5. Proper delivery of the packets after handover and temporary radio link failure.

### 8.1.1 Processing the Data Packet

The PDCP layer contains PDCP entities. As explained in Section 2.3, on the transmitting side, each PDCP entity receives PDCP SDUs from upper layers. It then performs the addition of the header and other necessary processing in order to generate PDCP PDUs. Thereafter, the PDCP entity passes the PDCP PDUs on to associated RLC entities as RLC SDUs for transmission over the air. On the receiving side, each PDCP entity receives PDCP PDUs from RLC entities. Then it removes the header and performs other necessary processing in order to retrieve PDCP SDUs. Thereafter, it passes the PDCP SDUs on to upper layers.

The PDCP layer processes user plane and control plane data for the following cases:

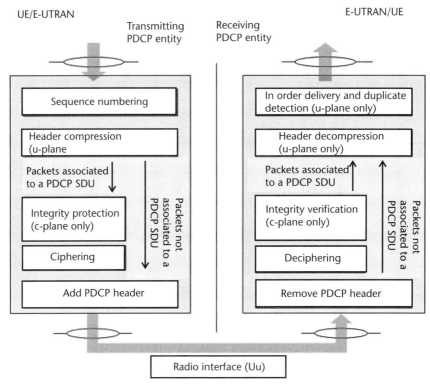

**Figure 8.1** Functional view of PDCP layer.

- SRB1 and SRB2 carrying RRC PDUs mapped on the DCCH logical channel. They use the RLC acknowledged mode.
- DRBs carrying IP packets mapped on the DTCH logical channels. They use the RLC acknowledged mode or the RLC unacknowledged mode.

Each radio bearer is associated with one PDCP entity as shown in Figure 8.2. The PDCP entities are associated with RLC entities as follows:

- *RLC unacknowledged mode:* The RLC unacknowledged mode provides a unidirectional data transfer. Therefore, two RLC entities are used for a bi-directional bearer, a transmitting UM RLC entity and a receiving UM RLC entity. In this case, each PDCP entity at UE or at eNodeB is associated with two RLC entities, one for transmission and one for reception.
- *RLC acknowledged mode:* The RLC acknowledged mode provides a bidirectional data transfer. A single AM RLC entity is used for a bidirectional bearer and it has two sides, a transmitting side and a receiving side. Thus, in this case, each PDCP entity at UE or at eNodeB is associated with one RLC entity.

### 8.1.2 Security Function

The PDCP layer performs security function. It applies ciphering to the RRC PDUs and user plane data. It also applies integrity protection to the RRC PDUs. This will be explained in Chapter 12. On the contrary, in case of HSPA, the RLC layer applies

## 8.1 PDCP Layer Functions

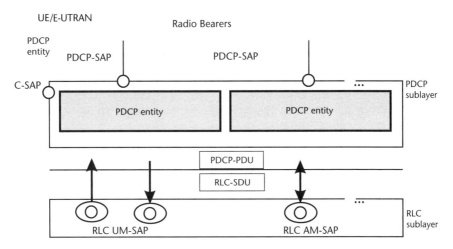

**Figure 8.2** Structural view of the PDCP layer.

ciphering to user data for the unacknowledged mode and the acknowledged mode and the RRC layer applies integrity protection.

### 8.1.3 Header Compression

The PDCP layer performs header compression of IP data at the transmitter side and header decompression of IP data at the receiver side in order to minimize protocol overheads transferred. The header compression also reduces transmission delay and packet loss rate. For this purpose, PDCP uses only the Robust Header Compression (ROHC) protocol, defined by the Internet Engineering Task Force (IETF); this protocol is a robust and efficient technique. IETF specifies a number of different sets of rules and parameters for header compression; these are known as different ROHC profiles. PDCP uses the ROHC profiles shown in Table 8.1. It may be noted that HSPA supports both IP header compression and ROHC.

The header compression is particularly useful for VoIP because there is no Circuit-Switched (CS) service for voice transfer. VoIP typically uses RTP/UDP/IP on Packet-Switched (PS) services and the header becomes very large compared to the

**Table 8.1** ROHC Profiles

| ROHC Profile Identifier | Usage | Specification |
|---|---|---|
| 0x0000 | No compression | RFC 4995 |
| 0x0001 | RTP/UDP/IP | RFC 3095, RFC 4815 |
| 0x0002 | UDP/IP | RFC 3095, RFC 4815 |
| 0x0003 | ESP/IP | RFC 3095, RFC 4815 |
| 0x0004 | IP | RFC 3843, RFC 4815 |
| 0x0006 | TCP/IP | RFC 4996 |
| 0x0101 | RTP/UDP/IP | RFC 5225 |
| 0x0102 | UDP/IP | RFC 5225 |
| 0x0103 | ESP/IP | RFC 5225 |
| 0x0104 | IP | RFC 5225 |

payload unless it is compressed. The typical size of the headers and the payload for VoIP without compression is shown here:

- RTP header: 12 octets;
- UDP header: 8 octets;
- IP header: IPv4 header, 20 octets; IPv6 header, 40 octets;
- Payload: 30 octets.

ROHC can compress headers for RTP/UDP/IP to 1–3 bytes and headers for TCP/IP to 8 bytes.

The ROHC parameters are reset during handover. In the case of HSPA, the header compression parameters are transferred between RNCs during handover for lossless Serving Radio Network Controller (SRNC) relocation.

#### 8.1.3.1 Basic ROHC Procedure

The ROHC procedure includes a compressor at the transmitting side and a decompressor at the receiving side. In order to compress the headers, the compressor explores redundancy between header fields in one IP frame (e.g., RTP and IP header) and also redundancy between header-to-header fields in one stream. In order to work efficiently with the redundancy, ROHC differentiates the static information of the header and the header parts that keep changing dynamically. For example, the static and dynamic fields may be determined as shown in Table 8.2 for RTP/UDP/IP.

The compressor and the decompressor can store a static context consisting of the static information, and thus, the compressor can exclude them from the header. The static information can be provided only when the information changes. Also, the compressor and the decompressor store a dynamic context to help the compression of the dynamic fields. The compressor, for example, can compress the RTP timestamp by transmitting only the difference from a reference clock that is maintained in the compressor and the decompressor.

The compressor and the decompressor recognize different sessions in the packet flow based on the IP version, source IP address, destination IP address, source port, destination port, and protocols. They identify each session or IP flow using a Context Identifier (CID) and then maintain static and dynamic contexts for each session (i.e., each session or IP flow is associated with an ROHC context). A radio bearer can include multiple IP flows. This information is checked for each

**Table 8.2** Static and Dynamic Information in the Headers

| Protocol | Static Fields | Dynamic Fields |
|---|---|---|
| IPv4 | Version, Protocol, Source Address, Destination Address | Type of Service, Time to Live, Identification, DF, RND, NBO, Generic Extension Header List |
| IPv6 | Version, Flow Label, Next Header, Source Address, Destination Address | Traffic Class, Hop Limit, Generic Extension Header List |
| UDP | Source Port, Destination Port | Checksum |
| RTP | SSRC | Version, P, CC, M, PT, Sequence Number, Timestamp, Generic CSRC List, X, TIS, TSS, TS Stride, Time Stride |

incoming packet to identify the CID of the session to which the packet belongs. An ROHC profile is associated with each ROHC context. Then compression is applied according to the ROHC profile of the corresponding session. A new application stream or session will have a new CID if the incoming packet contains the aforementioned information inconsistent with any of the existing sessions. The PDCP-Config IE includes the HeaderCompression IE. The HeaderCompression IE includes the ROHC IE, which contains the MaxCID field. The MaxCID field specifies the maximum value of the CID that can be used. The ROHC channels are unidirectional (i.e., there are separate channels for uplink and downlink).

The UE notifies the eNodeB about which ROHC profiles it supports. As will be shown in Section 15.1.7.4, the UE sends UECapabilityInformation message to the eNodeB to inform its radio access capabilities. This message contains the PDCP-Parameters IE, which specifies the ROHC profiles supported by the UE. The RRC layer at the eNodeB notifies the UE about which ROHC profiles can be used out of the supported ones. The PDCP-Config IE includes the HeaderCompression IE, which contains the ROHC IE. The ROHC IE specifies the ROHC profiles that can be used.

### 8.1.3.2 Modes of Operation

The compressor and the decompressor can operate in any of the following three modes:

1. *Unidirectional mode (U-mode):* In this mode of operation, the packets are transmitted from the compressor to the decompressor and there is no feedback from the decompressor to the compressor. This mode is used when a return path from the decompressor to the compressor is unavailable or undesirable. Since there is no feedback, the compressor is not aware of the condition of the decompressor. Thus, the compressor cannot use the information about the condition of the decompressor to change its state.
2. *Bidirectional optimistic mode (O-mode):* In this mode of operation, the feedback from the decompressor to the compressor is used to send error recovery requests. Optionally, the feedback is also used to send acknowledgments of significant context updates. Here the feedback channel is used only sparsely. This mode attempts to maximize compression efficiency through the sparse usage of the feedback channel.
3. *Bidirectional reliable mode (R-mode):* In this mode of operation, the robustness against loss and damage is maximized although this requires the intensive usage of the feedback channel. Attempts are made to prevent the loss of context synchronization between the compressor and the decompressor. The feedback channel is used to acknowledge all context changes. In R-mode, two metrics of compression performance are the best addressed. They are the robustness against damage propagation and the robustness against loss propagation. The damage propagation refers to incorrect decompressed headers because of error-prone headers or feedback information. The loss propagation refers to out-of-sync contexts because of lost packets.

### 8.1.3.3 States of Compressor and Decompressor

The compressor can acquire any of the following three states as shown in Figure 8.3:

1. *Initialization and Refresh State (IR state):* The compressor stays in this state in the beginning of transmission and attempts to initialize the decompressor with static and dynamic contexts. The compressor sends IR packets that initialize static and dynamic contexts. A feedback from the decompressor may indicate that the IR packet has been decompressed successfully. If feedback is not possible, then the compressor may send a few IR packets consecutively and consider it as successful initialization. Thereafter, the compressor can move to the first-order state (FO state).
2. *First-Order State(FO state):* The compressor sends the partially compressed dynamic context (i.e., the changes in the dynamic context in this state). The static context is stored and only a few numbers are updated from the static context. The compressor moves to the FO state from the IR state after transmitting the static context. Also, the compressor moves to the FO state from the second-order (SO) state when there are changes in the dynamic context. The compressor moves to the IR state from the FO state when parts of the static context change. The compressor moves to the SO state from the FO state when the compressor is confident enough that the decompressor is fully aware of the dynamic context and all parameters to derive the header fields.
3. *Second-Order State (SO state):* In general, the compressor compresses all the static fields and most of the dynamic fields in this state. Thus, it provides the optimum level of compression and it is the most efficient. The compressor suppresses all dynamic fields and sends only the sequence number and a few bits. The decompressor intuitively generates and verifies the headers of the next expected packet. The compressor moves to the SO state when the compressor is confident enough that the decompressor is fully aware of the dynamic context and all parameters to derive the header fields. The compressor moves to the SO state from the FO state when the compressor is confident enough that the decompressor is aware of the dynamic context.

The decompressor can acquire any of the following three states as shown in Figure 8.4:

1. *No context:* The decompressor does not possess any context in this state. The decompressor stays in this state in the beginning of transmission and it **the decompressor** is yet to be initialized with static and dynamic contexts. The decompressor receives only IR packets from the compressor.

**Figure 8.3** Compressor states.

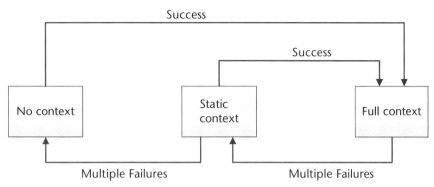

**Figure 8.4** Decompressor states.

2. *Static context:* The decompressor possesses the static context but not the dynamic context in this state. The decompressor can decompress certain type of packets. Once the decompressor acquires the dynamic context, it moves to the full context state. However, if decompression fails repeatedly while in the static context state, the decompressor will move to the no context state.

3. *Full context:* The decompressor has acquired the full context in this state. Thus, it can now decompress packets from the compressor that is in the SO state and provide maximum efficiency. If decompression fails repeatedly while in the full context state, the decompressor will move to the static context state.

### 8.1.4 Discard of Data Packets

In order to prevent from too much buffering at the transmitter when the data rate over the radio link gets worse, the PDCP layer can be configured to discard data packets after they have waited for transmission for certain time. This loss of packets is taken care of by upper layer protocols, TCP or RTP. TCP or RTP controls the rate of data flow considering the amount of packet losses as an indication of the available radio link capacity.

The transmitting side of PDCP layer in both uplink and downlink performs the discard of packets. The RRC layer at the eNodeB configures how long the PDCP layer at the UE would wait for transmission before discarding a PDCP SDU using a timer called DiscardTimer. The eNodeB includes the DiscardTimer field in the PDCP-Config IE. The DiscardTimer field can have the values of 50, 100, 150, 300, 500, 750, or 1,500 ms. The DiscardTimer field can also be infinity, which is interpreted as not configuring any discard of packets. The value of the DiscardTimer field can be set according to the QoS of the radio bearer. The UE starts the DiscardTimer when a PDCP SDU is received from upper layers. If the DiscardTimer expires without the transmission of the PDCP SDU, the UE discards the PDCP SDU along with its PDCP PDU. If the PDCP PDU has already been moved to lower layers of the UE, the PDCP layer notifies the lower layers that the PDCP PDU has been discarded.

### 8.1.5 Proper Delivery of Packets After Handover and Temporary Radio Link Failure

The PDCP layer is reestablished after handover and when the RRC connection reestablishment occurs after a temporary radio link failure. In this case, RLC and MAC layers are reset. Since the RLC layer is reset, it can no longer deliver PDCP SDUs in order. Also, there can be data missing due to the interruption of service. It is then the PDCP layer that offers the delivery of packets in order, duplicate detection, and necessary retransmission. In the case of the HSPA, PDCP layer cannot offer these services.

After the reestablishment of PDCP layer, the ROHC parameters are reset. In addition, the PDCP layer acts as follows:

- *Data Radio Bearer (DRB) of the RLC acknowledged mode:* SN and HFN values are maintained and not reset. SN and HFN values will be explained in Section 12.4.2. The PDCP SN is 12 bits long for the RLC AM and the PDCP SN cannot be 7 bits. This long PDCP SN is used so that the number of lost PDCP PDUs does not cause the PDCP SN to wrap around its range (i.e. the PDCP SN to reach its maximum limit, restart and get back to its previous value within the loss of PDCP PDUs). The transmitting side in both uplink and downlink retransmits all PDCP SDUs that are not yet acknowledged by the receiving RLC AM entity. The receiving side in both uplink and downlink reorders the PDCP PDUs using SN and HFN values. It detects duplicate PDCP PDUs and discards them. The reordering window size is 2,048. If there is any PDCP PDU missing and it does not arrive within the reordering window, the PDCP layer of the receiving side forwards the existing PDCP SDUs to the upper layer. It also sends a PDCP status report to the transmitting side indicating the missing PDCP PDUs, if configured, and thus requests for necessary retransmission. If a PDCP SDU is correctly received but its header decompression fails, then it is also indicated as missing in the PDCP status report. The PDCP status report is explained in Section 8.2.2. The transmitting side stores PDCP PDUs until the DiscardTimer expires, as explained in Section 8.1.4. When the transmitting side finds that the receiving side has any of the stored PDCP PDUs missing, it retransmits the missing ones. An additional advantage of the PDCP status report is that if a PDCP SDU is received successfully but its RLC acknowledgment is lost, then the PDCP SDU can be retransmitted unnecessarily. The PDCP status report precludes this unnecessary retransmission.
  
  The PDCP-Config IE configures whether the UE would send a PDCP status report upon reestablishment of the PDCP entity or not. The PDCP-Config IE includes the RLC-AM IE. The RLC-AM IE includes the StatusReportRequired field, which specifies if the UE will send a PDCP status report or not.

- *Data Radio Bearer (DRB) of the RLC unacknowledged mode:* SN and HFN values are reset. The transmitting side transmits all PDCP SDUs that have been buffered but not yet sent to the RLC layer. The receiving side does not send a PDCP status report.

- *Signaling Radio Bearer (SRB):* SN and HFN values are reset. All PDCP SDUs are discarded that have been received in the meantime.

## 8.2 PDCP Protocol Data Unit (PDU)

The PDCP PDU can be of two types:

1. *PDCP data PDU:* It contains user plane data or control plane data.
2. *PDCP control PDU:* It contains PDCP status report or header compression control information.

The PDCP PDUs includes D/C field in the beginning of the header of PDCP PDUs. The D/C field indicates if the PDCP PDU is PDCP data PDU or PDCP control PDU. Security is applied only to the PDCP data PDUs. The PDCP control PDUs are neither ciphered nor integrity protected. Therefore, the PDCP data PDUs include PDCP SN, but the PDCP control PDUs do not.

### 8.2.1 PDCP Data PDU

The PDCP layer generates PDCP data PDUs from PDCP SDUs. The PDCP data PDU is of two types, based on the content of the PDCP SDU:

1. User plane PDCP data PDU;
2. Control plane PDCP data PDU.

Ciphering is applied to both user plane PDCP data PDU and control plane PDCP data PDU. Integrity protection is applied only to the control plane PDCP data PDU.

#### 8.2.1.1 User Plane PDCP Data PDU

The user plane PDCP data PDU contains user plane data for the DRB. As will be shown in Section 12.4.2, two different sizes of PDCP Sequence Number (SN), 7 bits and 12 bits, are defined for DRBs with RLC UM. Therefore, the PDCP data PDU can have any of the following two formats, shown in Figure 8.5:

1. *PDCP data PDU with long PDCP SN:* This format of the PDCP data PDU includes 12-bit-long PDCP SN in the header. DRBs with RLC AM or RLC UM use this format. Thus, the range of the PDCP SN is from 0 to 4,095. In the case of HSPA, the SN ranges from 0 to 65,535.
2. *PDCP data PDU with short PDCP SN:* This format of the PDCP data PDU includes a 7-bit-long PDCP SN in the header. DRBs with only RLC UM use this format.

#### 8.2.1.2 Control Plane PDCP Data PDU

The control plane PDCP data PDU contains control plane data for SRBs (i.e., RRC signaling messages). It includes a 5-bit-long PDCP SN in the header, as shown in Figure 8.6. It also includes a MAC-I at the end of the PDU for integrity protection as will be explained in Section 12.4.4.

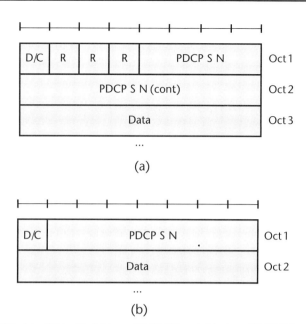

**Figure 8.5** PDCP data PDU with (a) long PDCP SN and (b) short PDCP SN.

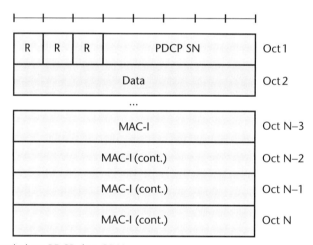

**Figure 8.6** Control plane PDCP data PDU.

### 8.2.2 PDCP Control PDU

The PDCP control PDU has two types:

1. *Interspersed ROHC feedback packet:* The PDCP layer sends this PDU to transmit feedback for the ROHC operation explained in Section 8.1.3. Its format is shown in Figure 8.7(a). The PDU type field indicates that the PDU type is interspersed ROHC feedback packet and not PDCP status report. The PDU contains no more than one ROHC feedback packet in order to avoid complexity.
2. *PDCP Status Report:* The format for the PDCP control PDU for the PDCP status report is shown in Figure 8.7(b). The PDU type field indicates that

## 8.2 PDCP Protocol Data Unit (PDU)

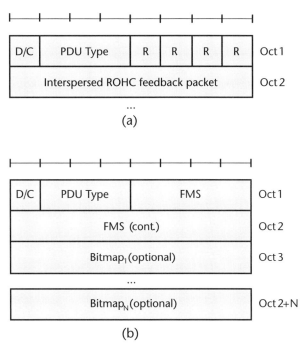

**Figure 8.7** PDCP control PDU for (a) the interspersed ROHC feedback packet and (b) the PDCP status report.

the PDU type is PDCP status report and not interspersed ROHC feedback packet. The PDU also includes the following fields:

- *First missing PDCP SN (FMS):* The FMS field is set to the PDCP SN of the first missing PDCP SDU.
- *Bitmap:* The bitmap indicates which PDCP SDUs requires retransmission considering FMS as the reference for the first missing PDCP SDU. The length of the bitmap field in bits is made equal to the number of PDCP SDUs from and not including the first missing PDCP SDU, up to and including the last out-of-sequence PDCP SDU. A bit in the bitmap is set to 0 if the corresponding PDCP SDU has not been received successfully.

If all PDCP SDUs are received successfully in sequence, then the bitmap field is not included and the FMS field indicates the next expected PDCP SN.

# CHAPTER 9

# Powering Up UE

When the UE is powered up, it first performs cell selection. The cell selection procedure is explained in Section 9.1. The UE camps on a cell at the end of successful cell selection and then it enters normal service of network. The network service types are explained in Section 9.2. Thereafter, the UE performs initial attach procedure in the cell in order to register with the network. The initial attach procedure will be explained in Section 15.1.

## 9.1 Cell Selection

The UE selects a suitable cell for services; this is known as *cell selection*. Apart from powering up, the UE may also perform cell selection after it has lost services. In RRC_IDLE state, if the UE fails to find any new suitable cell by searching and measuring intrafrequency, interfrequency, and inter-RAT cells based on information in the system information for 10 seconds, then the UE is considered to be out of service. In this case, the UE initiates the cell selection procedure for the selected PLMN.

The cell selection takes place in one of the two following ways.

- Initial cell selection;
- Stored information cell selection.

These procedures are explained in the following sections.

### 9.1.1 Initial Cell Selection

The UE performs initial cell selection when the UE does not have information stored about the E-UTRA carrier frequency or about the cell to be selected. The initial cell selection procedure includes the following steps.

1. *Cell search:* The UE performs a cell search procedure in order to find cells with strong signals on different E-UTRA carriers, and, in general, this procedure takes place as shown next. Apart from powering up, the UE also carries out cell search procedure for neighbor cell measurements as a part of cell reselection and handover procedure.

- *Scanning E-UTRA bands:* In all operating bands, the defined center carrier frequencies indicated by downlink EARFCNs have a 100-kHz step size. In addition, PSS, SSS, and PBCH are mapped on the central 72 subcarriers regardless of the size of the channel bandwidth. The UE may not use any prior knowledge of which RF channels are the E-UTRA carriers. The UE keeps a list of supported E-UTRA bands with priority. The UE sweeps through all supported E-UTRA bands one by one in priority order in steps of 100 kHz and searches for a strong signal on E-UTRA carriers within the bands. The UE checks for energy only in the central part of the channel bandwidth around downlink EARFCNs regardless of the size of the channel bandwidth. In essence, the UE performs a rough frequency synchronization to find out good candidates with strong signals on 72 subcarriers on any E-UTRA carrier. It is possible that the UE remembers the E-UTRA carrier on which it camped last time and this E-UTRA carrier is prioritized.
- *Downlink synchronization:* The UE performs synchronization procedure on strong cells using PSS and SSS as explained in Section 10.1. This allows the UE to attain the following:
  - Synchronization with downlink transmission;
  - Confirmation of exact carrier frequency or EARFCN;
  - Detection of system, if it is FDD or TDD;
  - Determination of the PCI of the cell;
  - Determination of the downlink cell-specific reference signal sequence;
  - Determination of the length of cyclic prefix, if it is normal or extended;
  - Estimation of downlink propagation channel.
- *Measurement of downlink channel quality:* The UE roughly measures a cell-specific reference signal, determines RSRP and RSRQ, and, based on RSRP and RSRQ, determines the strongest cell on each E-UTRA carrier.

2. *Reading MIB:* After downlink synchronization, the UE has both frequency and timing information about PBCH. The UE reads MIB on PBCH for the best cells on each E-UTRA carrier. The MIB message includes the DL-Bandwidth field, which gives the size of the system bandwidth in terms of the number of resource blocks. The size can be 6, 15, 25, 50, 75, or 100 resource blocks, which corresponds to the system bandwidth of 1.4, 3, 5, 10, 15 or 20 MHz, respectively. The MIB message also includes PHICH-Config IE, which contains configuration for PHICH as shown in Section 5.1.4 and so, this IE is required for reading PDCCH in the control region. In addition, the MIB message includes the SystemFrameNumber field, which gives the 8 most significant bits of the System Frame Number (SFN). The UE can determine the whole SFN from this information, as shown in Section 4.3. Moreover, the UE determines the number of transmit antennas by attempting to decode PBCH blindly for all antenna diversities. At this juncture, the UE can first detect PCFICH, then it can detect PDCCH from PCFICH, and then it can detect PDSCH from PDCCH.

3. *Reading SIB type 1:* The UE reads SIB type 1 on PDSCH. SIB type 1 message includes the following IEs:

- *FreqBandIndicator:* This field indicates the operating frequency band of the cell.
- *PLMN-IdentityList:* This IE contains a list of PLMNs. It includes the Mobile Country Code (MCC) and the Mobile Network Code (MNC), which uniquely identifies the PLMN. The first PLMN in the list is the primary PLMN to which the cell belongs. For each of the PLMNs in the list, the CellReservedForOperatorUse IE indicates whether the specific PLMN is reserved for operator use or not. This IE is normally set to "not reserved," allowing an ordinary UE to select the particular cell. However, if this IE is set to reserved, only the UEs assigned to particular access classes can select the cell. The access class is stored in the USIM.
- *CellBarred:* This field indicates if the cell is "barred" or "not barred." In the case of multiple PLMNs indicated in the PLMN-IdentityList, this IE is common for all PLMNs. This IE is normally set to "not barred," allowing the UE to select the particular cell. However, if this IE is set to "barred," the UE is not permitted to select the cell, not even for emergency calls.
- *CSG-Indication:* If this field is true, then the cell is a CSG cell. In this case, SIB type 1 also includes the CSG-Identity field. Then only the UEs who are the members of the CSG can access the cell.
- *CSG-Identity:* This field gives the CSG identity of the CSG cell. A UE is only allowed to access the CSG cell if the CSG identity stored in the UE matches with this CSG identity.
- *TrackingAreaCode:* This field gives the Tracking Area Code (TAC) to which the cell belongs. Since SIB type 1 also contains the PLMN identity of the cell, it thus actually contains the tracking area identity (TAI) of the cell, which is constructed from MCC, MNC, and TAC. The UE can select the cell if the tracking area does not belong to forbidden tracking areas.
- *CellIdentity:* This field gives the eNodeB Cell Identifier (ECI) of the cell that uniquely identifies the cell within the PLMN. Since SIB type 1 also contains the PLMN identity of the cell, it thus actually contains eNodeB Cell Global Identifier (ECGI) of the cell, which is constructed from MCC, MNC, and ECI.

4. *PLMN selection:* Out of the observed cells, the UE selects a cell with a particular PLMN. The PLMN selection can be automatic or manual. In automatic mode, the UE attempts to select PLMN in the following ways with the shown order of priority:
   - The UE first attempts to select the last registered PLMN (RPLMN) if it is available.
   - The UE attempts to select a higher-priority EHPLMN from the EHPLMN list if the EHPLMN list exists. If the EHPLMN list does not exist, then the UE attempts to select the HPLMN (i.e., the PLMN code derived from the IMSI). This EHPLMN list is expected to include the PLMN code derived from the IMSI (i.e., HPLMN) at the top of its list but it is not mandatory for the EHPLMN list to include HPLMN. If not included, then the PLMN code derived from the IMSI is treated as a VPLMN instead of HPLMN.

- The UE attempts to select a PLMN from the user- and operator-defined list of PLMNs in priority order.
- The UE attempts to select a PLMN from high-quality PLMNs randomly.
- The UE attempts to select a PLMN from other PLMNs with higher RSRP value.

If the UE operates on a VPLMN, it periodically searches for higher priority PLMN. The higher-priority PLMN is chosen the same way as mentioned above. The interval between the periodic searches is defined in the USIM and it can be between 6 minutes and 8 hours.

The selected PLMN becomes an RPLMN and it is stored in the USIM for subsequent PLMN selection. If a PLMN is selected by the UE but registration with the selected PLMN is rejected, then the PLMN is added to forbidden PLMN (FPLMN) list stored in the UE. A forbidden PLMN can later be selected only with the manual mode of the UE where the user selects the PLMN manually.

5. *Reading other SIB types:* The UE reads SIB type 2 and other SIB types. The UE is required to read at least MIB, SIB type 1, and SIB type 2 in order to camp on a cell.
6. *Suitability (S) criteria:* The UE checks if the best cell on the selected PLMN is suitable or not. The UE measures the cell-specific reference signal of the chosen cell and determines the RSRP and RSRQ. The UE considers the cell suitable if it fulfills the following S criteria:

[Srxlev of the cell > 0] according to Release 8
[Srxlev of the cell > 0] and [Squal of the cell > 0] according to Release 9

Here

Srxlev = Qrxlevmeas − (Qrxlevmin + Qrxlevminoffset) − Pcompensation

Pcompensation = MAX [$P_{EMAX} - P_{PowerClass}$, 0]
Squal = Qqualmeas − (Qqualmin + Qqualminoffset)

The parameters in these equations are explained next:
- *Qrxlevmeas:* It is the reference signal received power (RSRP) in dBm measured by the UE in the cell.
- *Qqualmeas:* It is the Reference Signal Received Quality (RSRQ) in dB measured by the UE in the cell.
- $P_{PowerClass}$: $P_{PowerClass}$ is the maximum RF output power of the UE according to the particular power class of the UE.
  The rest of the parameters in the equations are broadcast on SIB type 1 and they are explained next:
- *Qrxlevmin:* Qrxlevmin is the minimum required RSRP for selection of the cell. The Q-Rxlevmin field on SIB type 1 specifies the value of Qrxlevmin. The actual value of Qrxlevmin is, the Q-Rxlevmin field value × 2 [dBm]. The value of Q-Rxlevmin can be any integer between −70 and −22.

- *Qrxlevminoffset:* Qrxlevminoffset is an offset used to prioritize camping on a PLMN. The Q-RxLevMinOffset field on SIB type 1 specifies the value of Qrxlevminoffset. The actual value of Qrxlevminoffset is the Q-RxLevMinOffset field value × 2 [dBm]. The value of Q-RxLevMinOffset can be any integer up to 8. The Qrxlevminoffset is applied in S criteria while evaluating cells of different priority PLMN during the periodic search. The UE might have stored the value of QrxlevminOffset of a different cell earlier from the SIB type 1.
- $P_{EMAX}$: $P_{EMAX}$ is the maximum allowed uplink transmission power broadcast by the eNodeB. The P-Max field on SIB type 1 specifies the value of $P_{EMAX}$ to be used for the current cell. Its value can range between – 30 dBm and 33 dBm.
- *Qqualmin:* This IE is introduced in Release 9. Qqualmin is the minimum required RSRQ in the cell for selection of the cell. The Q-QualMinfield on SIB type 1 specifies the value of Qqualmin. The value of Q-QualMin can be any integer between – 34 dB and – 3 dB.
- *Qqualminoffset:* This IE is introduced in Release 9. Qqualminoffset is an offset used to prioritize camping on a PLMN the same way as a QrxlevminOffset IE is used. The Q-QualMinOffsetfield on SIB type 1 specifies the value of Qqualminoffset. The value of Q-QualMinOffset can be any integer up to 8 dB. The UE might have stored the value of Qqualminoffset of a different cell earlier from the SIB type 1.

7. *Camping:* If the cell is suitable, then the UE selects the cell for services. This is known as camping on the cell. The UE is now ready for initial attach procedure.

### 9.1.2 Stored Information Cell Selection

The UE performs stored information cell selection when the UE has information stored about the E-UTRA carrier frequency, PLMN, and/or cell for selection from previously received measurement report information or from previously detected cells. Then the UE attempts to find a suitable cell that matches the information stored. If such a cell is found, the UE camps on it. However, if no suitable cell is found matching the stored information, the UE initiates the initial cell selection procedure.

## 9.2 Network Services

After camping on a cell, the UE enters one of the following types of services:

1. *Normal service:* After camping on a suitable cell, the UE enters normal service. The cell is now called a *serving cell*. The next step for the UE is the initial attach, which registers the UE with the network. The initial attach procedure is explained in Section 15.1. In addition, the UE continues performing the following tasks:
   - The UE monitors system information messages. SIB type 2 provides uplink EARFCN, uplink bandwidth, and random access configuration

required to initiate uplink communication. The UE uses them to begin the attach procedure.
- The UE monitors the indicated paging channels according to information sent in the system information.
- The UE measures the RSRP and RSRQ from the cell-specific reference signals of the current cell and evaluates if the cell remains suitable at least once in every DRX cycle.
- The UE performs measurements of neighbor cells and also, performs cell reselection when certain conditions are satisfied.

2. *Limited service:* When a suitable cell is not found but a cell is found fulfilling the following requirements, the cell is called an *acceptable cell* and the UE enters limited service:
   - The cell is not barred. If a cell is barred, it is indicated in SIB type 1.
   - The S criteria are fulfilled.

   When the UE stays in limited service, it can only make emergency calls and receive ETWS notifications.

3. *Operator service:* The UEs assigned to particular access classes can select a reserved cell and operate on it for operator services purposes. In this case, the UE enters operator service. If a cell is reserved, it is indicated in SIB type 1.

# CHAPTER 10

# Synchronization

## 10.1 Downlink Synchronization

When the UE is powered up or after the UE has lost services, the UE is required to synchronize with the downlink transmission from the eNodeB. Thus, the UE performs downlink synchronization during the cell search as shown in Section 9.1.1. For downlink synchronization, the physical layer of the UE goes through a step with Primary Synchronization Signal (PSS) and then another step with Secondary Synchronization Signal (SSS). It may be noted that the UE performs two steps similarly in case of UMTS, with Primary Synchronization Channel (P-SCH) and Secondary Synchronization Channel (S-SCH).

### 10.1.1 Scheduling PSS and SSS

The eNodeB transmits the PSS and SSS, which are specific sequences. They are made available on the first 0.5-ms slot of the first and the sixth subframes (subframes 0 and 5) of each radio frame. On these slots, the last symbol (the sixth symbol for extended cyclic prefix and the seventh symbol for the normal cyclic prefix) carries the PSS. The symbol right before the last one (the fifth or sixth symbol) carries the SSS. In the frequency domain, the PSS and SSS sequences are placed on the central six resource blocks regardless of the system bandwidth (i.e., they are allocated 6 × 12 = 72 subcarriers around the DC subcarrier). It may be noted that the smallest system bandwidth consists of 72 subcarriers. The UE does not need to know the system bandwidth when it detects the PSS and SSS. A PSS or SSS instance is a 62-symbol-long sequence, although it is allocated 72 symbols on the 72 subcarriers. It is mapped on the central 62 subcarriers of the available 72 subcarriers. The remaining 10 subcarriers on the two sides are not used. This allows the UE to detect the PSS and SSS using a size-64 FFT, which is a lower sampling rate.

When the eNodeB uses multiple transmit antennas, the PSS and SSS are always transmitted from the same antenna port in any given subframe. However, various antenna ports may be used in different subframes in order to benefit from the time-switched antenna diversity.

### 10.1.2 Physical Cell Identifier (PCI)

There are 504 different Physical Cell Identifiers (PCIs) to identify different cells in a neighborhood (i.e., each cell in a neighborhood is assigned a unique PCI). The 504 PCIs are divided into 168 different cell identity groups. Each cell identity group consists of three PCIs. Typically, each eNodeB is allocated a cell identity group. Then the three sectors under the eNodeB, namely, alpha, beta, and gamma are allocated the three PCIs belonging to the cell identity group. Since there are as many as 504 possible PCIs, in usual cell planning, there is no chance of the reuse of the same PCI in the neighborhood. Thus, there should not be any conflict with PCIs. However, if too many Home eNodeBs are used in high-density areas, special management may be required with PCIs in order for the UE to avoid coming across the same PCI representing two different cells. Also, the available PCIs can be split between regular cells and home eNodeBs to allow the UE to limit the number of PCIs that it needs to be monitored.

As shown in Section 5.1.5.1, the downlink cell-specific reference signal sequence is derived using the PCI of the cell. There are 504 reference signal sequences defined corresponding to the 504 different PCIs. Once the UE detects the PCI of the cell, the UE can then also identify the reference signal sequence. When the reference signal sequence is known, the UE can perform estimation of downlink propagation channel and measurement of downlink channel quality from the cell-specific reference signals.

### 10.1.3 PSS Sequence

The PSS sequence is constructed from a frequency-domain Zadoff-Chu (ZC) sequence. The PSS uses roots $u = 25, 29$, and 34 for a ZC sequence as they provide good periodic autocorrelation and cross-correlation properties. Thus, three particular PSS sequences are used. These three PSS sequences are mapped to the three cell identities in the cell-identity group. Thus, typically, the three sectors under the eNodeB, namely, alpha, beta, and gamma, are assigned the three PSS sequences.

If the length of the ZC sequence, $N_{ZC}$, is a prime number, then it results in an optimal cyclic cross-correlation between any pair. The PSS uses the length, $N_{ZC} = 63$, which is a prime number fitting into the available resources appropriately. The ZC sequence for the PSS is as follows:

$$d_u(n) = \exp\left[-j\frac{\pi u n(n-1)}{63}\right] \quad n = 0, 1, 2, \ldots, 30$$

where the root $u = 25, 29$, or 34.

The middle element of the 63-element-long PSS sequence is punctured to suppress transmission on the DC subcarrier. The remaining 62-symbol-long sequence is mapped on the central 62 subcarriers of the available 72 subcarriers.

The UE detects the PSS by correlating with three possible PSS sequences. In this detection, the UE does not have any prior knowledge of the channel, so it uses noncoherent detector. Once the PSS is detected, the UE acquires the following information.

1. *Confirmation of downlink EARFCN:* With the detection of the PSS, the UE can confirm the location of central 72 subcarriers and thus can confirm the downlink E-UTRA carrier frequency or EARFCN. Such confirmation is more important for high-speed users because of the Doppler shift.
2. *Slot timing:* Since the PSS is located on the last symbols (sixth or seventh) of the slots, the UE can acquire the slot boundary regardless of the length of the cyclic prefix (CP). Thus, the UE is synchronized with the slot timing of the downlink transmission.
3. *Cell identity within the cell-identity group:* The UE identifies one of the three roots of the ZC sequence, and thus, the UE identifies the cell identity within the cell-identity group. However, the cell-identity group is still unknown. However, it is not clear yet where the radio frame starts because both first and sixth subframes have the same PSS.
4. *Estimation of propagation channel:* The UE acquires estimation of downlink propagation channel from PSS sequence detection. The UE can use this estimate for subsequent coherent detection.

### 10.1.4 SSS Sequence

The SSS sequences are based on maximum length sequences, known as *M-sequences*, which can be created by cycling through every possible state of a shift register of length 16. This results in a sequence of length 31. Two codes, SSC1 and SSC2, are first generated as two different cyclic shifts of a single length-31 M-sequence. Then they are interleaved and concatenated, giving rise to an SSS sequence of length 62. The concatenated sequence is scrambled with a scrambling sequence given by PSS. The 62-symbol-long sequence is mapped on the central 62 subcarriers of the available 72 subcarriers. There are 168 different pairs of sequences (S1, S2) used for the SSS where S1 and S2 are mapped on the first and the sixth subframes, respectively, in each radio frame. Thus, there are 168 specific SSS sequences. Each SSS sequence corresponds to a cell-identity group. Also, typically, each eNodeB in a neighborhood is assigned a particular SSS sequence such that the same SSS sequence does not repeat in the close vicinity.

The UE detects SSS by correlating with 168 possible SSS sequences. Since the UE has estimated downlink propagation channel from PSS sequence detection, it can use either coherent detector or non-coherent detector for SSS sequence detection. Once SSS is detected, the UE acquires the following information.

1. *Cell-identity group:* Each pair of SSS sequence (S1, S2) identifies the corresponding cell-identity group. Thus, after detection of both the PSS and the SSS, the UE is aware of both the cell-identity group and the cell identity within the group. Therefore, the UE can determine the PCI of the cell. Then the UE can determine the downlink cell-specific reference signal sequence. The UE can now measure the downlink channel quality from the cell-specific reference signal.
2. *Radio frame boundary:* The SSS has S1 and S2 sequences mapped on the first and the sixth subframes, respectively, in every radio frame, but a reverse sequence (S2, S1) does not exist. Therefore, the UE can determine

which one is the first subframe in a radio frame and which one is the sixth subframe. Thus, the boundary of the radio frame is detected.

3. *Length of the Cyclic Prefix (CP):* Since the PSS is located on the last symbol of the slot, the UE acquires the slot boundary from the PSS, regardless of the length of the CP in the downlink transmission. Now the precise position of the SSS in the slot depends on the length of the CP. The UE checks on the position of the SSS assuming that there could be either a normal CP or an extended CP. Thus, once the SSS is detected, the UE determines the length of the CP in the downlink transmission.

4. *Early detection of system type, FDD or TDD:* The UE identifies if the system type is FDD or TDD from the detection of the SSS, if required. In case of the TDD, the PSS is located in the third symbol of the third and thirteenth slots of a radio frame (the first slot of the second subframe and the first slot of the seventh subframe). This allows the PSS sequences to appear at every 5 ms the same way as they do in case of FDD. Thus, the detection of the PSS does not allow the UE to identify if the system type is FDD or TDD. However, the SSS is located three symbols earlier than the location of the PSS in the case of TDD. On the other hand, the SSS is located immediately before the PSS in case of FDD. Therefore, once the SSS is detected, the UE can determine if the system type is FDD or TDD.

The detection of the SSS completes the downlink synchronization of the UE. It may be noted that the UE may need to check a total of four possible positions for the SSS sequence: two positions for the normal CP and the extended CP for FDD and two positions for the normal CP and the extended CP for TDD.

## 10.2 Uplink Synchronization

The physical layer of the UE is required to have an uplink time alignment or synchronization with the eNodeB so that there is no interference among the uplink transmissions from different UEs. This requires the consideration of downlink and uplink propagation delays. The uplink transmissions from various UEs need to arrive at the eNodeB within the respective CP. The uplink time alignment is affected by the following primary factors:

1. The distance between the UE and the eNodeB varies because of the movement of the UE. Thus, the propagation delay between the UE and the eNodeB also varies.
2. The multipath propagation channel constantly varies with time. This causes a variation in the delay spread.
3. The velocity of the UE gives rise to a Doppler shift. This causes a shift in timing.
4. Oscillator drift in the UE can cause a shift in timing.

A timing control is required to establish uplink time alignment and update it constantly. The MAC layer at the eNodeB performs this timing control. The eNodeB sends a timing advance command to the UE with the necessary time adjustment information. The eNodeB first establishes the timing alignment and thereafter maintains it through regular updates. The timing alignment involves the following items:

- *Timing offset:* After the timing alignment, the UE sets up an appropriate timing offset between its transmitted radio frames in uplink and its received downlink radio frames, as shown in Figure 10.1. The timing offset accounts for both downlink and uplink propagation delays. The timing offset is denoted as $N_{TA}$ and expressed in terms of the basic time unit, $Ts = 1/(2{,}048 \times \Delta f)$ = 1/30,720,000 seconds ≈ 32.552 ns. The transmission of an uplink radio frame from the UE starts $N_{TA} \times Ts$ seconds before the start of the corresponding received downlink radio frame at the UE. The timing offset can support at least a 100-km cell range.

- *Timing advance command:* The eNodeB sends a timing advance command to the UE for both the initial establishment of the timing alignment and subsequent maintenance of the timing alignment. The timing advance command gives a value, $T_A$. The UE can calculate $N_{TA}$ based on $T_A$ and thus, the UE can also calculate the timing offset. Once the UE receives the timing advance command, it adjusts timing for the uplink transmission on PUCCH, PUSCH, and SRS. If the timing advance command is received on subframe $n$, the adjustment of the timing applies from the subframe $n + 6$.

### 10.2.1 Establishment of Timing Alignment

The UE must be already synchronized to the downlink transmission before it establishes the uplink synchronization. The uplink synchronization is established using the random access procedure explained in Chapter 11. The UE sends a random access preamble on the Physical Random Access Channel (PRACH), while the UE is unaware of its propagation delay in the uplink. This is the only allowed nonsynchronized transmission of the UE in the LTE. After the reception of the random access preamble, the eNodeB estimates the uplink transmission timing of the UE. Then the eNodeB sends a Random Access Response (RAR) message on the PDSCH and an indication of the RAR message on the PDCCH. As shown in Section 6.2.2.2, the MAC layer uses a special format of the PDU to transport RAR messages that contain one or more MAC Random Access Responses (MAC RAR). The MAC RAR contains a Timing Advance Command field. The Timing Advance Command field is 11 bits long and it gives the value, $T_A$. The timing offset between uplink and downlink radio frames is calculated as follows. The UE applies the timing offset in subsequent uplink transmissions.

1. $T_A$ is an integer between 0 and 1,282.
2. The UE calculates $N_{TA}$ as

$$N_{TA} = 16 \times T_A$$

   Thus, $N_{TA}$ can range between 0 and (16 × 1,282) = 20,512 with a granularity of 16.

3. The timing offset between uplink and downlink radio frames becomes

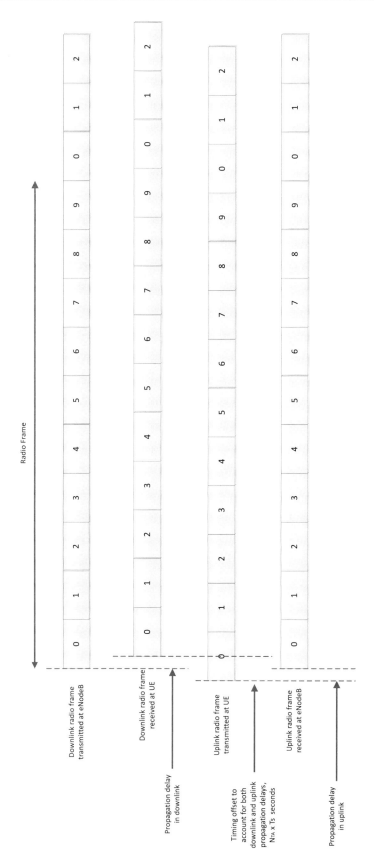

**Figure 10.1** Timing relation between uplink and downlink transmission.

## 10.2 Uplink Synchronization

$$N_{TA} \times T_s \text{ seconds} = 16 \times T_A/30{,}720{,}000 \text{ seconds}$$

Thus, the timing offset between uplink and downlink radio frames can range between 0 and $(16 \times 1{,}282/30{,}720{,}000) = 0.67$ ms with a granularity of $(16/30{,}720{,}000) = 0.52$ $\mu$s. The velocity of radio wave, $c \approx 3 \times 10^5$ km/sec. The two-way propagation delay covers a distance, $(3 \times 10^5 \times 0.67 \times 10^{-3})/2 \approx 100$ km for 0.67 ms. Thus, the maximum possible timing offset, 0.67 ms supports 100 km cell size.

### 10.2.2 Maintenance of Timing Alignment

After the initial establishment of the timing alignment, the eNodeB monitors uplink transmissions from the UE in order to maintain the uplink synchronization. The eNodeB may monitor the SRS, uplink the data transfer on the PUSCH, CQI/PMI/RI report, or HARQ ACK/NACK for this purpose. Thus, the eNodeB estimates the uplink transmission timing of the UE and determines the necessary correction.

From time to time, the eNodeB updates uplink synchronization by notifying the UE about the correction needed in the uplink transmission timing. To do so, the eNodeB adds timing advance command MAC control element to a downlink MAC PDU. The timing advance command MAC control element contains the Timing Advance Command field as shown in Section 6.2.1.2. The timing advance command indicates the necessary change in timing for an uplink transmission relative to the current timing. The Timing Advance Command field is 6 bits long and it gives the value, $T_A$. The new timing offset is calculated as follows:

1. $T_A$ is an integer between 0 and 63.
2. The UE updates $N_{TA}$ as

$$N_{TA,\ NEW} = N_{TA,\ OLD} + (T_A - 31) \times 16$$

Thus, $N_{TA}$ changes by adding –496, –480, ……………, –32, –16, 0, 16, 32, ……………, 480, 496, or 512, which has a granularity of 16.

3. The UE adjusts the timing offset between uplink and downlink radio frames. The new timing offset is

$$N_{TA,\ NEW} \times T_s \text{ seconds} = N_{TA,\ NEW}/30{,}720{,}000 \text{ seconds}$$

Thus, the timing offset changes by adding, –16.15 $\mu$s, –15.63 $\mu$s, ……………, –0.52 $\mu$s, 0, 0.52 $\mu$s, ……………, 15.63 $\mu$s, 16.15 $\mu$s, or 16.67 $\mu$s, which has a granularity of $(16/30{,}720{,}000) = 0.52$ $\mu$s.

The maintenance of uplink synchronization requires the consumption of resources for the transmission of a timing advance command. Also, the UE must be awake to read the timing advance command and perform the necessary adjustments; this consumes the battery power of the UE. Therefore, the uplink synchronization is maintained only when it is necessary. In the RRC_IDLE state, the uplink synchronization is not maintained. The UE always loses uplink synchronization as it enters the RRC_IDLE state. In the RRC_CONNECTED state, the uplink synchronization of the UE is maintained. However, the UE loses uplink synchronization in the following cases:

1. Radio link failure;
2. Nonsynchronized handover;
3. Expiration of TimeAlignmentTimer: Whenever the uplink time alignment is established or updated, the UE starts the timer called the *TimeAlignmentTimer*. If the TimeAlignmentTimer expires without reception of the timing advance command MAC control element, the UE is considered to have lost the uplink time alignment. Then the UE considers the latest uplink time alignment to be valid for a certain period. There can be a cell-specific or a UE-specific configuration for the TimeAlignmentTimer. If both the cell-specific and the UE-specific timers are configured, then the UE applies the UE-specific TimeAlignmentTimer. The eNodeB broadcasts TimeAlignmentTimerCommon IE for the configuration of the cell-specific TimeAlignmentTimer using the SIB type 2 as shown in Appendix A.1. The eNodeB sends the TimeAlignmentTimerDedicated IE to the UE for the configuration of the UE-specific TimeAlignmentTimer while setting up or modifying the radio bearers as shown in Appendix A.2. The values of both the TimeAlignmentTimerCommon IE and the TimeAlignmentTimerDedicated IE are given by the TimeAlignmentTimer IE. The TimeAlignmentTimer IE indicates a length of 500, 750, 1,280, 1,920, 2,560, 5,120, or 10,240 subframes. Its value can also be infinity, indicating that the UE would keep up the uplink time alignment indefinitely.

When the UE loses the uplink synchronization, it stops all uplink transmissions. It releases the PUCCH and SRS resources, flushes the HARQ buffers, and reinitializes HARQ processes. The UE can then only transmit on the PRACH in the uplink (i.e., the UE performs a contention-based random access procedure and obtains uplink synchronization when it needs any uplink transmission).

# CHAPTER 11

# Random Access

The UE attempts to gain access to the network and create a scope for communication with the network using the random access procedure. The MAC layer performs the random access procedure and establishes layer 1 connectivity between the UE and the eNodeB. It establishes uplink synchronization for the UE and also allocates resources for uplink transmission. The random access procedure has two forms: contention-based random access and noncontention-based random access.

## 11.1 Configuration of Random Access

The RRC layer at eNodeB sends the RACH-ConfigCommon IE to the UE for the configuration of the random access procedure. This takes place in two different ways, depending on whether the scenario of random access is handover or not:

1. *Cases other than handover:* The eNodeB broadcasts the SIB type 2 in the cell. As shown in Appendix A.1, the SIB type 2 includes the RadioResourceConfigCommonSIB IE. The RadioResourceConfigCommonSIB IE includes the RACH-ConfigCommon IE, which specifies how the UE would perform the random access procedure at the cell.
2. *Handover:* As shown in Section 23.3.3.2, when handover takes place, the source eNodeB sends an RRCConnectionReconfiguration message to the UE. As shown in Appendix A.3, the RRCConnectionReconfiguration message includes the RadioResourceConfigCommon IE. The RadioResourceConfigCommon IE includes the RACH-ConfigCommon IE, which specifies how the UE would perform random access procedure at the target eNodeB.

    Apart from the RACH-ConfigCommon IE, the MobilityControlInfo IE of the RRCConnectionReconfiguration message may or may not include the RACH-ConfigDedicated IE as shown in Appendix A.3. If the RACH-ConfigDedicated IE is included, then the UE performs the noncontention-based random access procedure at the target eNodeB using the parameters in the RACH-ConfigDedicated IE. On the other hand, if the RACH-ConfigDedicated IE is not included, then the UE performs the contention-based random access procedure at the target eNodeB.

## 11.2 Contention-Based Random Access

The contention-based random access procedure takes place in five cases:

1. *RRC connection establishment:* The UE performs the contention-based random access procedure when it initiates the RRC connection establishment procedure. This procedure will be explained in Section 14.1.1.
2. *RRC connection reestablishment:* The UE performs the contention-based random access procedure when it initiates the RRC connection reestablishment procedure. This procedure is explained in Section 14.1.4.
3. *Initiation of the downlink data transfer:* The UE performs the contention-based random access procedure when it is in the RRC_CONNECTED state and downlink data has become available for transmission but the UE does not have its uplink synchronized. As shown in Section 10.2.2, the UE loses uplink synchronization when the TimeAlignmentTimer expires.

    If downlink data becomes available for transmission when the UE is in the RRC_IDLE state, then the RRC connection needs to be established first by invoking the service request procedure. This is categorized here as the first case (i.e., the RRC connection establishment case).
4. *Initiation of the uplink data transfer:* The UE performs the contention-based random access procedure when it is in the RRC_CONNECTED state and it has the uplink data available but it does not have its uplink synchronized. As shown in Section 10.2.2, the UE loses uplink synchronization when the TimeAlignmentTimer expires. The UE also performs the contention-based random access procedure when it has its uplink synchronized and has uplink data available but there are neither PUCCH resources allocated to send a scheduling request (SR) nor PUSCH resources allocated to send a buffer status report (BSR).

    If the UE needs to send uplink data when it is in the RRC_IDLE state, then the RRC connection needs to be established first by invoking service request procedure. This is categorized here as the first case (i.e., the RRC connection establishment case).
5. *Handover:* The handover takes place while the UE is in the RRC_CONNECTED state. As mentioned earlier, the UE performs the contention-based random access procedure if the MobilityControlInfo IE does not include the RACH-ConfigDedicated IE.

The contention-based random access procedure includes four steps, explained in the following and shown schematically in Figure 11.1:

1. Transmission of the random access preamble on the PRACH;
2. Transmission of the RAR on the PDSCH with indication on the PDCCH;
3. Transmission of message 3 on the PUSCH;
4. Contention resolution.

## 11.2 Contention-Based Random Access

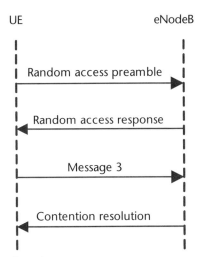

**Figure 11.1** Contention-based random access.

### 11.2.1 Transmission of the Random Access Preamble

The MAC layer of the UE picks a random access preamble from the available random access preambles and sends it on the RACH transport channel. The RACH is mapped on the PRACH physical channel. Thus, the physical layer of the UE transmits the random access preamble on the PRACH. The PRACH transmission is explained Section 5.2.3.

Each cell is allocated 64 random access preambles for use in random access. A particular number of preambles out of these 64 preambles, often called *nondedicated random access preambles*, are allocated for use in contention-based random access. The remaining available 64 preambles are dedicated for UEs and used in noncontention-based random access. The RACH-ConfigCommon IE includes the NumberOfRA-Preambles field, which gives the number of preambles for use in contention-based random access. Its value can be 4, 8, 12, 16, 20, 24, 28, 32, 36, 40, 44, 48, 52, 56, 60, or 64.

The eNodeB can optionally divide the nondedicated random access preambles into group A and group B in order to allow the UE to indicate whether it needs a small or large amount of resources for transmission at step 3 as shown in Figure 11.2. The UE selects a preamble from group A to indicate a small amount of resources and selects a preamble from group B to indicate a large amount of resources. This helps the eNodeB to allocate appropriate resources for message 3 when it sends a random access response at step 2. The RACH-ConfigCommon IE optionally includes the PreamblesGroupAConfig IE. If this IE is not included, then all nondedicated random access preambles belong to group A; group B does not exist. On the other hand, if this IE is included, then group B possesses preambles and this IE includes the following fields:

- *SizeOfRA-PreamblesGroupA:* This field gives the number of nondedicated random access preambles in group A. Its value can be 4, 8, 12, 16, 20, 24, 28, 32, 36, 40, 44, 48, 52, 56, or 60. The first SizeOfRA-PreamblesGroupA number of preambles from the NumberOfRA-Preambles number of nondedicated

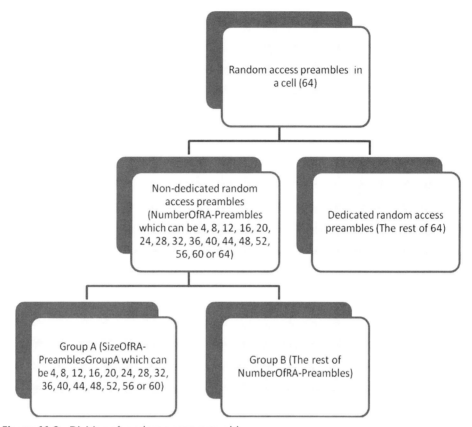

**Figure 11.2** Division of random access preambles.

preambles is selected for group A. The rest of the NumberOfRA-Preambles number of preambles are selected for group B.

- *MessageSizeGroupA:* This field gives the maximum potential resource size for message 3 in the number of bits for the selection of preambles from group A. Its value can be 56, 144, 208, or 256. If the potential size of message 3, including data available for transmission, MAC header, and MAC control elements, is greater than the indication of this field and also if the estimated path loss is less than certain limit, then the UE selects preambles from group B. Otherwise, the UE selects preambles from group A.

The UE transmits the initial random access preamble using a power level, PreambleInitialReceivedTargetPower + DELTA_PREAMBLE. Here PreambleInitialReceivedTargetPower is a cell-specific instruction from the eNodeB based on interferences in the cell. Its value is given by the PreambleInitialReceivedTargetPower field included in the RACH-ConfigCommon IE. Its value can be −120, −118, 116, −114, −112, −110, −108, −106, −104, −102, −100, −98, −96, −94, −92, or −90 dBm. The value of the DELTA_PREAMBLE depends on the preamble format, which can be chosen based on the cell size as shown in Table 5.9. Thus, the DELTA_PREAMBLE accounts for an open-loop estimation of the path loss.

There is really no good estimate of the path loss and the required uplink power level for the UE for the transmission of the initial random access preamble. Only as much power as deemed necessary is used in order to minimize the interference.

### 11.2.2 Random Access Response (RAR)

#### 11.2.2.1 Successful RAR

If the eNodeB receives the random access preamble, it sends the RAR message on the PDSCH and also sends an indication of the RAR message on the PDCCH. Thus, once the random access preamble is transmitted, the UE initiates monitoring the PDCCH for the indication of the RAR. The PDCCH instance has its CRC scrambled by the Random Access RNTI (RA-RNTI). The UE is aware of the RA-RNTI because it is computed from the indices corresponding to both time and frequency resources that were utilized by the UE to transmit the random access preamble. Thus, the UE can identify if the PDCCH instance is addressed to itself using the RA-RNTI. The UE checks the RAR on the PDSCH as indicated by the PDCCH instance if it is addressed to the UE itself. The value of the RA-RNTI is shown in Section 1.6.3.

The RACH-ConfigCommon IE includes the RA-ResponseWindowSize field, which indicates the duration of the RA Response Window in the number of subframes. It can be 2, 3, 4, 5, 6, 7, 8, or 10 subframes. The UE starts monitoring the PDCCH for the indication of the RAR three subframes after the transmission of the random access preamble (i.e., 2 ms after the end of the subframe with the preamble). Then the UE keeps monitoring the PDCCH for a period of the RA response window. The transmission of the RAR does not use Hybrid ARQ (HARQ).

As shown in Section 6.2.2.2, the MAC layer uses a PDU of a special format for RAR messages. The MAC PDU consists of a MAC header and one or more MAC random access responses (MAC RARs). The MAC header includes a MAC PDU subheader corresponding to each MAC RAR and the MAC PDU subheader contains Random Access Preamble Identifier (RAPID) field, as shown in Section 6.2.2.2. The RAPID field is an identifier of the one of the 64 random access preambles that the UE transmitted earlier. After detection of the PDCCH with the associated RA-RNTI, the UE checks the RAPID. If the RAPID matches the transmitted random access preamble, the UE applies the instructions of the corresponding MAC RAR and it may stop monitoring further PDCCH for the indication of RAR.

The MAC RAR allocates resources on the PUSCH for the UE to send message 3. This resource allocation is explained in Section 6.2.2.2. The MAC RAR also assigns a temporary C-RNTI for temporary use to the UE. After successful contention resolution, this temporary C-RNTI can be made permanent or a new C-RNTI can be assigned. However, if the UE already has a C-RNTI previously allocated, then that C-RNTI is used in the same field instead. Typically, if the UE has been in the RRC_IDLE state, the UE is assigned a temporary C-RNTI here and if the UE has been in the RRC_CONNECTED state, the UE already possesses a C-RNTI.

### 11.2.2.2 Unsuccessful RAR

As mentioned in Section 11.2.1, the UE may not have any good estimate of the required uplink power level while transmitting the random access preamble and therefore, the initial random access preamble may not reach the eNodeB properly. In addition, simultaneous transmission from multiple UEs using the same resources may interfere with each other. Consequently, the eNodeB may not send the RAR message. The UE considers the random access attempt to be unsuccessful if the UE does not receive any RAR within the RA response window, or if the UE receives the RAR but the RAPID does not match the transmitted random access preamble.

In the case of the unsuccessful random access attempt, the UE delays for back-off time and thereafter picks a random access preamble anew and transmits it with increased PRACH transmission power. Section 6.2.2.2 shows how the UE selects the back-off time. If no random access response is received in subframe $n$, where subframe $n$ is the last subframe of the RAR window, the physical layer of the UE becomes ready to transmit the new preamble by subframe $n + 4$. The UE keeps boosting the preamble power by a fixed step until it receives the RAR from the eNodeB or it reaches the maximum allowed number of transmissions. The RACH-ConfigCommon IE includes the PowerRampingStep field, which indicates how much the UE would boost the power level for every successive transmission of the random access preambles. Its value can be 0, 2, 4, or 6 dB. Thus, in the first attempt after the initial random access preamble, the UE adds the PowerRamping-Step field value to the power for initial transmission. In the next attempt, the UE adds the same value again and so on.

If the UE reaches the maximum allowed number of transmissions of random access preambles without the reception of the RAR, the physical layer reports the problem to higher layers. In this case, the UE waits for the random back-off period before attempting the transmission of preambles again. If the UE is in the RRC_CONNECTED state, it may come to a decision that Radio Link Failure (RLF) has occurred. The RACH-ConfigCommon IE includes the PreambleTransMax field to indicate the maximum allowed number for transmission. The value of Preamble-TransMax can be 3, 4, 5, 6, 7, 8, 10, 20, 50, 100, or 200.

### 11.2.3 Message 3

In the third step of random access procedure, the UE performs first scheduled uplink transmission on the PUSCH, and the message sent is known as message 3 (Msg3). The actual content of message 3 depends on the purpose of the random access.

The transmission of message 3 uses HARQ. When the UE reaches the maximum number of HARQ retransmissions, it initiates the random access procedure anew. The RACH-ConfigCommon IE includes the MaxHARQ-Msg3Txfield, which indicates the maximum allowed number of HARQ transmissions of message 3 in contention-based random access. Its value can range between 1 and 8.

If multiple UEs transmit random access preambles using the same resources resulting in a collision, the UEs will have the same RA-RNTI. Then they will accept the same RAR and the same temporary C-RNTI. Thus, the UEs can also collide while transmitting message 3. Since message 3 supports HARQ, the UEs will perform HARQ retransmissions in this case. When they reach the maximum number of HARQ retransmissions, the UEs will initiate random access procedure anew.

The transmission of message 3 can use either CCCH or DCCH as the logical channel. The use of CCCH and DCCH has scenarios as follows.

1. *Message 3 on CCCH:* If the UE was assigned a temporary C-RNTI in the second step, then message 3 is sent using the CCCH logical channel in this third step. The transport block on the PUSCH carrying message 3 is scrambled, and the scrambling is initialized using the temporary C-RNTI. Depending on the purpose of the random access, the content of message 3 varies as follows.
    - *RRC connection setup:* In the case of the RRC connection setup, the UE sends the RRCConnectionRequest message as message 3 to the eNodeB. The RRCConnectionRequest message typically includes S-TMSI as the UE identity, as will be shown in Section 14.1.1.1.
    - *RRC connection reestablishment:* In the case of the RRC connection reestablishment, the UE sends the RRCConnectionReestablishmentRequest message as message 3 to the eNodeB. The UE may possess a C-RNTI from the cell to which the UE was connected prior to the failure. Nevertheless, the UE is assigned a temporary C-RNTI as mentioned earlier. The RRCConnectionReestablishmentRequest message contains the C-RNTI and ShortMAC-I for identification of the UE as will be shown in Section 14.1.4.1.
2. *Message 3 on DCCH:* If the UE has been purely in the RRC_CONNECTED state, then the UE possesses the C-RNTI, which is a unique identity of the UE at the cell level identifying the RRC connection. In this case, message 3 is sent using the DCCH logical channel. The MAC layer of the UE adds the C-RNTI MAC control element to message 3. As shown in Section 6.2.1.2, the C-RNTI MAC control element contains the C-RNTI of the UE. Depending on the purpose of the random access, the content of message 3 varies as follows:
    - *Handover:* In case of handover, the UE sends the RRCConnectionReconfigurationComplete message as message 3 to the target eNodeB, as will be shown in Section 23.3.3.3. The UE earlier obtained the C-RNTI to be used at the target eNodeB via the RRCConnectionReconfiguration message from the source eNodeB.
    - *Transmission of uplink or downlink data:* If the UE needs to transfer uplink or downlink data in the RRC_CONNECTED state but it does not have its uplink synchronized, then this random access may take place and the UE may send the ULInformationTransfer message as message 3 to the eNodeB. The ULInformationTransfer message carries the SERVICE REQUEST message. The eNodeB forwards the SERVICE REQUEST message to the MME using the UPLINK NAS TRANSPORT message.

### 11.2.4 Contention Resolution

The UE randomly chooses the random access preamble signature. Therefore, it is possible that more than one UE simultaneously transmits the same random access preamble signature. As a result, the CRC of the PDCCH is scrambled by the same RA-RNTI and the same RAR message is sent on the PDSCH addressing multiple UEs. All these UEs receive the same resource allocation on the PUSCH for message 3 with the same timing adjustment and they all transmit accordingly. Therefore, at this juncture, a contention resolution process is required. The eNodeB receives message 3 from only one of the UEs for which it considers timing alignment most suitable. Then the steps shown in the following sections take place for successful and unsuccessful contention resolution.

A contention resolution timer specifies how long the UE would wait for contention resolution after sending Message 3. The RACH-ConfigCommon IE includes the MAC-ContentionResolutionTimer field, which gives the duration of contention resolution timer in number of subframes and it can be 8, 16, 24, 32, 40, 48, 56, or 64 subframes. The UE starts the contention resolution timer after the transmission of message 3 and it restarts the contention resolution timer anew at each HARQ retransmission of message 3. The UE keeps monitoring the PDCCH until the contention resolution becomes successful or until the contention resolution timer expires, regardless of the possible occurrence of a measurement gap. If the contention resolution timer expires without a successful contention resolution, the contention resolution is considered unsuccessful. In the case of either successful contention resolution or the expiry of the contention resolution timer, the UE stops monitoring the PDCCH as well as the contention resolution timer.

### 11.2.4.1 Successful Contention Resolution

The successful contention resolution occurs for the two different cases:

1. *Message 3 on CCCH:* If message 3 was sent over a CCCH logical channel, then in this fourth step, the eNodeB sends a PDCCH instance upon reception of message 3 and the PDCCH instance has its CRC scrambled by the temporary C-RNTI of the UE. The PDCCH instance indicates resource allocation for the downlink transmission on the DL-SCH. The eNodeB sends an RRC message to the UE on the allocated resources on the DL-SCH. The RRC message is the RRCConnectionSetup message in the case of RRC connection establishment and it is the RRCConnectionReestablishment message in the case of the RRC connection reestablishment. The MAC layer of the eNodeB adds the UE contention resolution identity MAC control element to the downlink MAC PDU on DL-SCH. As shown in Section 6.2.2.1, the UE contention resolution identity MAC control element contains the RRC message that was sent as message 3. Since message 3 contains the unique identity of the UE as shown earlier, the UE contention resolution identity MAC control element identifies the particular UE and acknowledges the reception of message 3 as well. Then the contention resolution is considered successful. After the successful contention resolution, the temporary C-RNTI is made permanent or a new C-RNTI is assigned.
2. *Message 3 on DCCH:* As shown earlier, when message 3 was sent over the DCCH logical channel, the UE added its C-RNTI using the C-RNTI MAC

control element. Now, in this fourth step, the eNodeB sends a PDCCH instance upon reception of message 3 and the PDCCH instance has its CRC scrambled by the C-RNTI of the UE. The PDCCH instance may carry uplink or downlink resource allocation for further activities. The PDCCH instance identifies the particular UE and acknowledges the reception of the C-RNTI MAC control element as well. Then the contention resolution is considered successful.

The contention resolution uses HARQ but using only ACK and no NACK. Thus, the UE sends ACK when the contention resolution is successful and it does not send anything when the contention resolution is unsuccessful.

### 11.2.4.2 Unsuccessful Contention Resolution

As mentioned earlier, the expiry of contention resolution timer without a successful contention resolution leads to the consideration of unsuccessful contention resolution. The contention resolution can be unsuccessful in several ways. If there is a collision, the UE may receive the PDCCH instance with the CRC scrambled by the C-RNTI or the temporary C-RNTI of a different UE. Also, the UE may miss the PDCCH instance or fail to decode it. When the contention resolution is unsuccessful, the UE delays for back-off time. Thereafter, the UE picks a random access preamble anew and transmits it with increased PRACH transmission power. This takes place in the same way as described in Section 11.2.2.2.

## 11.3 Noncontention-Based Random Access

The noncontention-based random access procedure takes place only for UEs in the RRC_CONNECTED state. It takes place in the following cases:

1. *Initiation of a downlink data transfer:* The UE performs noncontention-based random access procedure when downlink data has become available for transmission but the UE does not have its uplink synchronized. As shown in Section 10.2.2, the UE loses uplink synchronization when the TimeAlignmentTimer expires.
2. *Handover:* The UE performs the noncontention-based random access procedure at the target eNodeB if the MobilityControlInfo IE of the RRCConnectionReconfiguration message includes the RACH-ConfigDedicated IE.
3. *UE positioning:* The UE positioning may require that the UE performs random access procedure and, in this case, the random access must be noncontention based. This may occur, for example, when timing advance information is needed for UE positioning.

It may be noted that contention-based random access is also allowed in the case of the handover and the resumption of the downlink data transfer. However, the noncontention-based approach allows much lower latency access compared to the contention-based approach, and the noncontention-based random access may be adopted in these cases. The noncontention-based random access procedure includes three steps, as explained in the following sections and shown schematically in Figure 11.3:

1. Assignment of the random access preamble;
2. Transmission of the random access preamble on the PRACH;
3. The RAR on the PDSCH with indication on the PDCCH.

### 11.3.1 Assignment of the Random Access Preamble

The eNodeB assigns a dedicated random access preamble to the UE. This assignment takes place in two different ways depending on whether the purpose of random access is handover or not. Thus, the assignment procedure can be classified as:

1. The assignment of the preamble for downlink data transfer and UE positioning;
2. The assignment of the preamble for handover.

#### 11.3.1.1 Downlink Data Transfer and UE Positioning

In the case of downlink data transfer and UE positioning, the random access procedure is initiated by a PDCCH order that assigns a dedicated random access preamble to the UE. The PDCCH instance has its CRC scrambled by the C-RNTI, which allows the UE to identify its PDCCH order. PDCCH uses the DCI format 1A in this case.

The DCI format 1A includes a 6-bit-long Preamble Index field that can have any integer value from 0 through 63. The Preamble Index field indicates the assigned random access preamble. The preamble must be taken from the dedicated random access preambles available. The DCI format 1A also includes a 4-bit-long PRACH Mask Index field. The value of this field corresponds to the assigned PRACH resource index according to Table 11.1. The PRACH resource index indicates the subframe number in a radio frame that the UE can use for the transmission of the

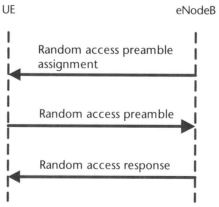

**Figure 11.3** Noncontention-based random access.

## 11.3 Noncontention-Based Random Access

**Table 11.1** PRACH Resource Index Allocation

| PRACH Mask Index | Allowed PRACH resource index |
|---|---|
| 0 | All |
| 1 | PRACH resource index 0 |
| 2 | PRACH resource index 1 |
| .. | ..................... |
| 9 | PRACH resource index 8 |
| 10 | PRACH resource index 9 |
| 11 | First PRACH resource index in subframe (even subframe) |
| 12 | First PRACH resource index in subframe (odd subframe) |

random access preamble. Section 5.2.2 shows the determination of the subframe number from the PRACH configuration index.

### 11.3.1.2 Handover

As mentioned in Section 11.1, the RACH-ConfigDedicated IE provides the UE with parameters for noncontention-based random access at the target eNodeB during handover. The RACH-ConfigDedicated IE contains the following fields:

- *RA-PreambleIndex:* This field indicates the random access preamble for dedicated use in the same way as the Preamble Index field does which is explained in Section 11.3.1.1. Its value also ranges between 0 and 63.
- *RA-PRACH-MaskIndex:* This field indicates the PRACH resource index in the same way as the PRACH Mask Index field does, which is explained in Section 11.3.1.1.

### 11.3.2 Transmission of the Random Access Preamble

The UE transmits the dedicated random access preamble assigned using PRACH in the same way as it does for contention-based random access. If a PDCCH order is sent indicating the downlink data arrival or UE positioning in subframe $n$, the UE transmits the random access preamble in the first subframe available for PRACH with subframe number $\geq n + 6$.

### 11.3.3 RAR

When the eNodeB receives the dedicated random access preamble, it is able to identify the particular UE. Then it sends the RAR on the PDSCH and the associated indication on the PDCCH. The transmission of the RAR was explained in Section 11.2.2. The PDCCH instance has its CRC scrambled by the RA-RNTI. The RAPID field in the MAC PDU subheader identifies the dedicated random access preamble that the UE transmitted. If the RAPID matches the transmitted random access preamble, the UE considers the random access procedure to be successfully completed. No further steps for UE identification or contention resolution are required here after the transmission of RAR. The MAC PDU header does not need to include a back-off indicator subheader because of the absence of contention resolution. The uplink grant is not present in MAC RAR if the random access is used to initiate the downlink data transfer, but it is present if the random access occurs during handover.

# CHAPTER 12
# NAS and AS Layer Security

In cellular communication, there is no dedicated path for information transfer between the UE and the network, and the information is rather launched into the air. Therefore, some measure must be taken to secure the information transferred. The following two types of measures are taken in the Nonaccess Stratum (NAS) layer and the Access Stratum (AS) layer for the sake of the security of information. The NAS layer and AS layer were defined in Section 1.3.1.

1. *Ciphering:* The ciphering of information is performed to prevent eavesdropping (i.e., prevent a third party from receiving the information transferred). Ciphering is performed using a secret key. Ciphering is applied to the following types of information:
   - NAS signaling messages.
   - RRC signaling messages on SRB1 and SRB2. The RRC signaling message may contain a NAS signaling message. SRB1 also includes messages that are not ciphered because security is not yet activated.
   - User plane data.
   
   Within the AS layer, the PDCP layer applies ciphering to the RRC signaling messages and the user plane data. On the other hand, the NAS layer independently applies ciphering to the NAS signaling messages.

2. *Integrity protection:* The integrity protection is performed to validate the content of the message and also to make sure that a third party has not modified the information through packet insertion, deletion, replacement, or replay. Integrity protection is applied only to control plane data. Integrity protection is performed using a secret key. Integrity protection is applied to the following types of information:
   - NAS signaling messages;
   - RRC signaling messages on SRB1 and SRB2. The RRC message may contain a NAS message. SRB1 also includes messages that are not integrity protected because security is not yet activated.
   
   The integrity protection is mandatory for both these cases, except that a few particular NAS and RRC signaling messages do not have to be integrity protected. Within the AS layer, the PDCP layer applies integrity protection to the RRC signaling messages. On the other hand, the NAS layer independently applies integrity protection to the NAS signaling messages.

Ciphering and integrity protection are applied after the activation of security for both the NAS layer and the AS layer. The activation of security can take place only when the UE is in the RRC_CONNECTED state. Also, they take place before the establishment of SRB2 or DRBs and when only SRB1 is established. SRB0 does not have security activated. Neither ciphering nor integrity protection is applied to SRB0.

## 12.1 EPS Security Context

The EPS security context is a state between the UE and the network. Once the EPS security context is established, both the UE and the network store EPS security context data. The EPS security context consists of the following parts:

1. *EPS NAS security context:* This exists in both the RRC_IDLE state and the RRC_CONNECTED state as long as the security is valid. Thus, the EPS NAS security context stays valid when the RRC connection is released. The EPS NAS security includes a key called $K_{ASME}$ that is the root for ciphering and integrity protection keys.
2. *EPS AS security context:* This exists only during the RRC_CONNECTED state. Thus, when the RRC connection is released, the EPS AS security context is deleted. When the RRC connection is established, a new EPS AS security context is set up.

## 12.2 EPS Authentication and Key Agreement (EPS AKA)

The EPS Authentication and Key Agreement (EPS AKA) procedure provides mutual authentication between the UE and the network and establishes the key, $K_{ASME}$. $K_{ASME}$ is used for the generation of the following keys:

1. The key for ciphering NAS messages;
2. The key for ciphering RRC messages;
3. The key for the integrity protection of NAS messages;
4. The key for the integrity protection of RRC messages;
5. The key for ciphering user data.

The EPS AKA procedure typically takes place during the attach procedure and the Tracking Area Update (TAU) procedure. It is always the network that initiates the EPS AKA procedure, and the UE can reject the EPS AKA procedure. After the successful completion of the EPS AKA procedure, an EPS NAS security context is established in the UE and the eNodeB.

The EPS AKA procedure uses a 128-bit-long permanent key, K, which is stored in the Universal Subscriber Identity Module (USIM) of the UE and in the Authentication Center (AuC). The AuC is typically a part of the home subscription server (HSS). The MME obtains authentication information from HSS. For this purpose, The MME sends an authentication information request message to HSS that includes the MCC, MNC, and IMSI of the UE. The HSS checks if the UE is known and the UE has a proper subscription. If HSS finds that the UE does, then it asks

## 12.2 EPS Authentication and Key Agreement (EPS AKA)

AuC to generate an EPS authentication vector. The authentication vector includes a Random Challenge (RAND), an Authentication Token (AUTN), XRES, and a Key Set Identifier ($KSI_{ASME}$). The HSS sends an authentication information answer message to the MME, which includes the EPS authentication vector. Then the MME sends the AUTHENTICATION REQUEST message to the UE as shown in Figure 12.1. The AUTHENTICATION REQUEST message includes a RAND, an AUTN, and $KSI_{ASME}$ for network authentication from the selected authentication vector. The USIM verifies whether the AUTN can be accepted. if the AUTN is acceptable, the USIM computes a signed response (RES) using K and RAND]. The USIM also computes CK and IK using K and RAND. Finally, the UE computes $K_{ASME}$ from the CK, IK, and Serving Network's identity (SN id) using Key Derivation Functions (KDF). The use of SN id implicitly authenticates the SN id when the keys derived from $K_{ASME}$ can be used successfully. Now the UE has generated an EPS NAS security context. The network also computes both RES and $K_{ASME}$.

At this stage, the UE sends the AUTHENTICATION RESPONSE message to the MME, which includes RES. The MME checks RES and if it is acceptable, then the UE is authenticated. The EPS NAS security contexts of both the eNodeB and the UE store $K_{ASME}$ and $KSI_{ASME}$. This completes the EPS AKA procedure.

The $KSI_{ASME}$ is used to indicate the validity of the EPS NAS security context. The $KSI_{ASME}$ identifies the stored $K_{ASME}$, and thus, the same $K_{ASME}$ can be used subsequently without invoking the EPS AKA procedure anew. Whenever a NAS signaling connection is established, the UE includes the $KSI_{ASME}$ in the initial NAS message to indicate the validity of the EPS NAS security context. The network checks on the $KSI_{ASME}$ and if it is acceptable, the old $K_{ASME}$ can be used again.

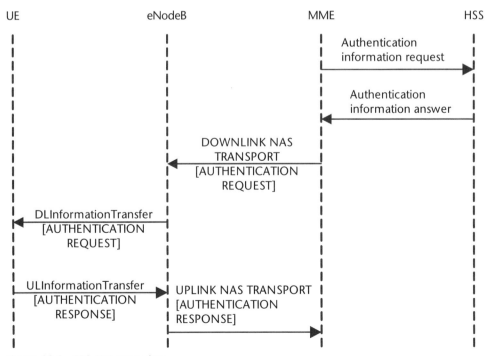

**Figure 12.1** EPS AKA procedure.

## 12.3 Security in the NAS Layer

The NAS layer of the UE and the MME generate the following keys from $K_{ASME}$:

1. The key for ciphering NAS messages;
2. The key for the integrity protection of NAS messages.

The MME selects a NAS integrity algorithm and a NAS ciphering algorithm. The UE has previously notified its supported ciphering and integrity protection algorithms. The MME selects algorithms from the notified ones and it may select the one with highest priority among the supported algorithms.

In order to activate or initialize security function in the NAS layer, the MME sends the SECURITY MODE COMMAND message to the UE, as shown in Figure 12.2. The SECURITY MODE COMMAND message includes $KSI_{ASME}$ as well as the selected NAS ciphering and NAS integrity algorithms. The SECURITY MODE COMMAND message is not ciphered, but the MME integrity protects this message with the NAS integrity key based on $K_{ASME}$ indicated by the $KSI_{ASME}$. The security header type of the message is set as "integrity protected with new EPS security context."

The UE performs an integrity check of the SECURITY MODE COMMAND message. If the SECURITY MODE COMMAND message is acceptable, the UE sends a SECURITY MODE COMPLETE message to the MME. The UE integrity protects the SECURITY MODE COMPLETE message with the selected NAS integrity algorithm and the NAS integrity key based on the $K_{ASME}$. The UE also ciphers the SECURITY MODE COMPLETE message with the selected NAS ciphering algorithm and the NAS ciphering key based on the $K_{ASME}$. The UE sets the security header type of the SECURITY MODE COMPLETE message to "integrity protected and ciphered with new EPS security context."

From this time onward, the security procedure is activated and the NAS layer independently applies ciphering and integrity protection to all NAS signaling

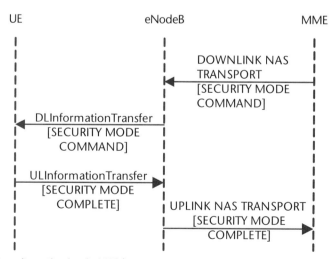

**Figure 12.2** Security activation in NAS layer.

messages with the selected NAS ciphering and NAS integrity algorithms. When the MME detects that the NAS COUNT values are about to wrap around its range (i.e., the NAS COUNT reaches its maximum limit and restarts) it initiates a new EPS AKA procedure and the entire key hierarchy is rekeyed.

## 12.4 Security in the AS Layer

### 12.4.1 Security Activation

The eNodeB receives the eNodeB specific key, $K_{eNB}$, for the UE from the MME, and then the eNodeB creates the AS security context for the UE. In order to activate or initialize the AS security function, the RRC layer of the eNodeB sends the SecurityModeCommand message to the UE as shown in Figure 12.3. The SecurityModeCommand message is integrity protected but not ciphered. The SecurityModeCommand message indicates the selected ciphering and integrity algorithms to be used at PDCP layer. The UE performs integrity check of the SecurityModeCommand message. If the SecurityModeCommand message is acceptable, then the AS layer of the UE and the eNodeB derives an eNodeB specific key, $K_{eNB}$, from $K_{ASME}$. Here $K_{ASME}$ is handled by the NAS layer. Thereafter, $K_{eNB}$ is used to generate the following security keys known as the AS derived keys:

1. The key for ciphering RRC messages;
2. The key for the integrity protection of RRC messages;
3. The key for ciphering user data.

The RRC layer of the UE sends the SecurityModeComplete message to the eNodeB. The SecurityModeComplete message is integrity protected as well as ciphered. From this time onward, the security procedure is activated in the AS layer. The PDCP layer performs ciphering and integrity protection.

The eNodeB may send the RRCConnectionReconfiguration message to establish SRB2 and DRBs before reception of the SecurityModeComplete message. However, the RRCConnectionReconfiguration message is both ciphered and integrity protected whether it is sent before or after the reception of the SecurityModeComplete message. The RRC connection is released if any or both of these two simultaneous attempts, initial security activation and radio bearer establishment, fail.

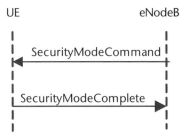

**Figure 12.3** Security activation in the AS layer.

## 12.4.2 COUNT

For each radio bearer an independent counter value, COUNT, is maintained in each direction, uplink and downlink. For each SRB, the COUNT is used as input for both ciphering and integrity protection algorithms. For each DRB, the COUNT is used as input for the ciphering algorithm. The COUNT is 32 bits long. The COUNT value is incremented for each PDCP data PDU during an RRC connection. Despite the COUNT value being 32 bits long, the UE takes into consideration that the COUNT value may wrap around its range (i.e., the COUNT reaches its maximum limit and restarts) during an RRC connection. The COUNT consists of two parts:

1. *PDCP sequence number (SN):* The PDCP SN is the LSB part of the COUNT value as shown in Figure 12.4. For SRBs, PDCP SN is 5 bits long. For DRBs with RLC AM., PDCP SN is always 12 bits long. For DRBs with RLC UM., PDCP SN is either 7 bits long or 12 bits long as configured by the PDCP-Config IE.
2. *Hyper frame number (HFN):* The HFN is the MSB part of the COUNT value. Since the COUNT is 32 bits long, the HFN is 27 bits long for SRBs and the HFN is either 25 bits long or 20 bits long for DRBs.

In order to limit the signaling overhead, the whole COUNT is not transmitted over the air and only the PDCP SN is signaled. Thus, as shown in Section 8.2.1, each PDCP data PDU includes only the PDCP SN in the header. Then the HFN is derived from counting overflows of the PDCP SN (i.e., when the PDCP SN wraps around its range every time, the HFN is incremented by 1). Thus, the whole COUNT value is derived.

If any PDCP data PDUs are lost, the receiver uses the PDCP SN of available PDCP data PDUs to determine the COUNT value. This requires the PDCP SN to be enough long such that the number of lost PDCP data PDUs does not cause the PDCP SN to wrap around. However, the long PDCP SN increases the overhead. Therefore, two different sizes of PDCP SN, 7 bits and 12 bits, are defined for DRBs with the RLC UM. The eNodeB configures the length of the PDCP SN in the case of the RLC UM using the PDCP-Config IE. Appendix A.2 shows how the eNodeB sends the PDCP-Config IE to the UE. The PDCP-Config IE includes the RLC-UM IE. The RLC-UM IE includes the PDCP-SN-Size field, which specifies if the size of the PDCP SN would be 7 bits or 12 bits.

If a third party attempts to retransmit a packet, the COUNT value helps to recognize it and thus protects replay. When the eNodeB detects that the PDCP COUNT values are about to wrap around its range, it refreshes the AS derived keys.

**Figure 12.4** Structure of the COUNT.

### 12.4.3 Ciphering

The PDCP layer on the transmitting side performs ciphering of RRC messages and user data using the EPS Encryption Algorithm (EEA). The EEA executes an XOR operation between the data and a ciphering key stream. The EEA allows the selection from several ciphering protection algorithms [e.g., SNOW 3G and Advanced Encryption Standard (AES)]. In the PDCP layer on the transmitting side, the ciphering algorithm generates the ciphering key stream using the following input parameters:

1. The 128-bit key for ciphering the RRC messages or user data from Access Stratum (AS)–derived keys explained in Section 12.4.1;
2. The radio bearer ID given by the SRB-Identity field and the DRB-Identity field explained in Sections 2.2.1 and 2.2.2;
3. The 1-bit direction of transmission indicating uplink or downlink;
4. The 32-bit COUNT value explained in Section 12.4.2;
5. The length of the ciphering key stream.

The PDCP layer on the receiving side uses EEA and generates the ciphering key stream using the same input parameters. Then from the received ciphered data and the ciphering key stream, it produces the original deciphered data.

### 12.4.4 Integrity Protection

The PDCP layer on the transmitting side performs the integrity protection of the RRC message using the EPS Integrity Algorithm (EIA). The EIA allows the selection from several integrity protection algorithms (e.g., SNOW 3G and AES).

In the PDCP layer on the transmitting side, the integrity protection algorithm derives a code called the Message Authentication Code for Integrity (MAC-I). The MAC-I is appended at the end of the control plane PDCP data PDU as shown in Section 8.2.1.2. The integrity protection algorithm generates the MAC-I using the following input parameters:

1. The 128-bit key for the integrity protection of RRC messages from the AS-derived keys explained in Section 12.4.1;
2. The radio bearer ID given by the SRB-Identity field explained in Section 2.2.1;
3. The 1-bit direction of transmission indicating uplink or downlink;
4. The 32-bit COUNT value explained in Section 12.4.2;
5. The RRC message as well as the header of the PDU.

To verify the integrity, the PDCP layer on the receiving side uses EIA and determines the computed MAC-I (X-MAC) based on the same input parameters and algorithm. If the X-MAC conforms to the received MAC-I, then the integrity is assumed to be retained perfectly.

# CHAPTER 13

# Paging

The MME sends a message to the UE, an action that is called *paging*, for the following purposes:

1. To inform a UE in the RRC_IDLE state about an incoming call or a downlink data packet arrival. The UE, in response, may initiate the RRC connection establishment.
2. To inform UEs in the RRC_IDLE state or in the RRC_CONNECTED state about a system information change.
3. To inform the Earthquake and Tsunami Warning System (ETWS)-capable UEs in the RRC_IDLE state or in RRC_CONNECTED state about ETWS primary and/or secondary notification. The ETWS is explained in Chapter 26.
4. To inform the Commercial Mobile Alert Service (CMAS)-capable UEs in the RRC_IDLE state or in the RRC_CONNECTED state about CMAS notification. The CMAS is explained in Chapter 26.

The paging message is transmitted within a particular region depending on the RRC state of the UE, as shown next. The paging message includes an UE identity that allows the UE to identify that the paging message is addressed to itself.

- *RRC_IDLE state:* The UE can move around within the tracking areas of its TAI list without performing any tracking area updating procedure. Thus, the MME is aware of the TAI list of the UE, but it is unaware of the current serving cell. Therefore, the MME sends the PAGING message to all eNodeBs that belong to the area covered by the current TAI list of the UE. These eNodeBs transmit the paging message in their cells.
- *RRC_CONNECTED state:* The MME is aware of the current serving cell of the UE. Therefore, the MME sends the PAGING message to the serving eNodeB, which transmits the paging message only in the current serving cell.

## 13.1 Mapping Among Layers

The mapping of a paging message among different layers is shown here:

1. The RRC layer at the eNodeB generates a paging message for the UE and forms an RRC PDU. The RRC PDU is directly mapped onto an RLC SDU without using the PDCP layer. The RLC layer applies the Transparent Mode (TM), so the RLC SDU is mapped straight onto an RLC PDU without any segmentation or concatenation and even without the addition of any header. The RLC layer delivers the RLC PDU to the PCCH logical channel.
2. The RLC PDU of paging message is treated as the MAC SDU. The MAC layer uses transparent MAC. The MAC SDU is mapped straight to a MAC PDU without any segmentation and even without addition of any header. The MAC PDU is delivered to the PCH transport channel as a transport block. The MAC PDU includes data from only one logical channel and no multiplexing is applied. There is one-to-one mapping between the PCCH and the PCH.
   Thus, the whole layer 2 does not segment the paging message or add any header to it.
3. The physical layer receives the transport block of paging message, performs the necessary processing, and transmits the paging message over physical channel, PDSCH. There is no dedicated physical channel for paging such as GSM or UMTS. The modulation scheme used is always QPSK. Paging does not use HARQ.

## 13.2 Indication of Paging Messages

A PDCCH instance indicates the presence of a paging message on the PDSCH and this particular PDCCH instance has its CRC is scrambled by the Paging Radio Network Temporary Identifier (P-RNTI). The P-RNTI has a fixed value that is FFFE in hexadecimals. The UE monitors the PDCCH channel for indication of paging messages and identifies it using P-RNTI. PDCCH signaling is short in duration and does not lead to significant power consumption of the UE. Therefore, unlike UMTS, a separate paging indicator channel is not required in LTE. The PDCCH instance uses the DCI format 1A or the DCI format 1C, which schedules only one PDSCH code word using resource allocation type 2.

When the eNodeB needs to send a paging message to a particular UE, the eNodeB is allowed to send the PDCCH indication for the paging message only at specific subframes within specific radio frames. These radio frames and subframes appear periodically at regular intervals and they are defined next. The UE needs to monitor the PDCCH channel only at these radio frames and subframes.

- *Paging Occasion (PO):* The subframe that can carry the PDCCH instance indicating a paging message;
- *Paging Frame (PF):* The radio frame that can carry one or more paging occasions;

- *Paging cycle:* The number of radio frames within which there can be one PO.

In the RRC_IDLE state, the UE may use Discontinuous Reception (DRX) in order to reduce power consumption, as will be shown in Section 16.1. In this case, the DRX cycle is the same as the paging cycle. The UE has one PO per DRX cycle or paging cycle (i.e., the UE checks the PDCCH indication for the paging message once in every paging cycle). The eNodeB configures the PF and PO and thus transmits the paging cycle information. This can take place in two ways:

1. *Default paging cycle:* In each cell, the eNodeB broadcasts a default paging cycle. This is the common type of paging cycle configuration and is explained in Section 13.3.
2. *UE-specific paging cycle:* The upper layers may use dedicated signaling to configure a UE-specific paging cycle.

When both the default paging cycle and the UE-specific paging cycle are configured, the UE applies the lowest value of paging cycle.

In the RRC_CONNECTED state, the UE looks for a paging message in order to check on the system information change, ETWS, and/or CMAS notification. The UE looks for a paging message at least once in every default paging cycle if the UE is ETWS and/or CMAS capable.

## 13.3 Default Paging Cycle

The timings for paging occasions are distributed among the UEs in a cell in order to allow each UE to monitor fewer paging occasions. The UEs are grouped in a cell based on their IMSIs so that the paging frames can be allotted at different radio frames within the paging cycle for different groups of UEs. Also, the paging occasions are allotted at different subframes within the paging frame for different groups of UEs.

The eNodeB configures the paging frame and paging occasion using the PCCH-Config IE. Appendix A.1 will show how the PCCH-Config IE is transmitted. The PCCH-Config IE contains the following fields:

- *DefaultPagingCycle:* This field gives the length of default paging or DRX cycle in number of radio frames. It can be 32, 64, 128 or 256 radio frames.
- *NB:* This field is used to calculate the PF and the PO. Its value can be 4, 2, 1, 1/2, 1/4, 1/8, 1/16, or 1/32 and NB is considered as these values times of the DefaultPagingCycle.

To determine the PF and the PO, the IMSI of the UE is used in decimal number. Also, a parameter $N$ is used as follows.

$N$ = DefaultPagingCycle if $NB$ = 4, 2, or 1 times of the DefaultPagingCycle

$N = NB$ if $NB$ = 1/2, 1/4, 1/8, 1/16, or 1/32 times of the DefaultPagingCycle

### 13.3.1 Allocation of the Paging Frame

The SFN of the PF satisfies the following relation, and thus, the PFs are distributed among UEs in a cell.

$$\text{SFN mod (DRX cycle)} = [\text{DRX cycle}/N] \times [(\text{IMSI mod } 1{,}024) \text{ mod } N]$$

According to this relation, the valid SFN for paging frame appears at every DRX cycle/$N$ radio frames within the DRX cycle for all UEs in the cell and a particular UE needs to monitor only one of these PFs within the DRX cycle. Thus, a particular UE monitors the PF once in every DRX cycle, which can be 32, 64, 128, or 256 radio frames long depending on the value of the DefaultPagingCycle field. On the other hand, the eNodeB in the cell may transmit a PF at every radio frame or at every 2, 4, 8, 16, or 32 radio frames within the DRX cycle depending on the value of the NB field.

### 13.3.1 Allocation of the Paging Occasion

The eNodeB can use 1, 2, or 4 POs in a PF, but a particular UE needs to monitor only one PO in the PF. The allowed subframe numbers of the POs in a PF are 0, 4, 5, and 9. These subframes are not used for MBMS related information as shown in Section 24.3. Ns denotes the number of POs in a PF (i.e., the number of subframes used for paging in a PF). Ns is calculated as

Ns = 1 if NB = 1, 1/2, 1/4, 1/8, 1/16, or 1/32 times of the DefaultPagingCycle
   = 2 if NB = 2 times of DefaultPagingCycle
   = 4 if NB = 4 times of DefaultPagingCycle

In order to distribute the POs among the UEs within the PF in a cell, the PO is determined using an index, $i\_s$, which is calculated as

$$i\_s = \text{floor}((\text{IMSI mod } 1{,}024)/N) \text{ mod Ns}$$

Thus,

$i\_s$ must be 0 when Ns = 1;
$i\_s$ must be 0 or 1 when Ns = 2;
$i\_s$ must be 0, 1, 2, or 3 when Ns = 4.

Then the subframe number of the PO in a PF is determined according to Table 13.1.

## 13.4 Paging Message

A paging message can address one UE or multiple UEs. The paging message includes the following IEs:

## 13.4 Paging Message

**Table 13.1** Subframe Number of POs

| Ns | Subframe Number | | | |
|---|---|---|---|---|
| | $i\_s = 0$ | $i\_s = 1$ | $i\_s = 2$ | $i\_s = 3$ |
| 1 | 9 | — | — | — |
| 2 | 4 | 9 | — | — |
| 4 | 0 | 4 | 5 | 9 |

1. *PagingRecordList:* This IE includes the PagingRecordList IE for each of the UEs addressed by the paging message. It can include 16 PagingRecord IEs at the maximum. The PagingRecord IE for a particular UE includes the following fields:
    - *UE-Identity:* This field contains the identity of the UE. The UE identifies that the paging message is addressed to itself using this identity. The following identities can be used:
        - *S-TMSI:* S-TMSI is normally used to page the UE for initiation of transfer of NAS signaling messages or user data.
        - *IMSI:* IMSI is used to page the UE when S-TMSI is not available because of some network failure. The network may send paging message using this IMSI in its attempt to recover error. When the UE receives paging message including IMSI instead of S-TMSI, it initiates the detach procedure and thereafter initiates the attach procedure.
    - *CN-Domain:* This field indicates if the domain is PS or CS. It is set to CS when the paging is performed for the CS fallback and it is set to PS otherwise.
3. *SystemInfoModification:* If this field is present in the paging message, then it indicates a change in the system information (SI) messages at the next modification period, explained in Section 4.5.2.2.
4. *ETWS-Indication:* If this field is present in the paging message, then it indicates the transmission of the ETWS primary notification on SIB type 10 and/or the ETWS secondary notification on SIB type 11. The transmission of such a notification can occur at any point in time regardless of any modification period.
5. *CMAS-Indication:* If this field is present in the paging message, then it indicates transmission of one or more CMAS notifications on SIB type 12. The transmission of such a notification can occur at any point in time regardless of any modification period.

# CHAPTER 14
# NAS Signaling Connection

The Non-Access Stratum (NAS) signaling connection is established between the UE and the MME paving the way for communication between the NAS layer of the UE and the MME. The UE and the MME move from the ECM-IDLE state to the ECM-CONNECTED state when the NAS signaling connection is established. The NAS signaling connection consists of an RRC connection over the LTE-Uu interface and a UE-associated logical S1-connection over the S1-MME connection. The establishment of the NAS signaling connection is generally comprised of the following steps:

1. *RRC connection establishment:* The connection between the RRC layers of the UE and the eNodeB is established. This is explained in Section 14.1.1.
2. *MME selection:* An appropriate MME is selected for services, if needed. This is explained in Section 14.2.
3. *UE-associated logical S1-connection establishment:* The UE-associated logical S1-connection provides for the exchange of signaling information dedicated for a particular UE between the eNodeB and the MME. The establishment of this connection is explained in Section 14.3.2.

The RRC connection establishment is completed before the UE-associated logical S1-connection completes. The establishment of NAS signaling connection is generally triggered when the UE wants to send the initial NAS message. The initial NAS messages are as follows:

1. ATTACH REQUEST;
2. TRACKING AREA UPDATE REQUEST;
3. SERVICE REQUEST;
4. EXTENDED SERVICE REQUEST;
5. DETACH REQUEST.

## 14.1 RRC Connection

The connection between the RRC layers of the UE and the eNodeB is referred to as an *RRC connection*. The RRC connection is required for exchange of dedicated signaling information or user data between UE and EPS. Therefore, if uplink or downlink data transfer is required and no RRC connection exists, then RRC connection needs to be established first. On the other hand, if there is a long period of inactivity in the data transfer, the eNodeB can release the RRC connection. The UE is in the RRC_IDLE state when there is no RRC connection between the UE and the eNodeB. Conversely, the UE is in the RRC_CONNECTED state if the RRC connection has been established.

The eNodeB does not store any information about the UE in the RRC_IDLE state.

### 14.1.1 RRC Connection Establishment

The RRC connection establishment procedure generally takes place in the following cases:

1. Initial access to the network;
2. Tracking area update;
3. Initiation of uplink or downlink data transfer;
4. Detachment from the network.

The RRC connection establishment procedure is generally used to establish SRB1 and to establish the NAS signaling connection between the UE and the MME. The NAS layer of the UE asks the RRC layer of the UE to establish an RRC connection and the RRC layer initiates the RRC connection establishment procedure. The eNodeB performs admission control based on its Radio Resource Management (RRM) algorithm to check if the RRC connection can be established. The RRC connection establishment procedure consists of the steps explained in the following sections. The exchange of messages in the RRC connection establishment procedure is shown in Figure 14.1.

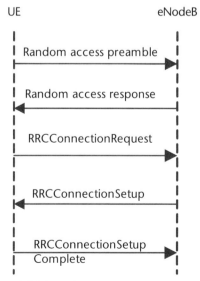

**Figure 14.1** RRC connection establishment.

### 14.1.1.1 Contention-Based Random Access

The UE performs contention-based random access in the beginning of the RRC connection establishment. This random access procedure is explained in Section 11.2. The UE sends a request to the eNodeB for RRC connection establishment in the third step of the random access procedure. To do so, the UE sends the RRCConnectionRequest message as message 3 using SRB0 on a CCCH logical channel. The RRCConnectionRequest message contains the following fields.

1. *UE-Identity:* If the UE is registered in the current tracking area, it already possesses an S-TMSI. The NAS layer of the UE provides the RRC layer with S-TMSI and the RRC layer uses the 40-bit-long S-TMSI for this UE-Identity field. This is typically the case when the random access occurs for the purpose of a periodic tracking area update, an UE-initiated dedicated EPS bearer setup, a service request, a default EPS bearer setup with a new PDN GW, and the detach procedure. Conversely, if the UE is not registered in the current tracking area, then it does not have an S-TMSI and the RRC layer of the UE uses a 40-bit-long random integer value between 0 and $2^{40} - 1$ for the UE-Identity field. Typically, the random integer value is used when the random access occurs for initial attach and tracking area update procedure for the sake of MME load rebalancing procedure.
2. *EstablishmentCause:* It depicts the cause for RRC connection establishment. The cause can be any of the following:
   - Emergency calls;
   - Mobile originated signaling;
   - Mobile originated data transfer;
   - Mobile terminated access;
   - High-priority access for UEs that have an access class between 11 and 15.

### 14.1.1.2 RRC Connection Setup

As shown in Section 11.2.4.1, the eNodeB sends a PDCCH instance indicating resource allocation on the DL-SCH in the fourth step of the random access procedure. The eNodeB sends an RRCConnectionSetup message to the UE on this allocated resource using SRB0 on the CCCH logical channel. The RRCConnectionSetup message includes the RadioResourceConfigDedicated IE. The RadioResourceConfigDedicated IE includes the SRB-Identity field. The value of the SRB-Identity is set to 1 indicating that SRB1 should be established. The RadioResourceConfigDedicated IE also includes detailed configuration settings of layer 1 and layer 2 to be used for the SRB1. Appendix A.2 shows the content of RadioResourceConfigDedicated IE. There are also default configuration parameter values specified that the UE would use if the eNodeB skip explicitly configures any of them.

SRB1 is set up after the reception of the RRCConnectionSetup message. Now the RRC layers of the UE and the eNodeB become connected and the UE enters the RRC_CONNECTED state. From now on, further signaling occurs on SRB1 using the DCCH logical channel.

#### 14.1.1.3 RRC Connection Setup Complete

The UE sends RRCConnectionSetupComplete message to the eNodeB using SRB1 on the DCCH and thus, it indicates successful completion of RRC connection establishment. The RRCConnectionSetupComplete message contains the following fields:

- *SelectedPLMN-Identity:* This field indicates the PLMN selected by the UE. It gives an index number that matches the position of the selected PLMN in the PLMN-identity list on SIB type 1. When an RRC connection is established for initial access of the UE, the UE selects a PLMN and indicates in this field.
- *RegisteredMME:* This field gives the GUMMEI of the MME that has registered the UE recently. If the RRC connection is established during tracking area update procedure for the sake of MME load rebalancing, the UE does not include the GUMMEI of the previously registered MME in this field, as will be explained in Section 25.3.
- *DedicatedInfoNAS:* The UE may send the initial NAS message piggybacked on the RRCConnectionSetupComplete message. In this case, this IE contains the initial NAS message. The eNodeB forwards the NAS message to the MME over the S1 connection.

#### 14.1.1.4 RRC Connection Setup Failure

The eNodeB may reject the request of RRC connection establishment based on its RRM algorithm. In this case, instead of the RRCConnectionSetup message shown in Section 14.1.1.2, the eNodeB sends an RRCConnectionReject message to the UE using SRB0 on the CCCH logical channel as shown in Figure 14.2.

The UE starts the timer T300 when it sends the RRCConnectionRequest message. If the UE receives neither an RRCConnectionSetup message nor an RRCConnectionReject message before the expiry of T300, the RRC connection procedure is terminated. The UE-TimersAndConstants IE contains the value of T300. Appendix A.1 shows that the eNodeB broadcasts the UE-TimersAndConstants IE on SIB type 2. The value of T300 can be 100, 200, 300, 400, 600, 1,000, 1,500, or 2,000 ms.

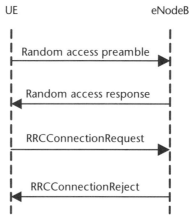

**Figure 14.2** RRC connection reject.

### 14.1.2 RRC Connection Release

The S1 release procedure generally causes the release of RRC connection between the eNodeB and the UE as mentioned in Section 14.4. The eNodeB sends an RRC-ConnectionRelease message to the UE, which triggers the release of the RRC connection. The UE cannot initiate the RRC connection release, so the UE cannot release the RRC connection itself to save power. The RRC connection release causes the release of all the established radio bearers and stops the use of dedicated radio resources. The eNodeB deletes the UE context. The UE leaves the RRC_CONNECTED state and enters the RRC_IDLE state. The RRCConnectionRelease message includes the ReleaseCause IE, which indicates the reason for releasing the RRC connection. The RRCConnectionRelease message can optionally include the cell reselection priority information to be used in the RRC_IDLE state, as will be explained in Section 23.2. The RRCConnectionRelease message can also include UTRAN/GERAN-related information for redirection to UTRAN/GERAN for the purpose of circuit-switched fallback (CSFB) as will be explained in Section 21.3.2.

### 14.1.3 Radio Link Failure (RLF)

While the UE is in the RRC_CONNECTED state, it may reach a coverage hole and the communication between the UE and the network can be lost because of poor signal. This is known as radio link failure (RLF).

#### 14.1.3.1 Detection of RLF

The RRC layer of the UE comes to a decision that RLF has occurred based on indications from layer 1 and layer 2:

1. *Poor signal quality detection:* The UE determines if a RLF has occurred based on out-of-sync and in-sync indications from the physical layer:
    - *Out-of-sync indication:* $Q_{out}$ is an indicator of a threshold level of unacceptably poor downlink radio link quality. $Q_{out}$ corresponds to a 10% block error rate of the PDCCH transmission. If the radio link quality estimated over the last $T_{out}$ duration is found to be worse than $Q_{out}$, the physical layer sends an out-of-sync indication to the higher layers in the $T_{out}$ duration.
    - *In-sync indication:* $Q_{in}$ is an indicator of a threshold level of acceptable downlink radio link quality. $Q_{out}$ corresponds to a 2% block error rate of the PDCCH transmission. If the radio link quality estimated over the last $T_{in}$ duration is found to be better than $Q_{in}$, the physical layer sends an in-sync indication to the higher layers in the $T_{in}$ duration.

    Two successive indications from the physical layer must be separated by at least 10 ms. If the DRX is not configured, $T_{out}$ and $T_{in}$ are 200 ms and 100 ms, respectively. If the DRX is configured, $T_{out}$ and $T_{in}$ depends on the DRX cycle length as shown in Table 14.1.

    If the physical layer sends N310 number of out-of-sync indications, then the T310 timer is started. If the T310 timer expires before the reception of N311 number of in-sync indications, then the UE comes to a decision

**Table 14.1** $T_{out}$ and $T_{in}$ for Different DRX Cycle Lengths

| DRX Cycle Length [sec] | $T_{out}$ and $T_{in}$ [sec] |
| --- | --- |
| Less than 0.04 | 20 × DRX cycle length |
| 0.08 | 0.8 [10× DRX cycle length] |
| 0.16 | 1.6 [10× DRX cycle length] |
| 0.32 | 3.2 [10× DRX cycle length] |
| 0.64 | 6.4 [10× DRX cycle length] |
| 1.28 | 6.4 [5× DRX cycle length] |
| 2.56 | 12.8 [5× DRX cycle length] |

that radio link failure has occurred. The UE turns off its transmitter power within 40 ms after the expiry of T310 timer. Conversely, if N311 number of in-sync indications are received while the T310 timer is running, the T310 timer is stopped and no RLF is considered.

The UE-TimersAndConstants IE contains the values of N310, N311, and T310. Appendix A.1 shows that the eNodeB broadcasts the UE-TimersAndConstants IE on SIB type 2. N310 can be 1, 2, 3, 4, 6, 8, 10, or 20. N311 can be 1, 2, 3, 4, 5, 6, 8, or 10. Typically, N311 is lower than N310. T310 can be 0, 50, 100, 200, 500, 1,000, or 2,000 ms.

2. *Random access failure:* As shown in Section 11.2, the UE must perform contention-based random access procedure if it has uplink data available but it does not have its uplink synchronized (i.e., the TimeAlignmentTimer has expired). In this contention-based random access procedure, if the UE reaches the maximum allowed number of transmissions without reception of Random Access Response (RAR), the physical layer reports the problem to RRC layer and it comes to a decision that radio link failure has occurred.

3. *Maximum RLC retransmissions:* The transmitting side of the AM RLC stops retransmission of an RLC PDU or a portion of an RLC PDU when it reaches a maximum limit specified by MaxRetxThreshold. For uplink data transfer, the eNodeB provides the UE with the value of the MaxRetxThreshold field as shown in Section 7.3.1.4. When the maximum number of retransmissions is reached, the RLC layer reports it to RRC layer and the RRC layer determines that a RLF has occurred.

4. *Handover failure:* The RLF can occur during handover action leading to handover failure as well. This occurs due to the failure in signaling related to handover with the source eNodeB because of a poor radio link. Such radio link failure is not too unlikely since soft handover is not supported.

The handover procedure that will be explained in Section 23.3.3 shows that the source eNodeB sends an RRCConnectionReconfiguration message to the UE as an instruction for handover action and then the UE starts the timer T304. Later in the handover procedure, the UE performs either noncontention-based random access or contention-based random access at the target eNodeB. If the random access procedure completes successfully, the UE stops the timer T304, but if T304 expires with no successful random access at the target eNodeB, the UE determines an RLF as well as a handover failure.

### 14.1.3.2 Actions to Overcome RLF

When the RRC layer of the UE determines that RLF has occurred, it starts the timer T311 and it attempts to select a cell for service. This can take place in one of the following ways:

1. *RRC connection reestablishment:* If the UE can select a suitable E-UTRA cell before the expiry of T311 and if AS security was activated prior to RLF, then the UE initiates the RRC connection reestablishment procedure explained in Section 14.1.4. The UE suspends SRB1 and releases SRB2 and all DRBs. The RRC connection reestablishment procedure resumes SRB1. Later, an RRC connection reconfiguration will be used to resume SRB2 and DRBs. The UE may select the same cell to which it was connected, and typically, this will be the case when the UE passes a coverage hole or shadowing in a cell. If the selected cell is different from the previously serving cell, then the new eNodeB can obtain the UE context from the old eNodeB. The UE may also select a cell using another RAT. If the RRC connection reestablishment procedure fails at the selected cell, the UE enters the RRC_IDLE state as explained in Section 14.1.4.4.

    If the RRC connection reestablishment occurs after handover failure explained as the fourth indication of RLF in Section 14.1.3.1, then this new access to a cell is referred to as an *RLF handover*.

2. *Entering the RRC_IDLE state due to the lack of security:* If the UE selects a suitable E-UTRA cell before the expiry of T311 but the AS security was not activated prior to RLF, then the UE enters the RRC_IDLE state at the same cell.

3. *Entering the RRC_IDLE state due to the expiry of T311:* If the UE fails to select a suitable cell before the expiry of T311, the UE enters the RRC_IDLE state at the same cell.

### 14.1.4 RRC Connection Reestablishment

The RRC connection between the UE and the network may be interrupted in case of different types of failures and then the UE may attempt to recover through the reestablishment of the RRC connection. The RRC connection reestablishment procedure is invoked for resumption of the function of SRB1 and for the reactivation of security. An RRC connection reconfiguration procedure is required subsequently in order to resume operation of SRB2 and DRBs.

The following events trigger the RRC connection reestablishment procedure provided that AS security has been activated. If AS security has not been activated, the UE does not attempt RRC connection reestablishment and rather moves to the RRC_IDLE state.

1. *RLF:* The RRC connection reestablishment procedure is triggered after RLF as explained in Section 14.1.3.2.

2. *Integrity check failure:* The RRC connection reestablishment procedure can be triggered when the PDCP layer indicates integrity check failure.
3. *RRC connection reconfiguration failure:* As shown in Section 14.1.5, the eNodeB sends an RRCConnectionReconfiguration message to the UE during the RRC connection reconfiguration procedure. The RRC connection reconfiguration failure may occur when the UE cannot comply with the configuration that the eNodeB has instructed in the RRCConnectionReconfiguration message. It may also occur when the RRCConnectionReconfiguration message causes a protocol error in the UE. In case of such failures, the UE does not apply any part of the configuration from the RRCConnectionReconfiguration message and the RRC connection reestablishment procedure is triggered.

The RRC layer of the UE initiates the RRC connection reestablishment procedure. The RRC connection reestablishment procedure consists of the steps explained in the following sections.

### 14.1.4.1 Contention-Based Random Access

The UE performs contention-based random access in the beginning of the RRC connection reestablishment as shown in Figure 14.3. The UE sends a request to the eNodeB for RRC connection reestablishment in the third step of the random access procedure. To do so, the UE sends the RRCConnectionReestablishmentRequest message as message 3 using SRB0 on the CCCH logical channel. The RRCConnectionReestablishmentRequest message contains the following fields:

1. *ReestabUE-Identity:* This IE contains the following UE identities to help the eNodeB retrieve the UE context and also to facilitate contention resolution.
   - *C-RNTI:* In the case of handover failure, this field gives the C-RNTI of the UE in the source cell. In case of other types of failures, this field gives

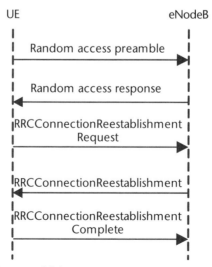

**Figure 14.3** RRC connection reestablishment.

the C-RNTI of the UE in the cell to which the UE was connected prior to the failure.
- *PhysCellId:* In the case of handover failure, this field gives the Physical Cell Identifier (PCI) of the source cell. In case of other types of failures, this field gives the PCI of the cell to which the UE was connected prior to the failure.
- *ShortMAC-I:* This field is used to identify and verify the UE at RRC connection reestablishment. This field is set to the 16 least significant bits of the MAC-I. The MAC-I is calculated based on the key for integrity protection of RRC messages and the integrity protection algorithm. In the case of handover failure, these key and integrity protection algorithms were used in the source cell. In the case of other types of failures, they were used in the cell to which the UE was connected prior to the failure.

2. *ReestablishmentCause:* This field depicts the type of failure that triggered the RRC connection reestablishment procedure. It can be any of the following:
   - Reconfiguration failure;
   - Handover failure;
   - Other failure.

#### 14.1.4.2 RRC Connection Setup

The eNodeB sends a PDCCH instance indicating resource allocation on DL-SCH in the fourth step of the random access procedure. The eNodeB sends RRCConnectionReestablishment message to the UE on this allocated resources using SRB0 on CCCH logical channel. SRB1 is set up after the reception of the RRCConnectionReestablishment message in the same way as explained in Section 14.1.1.2. Now the RRC layers of the UE and the eNodeB become connected and the UE enters the RRC_CONNECTED state. From now on, further signaling occurs on SRB1 using the DCCH logical channel.

#### 14.1.4.3 RRC Connection Reestablishment Complete

The UE sends an RRCConnectionReestablishmentComplete message to the eNodeB using SRB1 on DCCH and thus indicates the successful completion of the RRC connection reestablishment.

#### 14.1.4.4 RRC Connection Reestablishment Failure

The eNodeB may reject the request of the RRC connection reestablishment. This may happen, for example, if the cell does not have a valid UE context for the UE. In this case, instead of the RRCConnectionReestablishment message shown in Section 14.1.4.2, the eNodeB sends the RRCConnectionReestablishmentReject message to the UE using SRB0 on CCCH logical channel as shown in Figure 14.4. Then the UE enters the RRC_IDLE state.

The UE starts timer T301 when it sends the RRCConnectionReestablishmentRequest message. If the UE receives neither an RRCConnectionReestablishment message nor an RRCConnectionReestablishmentReject message before the expiry

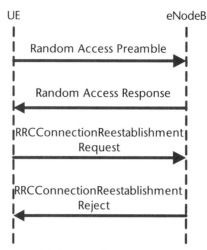

**Figure 14.4** RRC connection reestablishment reject.

of T301, the RRC connection reestablishment procedure is terminated and the UE enters the RRC_IDLE state. The UE-TimersAndConstants IE contains the value of T301. Appendix A.1 shows that the eNodeB broadcasts the UE-TimersAndConstants IE on SIB type 2. The value of T301 can be 100, 200, 300, 400, 600, 1,000, 1,500, or 2,000 ms.

### 14.1.5 RRC Connection Reconfiguration

The eNodeB initiates the RRC connection reconfiguration procedure by sending an RRCConnectionReconfiguration message to the UE. The RRCConnectionReconfiguration message is both ciphered and integrity protected. In response, the UE applies the instructions in the RRCConnectionReconfiguration message and sends an RRCConnectionReconfigurationComplete message to the eNodeB. The RRC connection reconfiguration procedure is used for various purposes:

- It can be used to set up SRB2 and DRBs.
- It can be used to modify or release radio bearers.
- It can be used to change C-RNTI.
- It can be used to command handover action.
- It can be used to configure or modify the measurement procedure.
- The RRCConnectionReconfiguration message can include a DedicatedInfoNAS IE that contains a NAS message. Thus, the RRC connection reconfiguration procedure can be used to carry a NAS message.

## 14.2 MME Selection

After the RRC connection establishment, the eNodeB selects an MME in order to establish an S1-MME connection for the particular UE. If the serving eNodeB is connected to a single MME, then it does not need to perform any MME selection.

However, in an MME pool area, the eNodeB can be connected to multiple MMEs. There is a NAS Node Selection Function (NNSF) located in the eNodeB to determine which MME would be associated with the UE. The selection of an appropriate MME accounts for the following factors:

1. The eNodeB may prefer an MME with service area that reduces the probability of changing the MME.
2. At the last stage of RRC connection establishment, the UE sends an RRC-ConnectionSetupComplete message to the eNodeB. This message indicates the MME that registered the UE most recently. The eNodeB may select this particular MME. If the information of this old MME is not available or this old MME does not serve the current eNodeB, then the eNodeB may select a new MME.
3. The eNodeB prioritizes an MME higher in the selection if it has a higher weight factor for the purpose of proper MME load balancing as explained in Section 25.3.

## 14.3 S1-MME Connection

The eNodeB and the selected MME communicate over S1-MME interface. The signaling service between the eNodeB and the selected MME uses S1AP protocol over S1-MME interface. These signaling services are classified as follows:

1. *Non-UE-associated services:* These signaling services are related to the whole S1 interface instance between the eNodeB and the selected MME and they are not related to a particular UE.
2. *UE-associated services:* These signaling services are related to a particular UE. They use a UE-associated logical S1-connection. The UE-associated logical S1-connection is identified by a unique pair of eNB UE S1AP ID and MME UE S1AP ID (i.e., each UE is assigned a unique pair of eNB UE S1AP ID and MME UE S1AP ID). The UE-associated logical S1-connection may exist before the S1 UE context is setup at eNodeB. When the MME sends any signaling message to the UE, it includes the assigned eNB UE S1AP ID allowing the eNodeB to forward the message to the right UE. When the eNodeB forwards any signaling message from the UE to the MME, it includes the assigned MME UE S1AP ID allowing the MME to identify the right UE.

### 14.3.1 S1 Setup

The S1 setup procedure is invoked during initial provisioning for initial S1-MME interface setup. This procedure uses non-UE associated signaling. The S1 setup procedure performs the exchange of configuration information needed for the eNodeB and the MME to interoperate properly on the S1-MME interface. The S1 interface uses SCTP/IP in the Transport Network Layer (TNL). The eNodeB first sets up SCTP association with the MME. For this purpose, the eNodeB may be initially configured with the IP address of the MME. A pair of SCTP stream is reserved for the exchange of signaling information of the common procedures that are not related to a particular UE. The S1 setup procedure is the first S1AP procedure after the establishment of SCTP association (i.e., after transport network layer has become operational). The S1 setup procedure includes the following steps:

1. The eNodeB requests the MME to set up S1 by sending S1 SETUP REQUEST message. The S1 SETUP REQUEST message includes the following fields:
   - *eNodeB Name:* This is a human readable name of the eNodeB.
   - *Global eNB ID:* This is the Global eNB ID of the eNodeB, explained in Section 1.6.2.
   - *Supported tracking areas:* This includes TAC and PLMN of the tracking areas that the eNodeB supports.
2. The S1 connection is set up after the reception of the S1 SETUP REQUEST message. Then the MME sends an S1 SETUP RESPONSE message to the eNodeB indicating a successful S1 setup. The S1 SETUP RESPONSE message contains the following fields:
   - *MME Name:* This is a human readable name of the MME.
   - *PLMNs:* A list of served PLMNs.
   - *Relative MME capacity:* This IE indicates the relative processing capacity of the MME with respect to the other MMEs in the MME pool. This is used in load balancing among MMEs in the MME pool area. Its value can be any integer up to 255.

If the MME is unable to accept the S1 setup request, then it sends S1 SETUP FAILURE message instead of the S1 SETUP RESPONSE message. This indicates an unsuccessful S1 setup. The S1 SETUP FAILURE message includes the reason of denial.

### 14.3.2 UE-Associated Logical S1-Connection Setup

The UE-associated logical S1-connection supports the exchange of signaling information dedicated for a UE between the eNodeB and the selected MME. The UE-associated logical S1-connection can be established any time after the initial provisioning using S1 setup procedure. It is established in the following ways:

1. *Transmission of INITIAL UE MESSAGE message:* When the eNodeB receives the initial NAS message from the UE, the eNodeB needs to forward the NAS message to the MME. Then the eNodeB sends INITIAL UE MESSAGE message to the MME. The INITIAL UE MESSAGE contains the following major IEs:
   - *eNB UE S1AP ID:* The eNodeB assigns a unique eNB UE S1AP ID for the UE and this IE contains the eNB UE S1AP ID.
   - *NAS-PDU:* This IE contains the initial NAS message from the UE.
   - *TAI:* This IE contains the current tracking area of the UE.
   - *E-UTRAN CGI:* This IE contains the ECGI of the cell currently serving the UE.

- *S-TMSI:* This IE contains the S-TMSI of the UE only if the UE has sent S-TMSI to the eNodeB.
- *GUMMEI:* If the eNodeB does not support the NNSF and the UE has sent eNodeB the GUMMEI from its most recent registration, then that GUMMEI is included in this IE.

  The MME will use the assigned eNB UE S1AP ID subsequently when it sends any signaling message to the UE. The UE-associated logical S1-connection is now established.

2. *Transmission of the DOWNLINK NAS TRANSPORT message:* When the MME needs to send a NAS message to the UE, it sends a DOWNLINK NAS TRANSPORT message to the eNodeB. The NAS-PDU IE of the DOWNLINK NAS TRANSPORT message contains the NAS message. If the UE-associated logical S1-connection already exists for the UE, the MME uses this connection. However, if the UE-associated logical S1-connection does not exist, the MME assigns a unique MME UE S1AP ID for the UE and includes it in the DOWNLINK NAS TRANSPORT message. The eNodeB will use the assigned MME UE S1AP ID subsequently when it forwards any message from the UE. The UE-associated logical S1-connection is now established.

3. *Transmission of the INITIAL CONTEXT SETUP REQUEST message:* When the MME sends an INITIAL CONTEXT SETUP REQUEST message, the UE-associated logical S1-connection is established if the UE-associated logical S1-connection did not exist.

4. *Path switch:* The UE-associated logical S1-connection can be established during the switching of a downlink GTP tunnel towards a new GTP tunnel endpoint. The UE-associated logical S1-connection is established when the MME sends PATH SWITCH REQUEST ACKNOWLEDGE message.

5. *Trace start:* The UE-associated logical S1-connection can be established when a trace session is initiated. The UE-associated logical S1-connection is established when the MME sends TRACE START message.

## 14.4 NAS Signaling Connection Release

The release of NAS signaling connection releases both RRC connection and UE-associated logical S1-connection. When RRC connection has been released, the UE considers that the NAS signaling connection is also released. When NAS signaling connection is released, the UE and the MME move from the ECM-CONNECTED state to the ECM-IDLE state. Additionally, all established radio bearers are released including S1-U bearers between the eNodeB and the serving gateway. The eNodeB deletes all UE related information and releases all signaling and user data transport resources. However, the MME still stores the UE context and information about the established bearers although it is in the ECM-IDLE state. The MME and the UE retain the GUTI.

The S1 release procedure generally causes the release of NAS signaling connection. The eNodeB or the MME initiates the signaling for S1 release procedure but the eNodeB or the MME may do so based on indications from the UE. The S1 release procedure may be invoked in the following cases:

- During the detach procedure.
- When there is long period of inactivity in data transfer. The release of NAS signaling connection in this case allows reduction of processing in the network and in the UE.
- If there is repeated RRC signaling integrity check failure, authentication failure, and so forth.
- For the purpose of load balancing. Then the ReleaseCause field in the RRC-ConnectionRelease message takes on the value LoadBalancingTAURequired This is explained in Section 25.3.

The S1 release procedure is comprised of the following steps as shown in Figure 14.5:

1. If the eNodeB initiates the signaling for S1 release procedure, then the eNodeB sends a UE CONTEXT RELEASE REQUEST message to the MME. However, if it is the MME that initiates the signaling for an S1 release procedure, then this step is not performed and the S1 release procedure starts from the next step.
2. The MME sends RELEASE ACCESS BEARERS REQUEST message to the serving GW and thus requests the release of all S1-U bearers for the UE.
3. The serving GW releases all eNodeB-related information for the UE and sends a RELEASE ACCESS BEARERS RESPONSE message to the MME. The serving GW keeps the S1-U bearer configuration of EPS bearers of the UE. If downlink packets arrive for the UE, the serving GW starts buffering them, and later the serving GW may initiate the network-triggered service request procedure.
4. The MME sends a UE CONTEXT RELEASE COMMAND message to the eNodeB.

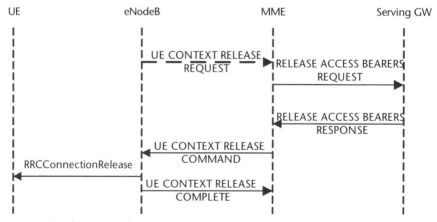

**Figure 14.5** S1 release procedure.

5. The eNodeB sends an RRCConnectionRelease message to the UE. This causes the release of the RRC connection between the eNodeB and the UE.
6. The eNodeB releases all signaling and user data transport resources. It also sends the UE CONTEXT RELEASE COMPLETE message to the MME to confirm the S1 release.

# CHAPTER 15

# Attach and Detach

The attach procedure is used to register the UE with the EPS (i.e., to establish an approved connection between the UE and the EPS). On the other hand, the detach procedure is used to disconnect the UE from the EPS.

## 15.1 Initial Attach Procedure

When the UE is powered up, it first performs cell selection, explained in Section 9.1. Thereafter, the UE performs an initial attach procedure in the selected cell in order to register with the EPS. A successful completion of the attach procedure causes the following changes:

- The UE and the MME move from an EMM-DEREGISTERED state to an EMM-REGISTERED state. An EPS Mobility Management (EMM) context is established both in the UE and in the MME.
- The selected PLMN becomes the Registered PLMN (RPLMN) of the UE.
- The Globally Unique Temporary UE Identity (GUTI) reallocation is performed as a part of the attach procedure. The GUTI reallocation was explained in Section 1.6.3.
- The MME becomes aware of the location of the UE and the MME can communicate with the UE.
- The UE starts reading paging and notification messages from the network.
- A default EPS bearer that enables always-on IP connectivity to the UE is established between the UE and the PDN GW. The connectivity allows the transportation of user data. The default EPS bearer, explained in Section 2.4, remains established throughout the lifetime of the PDN connection. The UE can establish connections with multiple PDNs. Then the PDN connectivity procedure is used to establish the first default EPS bearer with the first PDN GW during the initial attach procedure. Later, the UE requested PDN connectivity procedure can be invoked to establish additional default EPS bearers with additional PDN GWs.

In the case of the earlier 3GPP implementations, the UE attaches to the network and is authenticated initially. There is no always-on connectivity and so, for establishment of data services, the connectivity between the UE and the network needs to be set up using a second procedure. This introduces a delay in access and reception of services and consequently limits network performance. However, LTE allows having a default EPS bearer and IP connectivity to remain established all the time. This simplifies and speeds up the connectivity.

The UE and the network carry out the attach procedure and the default EPS bearer context activation procedure in parallel. The success of the attach procedure is dependent on the success of the default EPS bearer context activation procedure. The network may initiate the activation of dedicated bearers with the same PDN GW as a part of the attach procedure.

The attach procedure generally includes the steps one after another explained in the following sections and shown schematically in Figure 15.1.

### 15.1.1 NAS Signaling Connection Establishment and Attach Request

The UE needs a way to communicate with the network in order to send a request for the attachment. The UE and the network will also use the communication path for further negotiation in the attach procedure. Thus, in the beginning, the UE initiates the establishment of the NAS signaling connection between the UE and the MME. The NAS signaling connection establishment procedure includes the steps shown in the following sections.

#### 15.1.1.1 RRC Connection Establishment

The RRC connection establishment procedure was explained in Section 14.1.1. During this procedure, the following settings are used for different messages.

*RRCConnectionRequest Message*
- *UE-Identity:* The UE does not have an S-TMSI. A random integer value between 0 and $2^{40} - 1$ is used in this field.
- *EstablishmentCause:* This field indicates the cause for RRC connection establishment. This field is set to Mobile Originated Signaling.

*RRCConnectionSetup Message*
This message includes RadioResourceConfigDedicated IE, which configures the settings of layer 1 and layer 2 to be used for the SRB1. At this stage, the eNodeB does not know about the capabilities of the UE, so the eNodeB may use a minimum configuration in RadioResourceConfigDedicated IE that any UE may be able to support. The eNodeB can also skip configuring and allow the UE to use default parameter values.

*RRCConnectionSetupComplete Message*
1. *SelectedPLMN-Identity:* The UE selects a PLMN from the PLMN-identity list on SIB type 1 and indicates the PLMN in this field as explained in Section 14.1.1.3.

## 15.1 Initial Attach Procedure

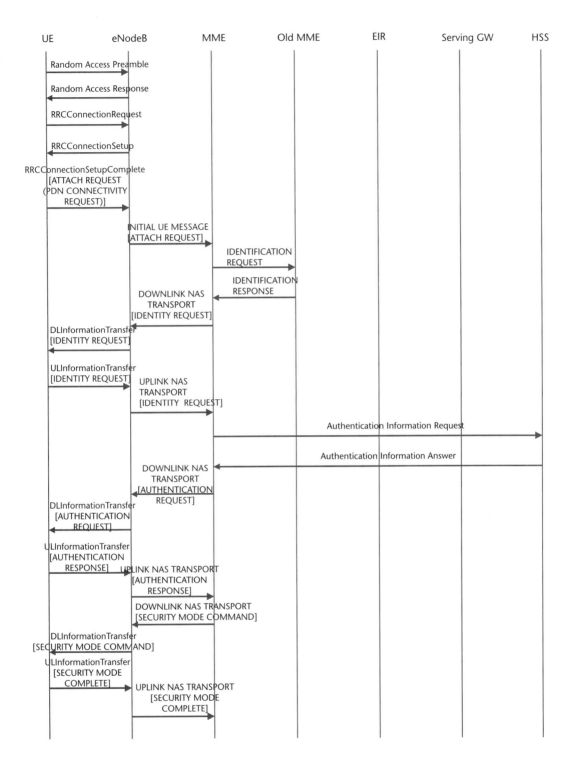

**Figure 15.1** Initial attach procedure.

2. *RegisteredMME:* This IE gives the GUMMEI of the MME that has registered the UE recently. The UE may not have any MME information to

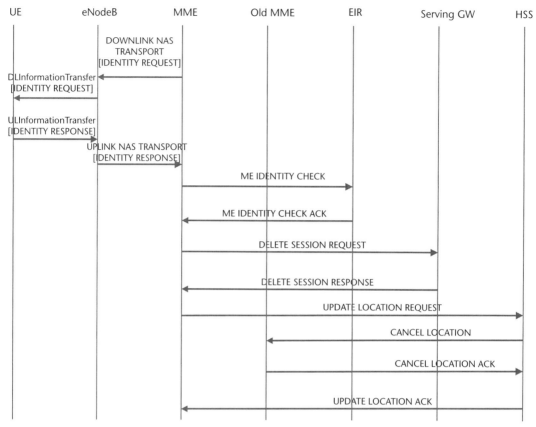

**Figure 15.1** (continued)

provide here as it is not yet registered with the network. However, the UE may still have valid MME information from the last registration and then the UE includes the GUMMEI of that MME.

3. *DedicatedInfoNAS:* The DedicatedInfoNAS IE in this message contains ATTACH REQUEST as the initial NAS message. Thus, the UE sends a request for attach to the network. With this request, the UE starts timer T3410. The UE stops the timer T3410 when the MME responds by sending ATTACH ACCEPT message as shown in Section 15.1.7.4. The ATTACH REQUEST message includes the following major IEs:
    - *EPS Mobile Identity:* This IE includes the IMSI of the UE because the UE has not been assigned a GUTI yet. However, the UE may possess the GUTI assigned when it was registered with the network most recently. In this case, this IE contains the GUTI of the UE instead. The inclusion of the old GUTI instead of IMSI at this stage helps prevent eavesdropping, which is important because security is not yet activated for over-the-air messages. The use of this IE is explained in Section 15.1.2.
    - *TAI:* This IE contains the TAI of the tracking area that the UE visited last. This information may or may not be available.
    - *EPS Attach Type:* This IE contains the type of the requested attach procedure. This IE can be set to any of the following values:

## 15.1 Initial Attach Procedure

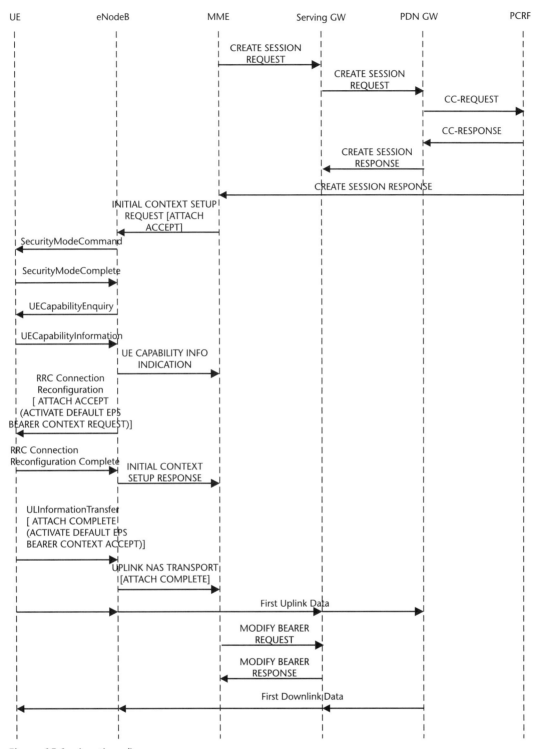

**Figure 15.1** (continued)

- *EPS attach:* This value indicates request for PS mode of operation, i.e. request for attach to EPS.

- *Combined EPS/IMSI attach:* It indicates request for attach to both EPS and 2G/3G (i.e., request for registration with both CS and PS domain). Setting to this value is required for the circuit-switched fallback (CSFB) operation explained in Section 21.3.2. Also, setting to this value indicates that the UE is CSFB capable.
- *EPS Emergency Attach:* This value indicates request for emergency bearer. This value is used when the UE cannot gain normal services from the network and it attempts to gain emergency services.
  - *Additional update type:* This IE is used to indicate if the UE requests for SMS service over the SGs' interface but not the CSFB operation. If this request is accepted, then the UE does not fall back to UTRAN/GERAN but the UE registers only for the exchange of SMS over the SGs' interface as explained in Section 21.4.2.
  - *Voice domain preference and UE's usage setting:* This IE is used to indicate if the UE supports CS fallback and SMS over the SGs' interface or if the UE supports IMS for voice.
  - *NAS Key Set Identifier, $KSI_{ASME}$:* If the UE has a valid NAS key set identifier, $KSI_{ASME}$, then it is included in this IE. The NAS key set identifier is composed of the type of security context and NAS key set ID. The type of security context is set to native if the UE operates only on EPS services. The UE does not have a security context during the initial attach.
  - *UE network capability:* This IE contains ciphering and integrity protection algorithms supported by the UE.
  - *ESM message container:* This IE contains an ESM message, PDN CONNECTIVITY REQUEST. The PDN CONNECTIVITY REQUEST message is used to request PDN connectivity, as explained in Section 15.1.7.1.

  A stored NAS security context may or may not exist. If a valid NAS security context does not exist, the ATTACH REQUEST message combined with the PDN CONNECTIVITY REQUEST message is not integrity protected. However, if a valid NAS security context exists, then the UE integrity protects the ATTACH REQUEST message combined with the PDN CONNECTIVITY REQUEST message.

### 15.1.1.2 MME Selection

The MME selection takes place as explained in Section 14.2.

### 15.1.1.3 UE-Associated Logical S1-Connection Setup

The UE-associated logical S1-connection establishment was explained in Section 14.3.2. The eNodeB sends an INITIAL UE MESSAGE to the MME, and thus, the UE-associated logical S1-connection is established. The INITIAL UE MESSAGE uses the following settings:

- *eNB UE S1AP ID:* The eNodeB assigns eNB UE S1AP ID uniquely to the UE. This IE contains the eNB UE S1AP ID. The MME will use this eNB UE S1AP ID subsequently.

- *NAS-PDU:* This IE contains the ATTACH REQUEST message originally sent by the UE, and thus, the eNodeB forwards the ATTACH REQUEST message to the MME.
- *TAI:* This IE contains the current tracking area of the UE.
- *E-UTRAN CGI:* This IE contains the ECGI of the eNodeB serving the UE currently.
- *S-TMSI:* This IE contains the S-TMSI of the UE only if the UE was able to include S-TMSI in the RRCConnectionRequest message, so this IE contains no information in this initial attach procedure.
- *GUMMEI:* If the UE has included GUMMEI from its last registration in the RegisteredMME IE of RRCConnectionSetupComplete message, then that GUMMEI is included in this IE.

### 15.1.2 User Identification

If the UE has included its IMSI in the ATTACH REQUEST message, then the MME checks if there is any UE context existing for the UE based on the IMSI. In this initial attach procedure, the MME may not have any UE context and then the MME starts building up an UE context based on the information of the UE in the ATTACH REQUEST message. If any UE context is available, it can be updated.

On the other hand, if the UE has included its GUTI in the ATTACH REQUEST message, then the MME identifies the user in the following ways:

1. If the UE was registered with the same MME last time, then the MME can identify the UE completely from the GUTI. The MME can also derive the IMSI of the UE.
2. If the UE was registered with a different MME last time, then the new MME first detects the old MME from the GUMMEI part of the GUTI. Thereafter, the new MME derives the IMSI of the UE from the old MME. To do so, the new MME sends IDENTIFICATION REQUEST message to the old MME. The IDENTIFICATION REQUEST message includes the GUTI of the UE and the ATTACH REQUEST message. The old MME first verifies the ATTACH REQUEST message and then responds with IDENTIFICATION RESPONSE message. The IDENTIFICATION RESPONSE message includes the IMSI of the UE.
3. If it happens that both the new MME and the old MME fail to identify the UE based on the GUTI, then the new MME sends an IDENTITY REQUEST message to the UE that requests for IMSI. In response, the UE sends an IDENTITY RESPONSE message to the MME including its IMSI.

### 15.1.3 Authentication and NAS Security Activation

The EPS Authentication and Key Agreement (EPS AKA) procedure takes place as explained in Section 12.2. Thereafter, the security function is activated in NAS layer as explained in Section 12.3. The EPS security context can be stored for future use and it is then known as the EPS-cached security context. If the EPS-cached security context is available, then the EPS AKA procedure is optional and it is mandatory otherwise. IIf the EPS AKA procedure is not performed, then the activation of NAS security function is not required. The activation of AS security function is later performed as shown in Section 15.1.7.4 whether EPS AKA procedure is performed or not.

### 15.1.4 ME Identity Check

The network checks if the mobile equipment is stolen or if it has faults. If the mobile equipment is not allowable, the network may send the Attach Reject message to the UE rejecting the attach procedure. In order to check the validity of the mobile equipment, the MME sends an IDENTITY REQUEST message to the UE. In response, the UE sends an IDENTITY RESPONSE message to the MME, which includes the IMEI of the mobile equipment.

The Equipment Identity Register (EIR) keeps records about mobile equipment and the MME may check with the EIR via S13 interface. If the MME is configured to check the IMEI with the EIR, it sends an ME IDENTITY CHECK message to the EIR, which includes the IMEI and IMSI of the UE. The EIR responds with ME IDENTITY CHECK ACK message indicating whether the mobile equipment is permitted for use or not.

In order to reduce signaling delays, the ME identity check procedure may be combined with NAS security activation step. In this case, the MME includes the ME Identity Request field in the SECURITY MODE COMMAND message and the UE includes the IMEI in the SECURITY MODE COMPLETE message.

### 15.1.5 Old EPS Bearer Context Delete

When the UE detaches from the network, the network deletes EPS bearer contexts, but it may happen sometimes that the detachment is not properly performed and some EPS bearer contexts are not deactivated in the network. In this case, the old EPS bearer contexts are deactivated while the UE attaches to the network again. Thus, if there is any old EPS bearer contexts left, they are deleted at this stage of the attach procedure. The MME sends DELETE SESSION REQUEST message to the Serving GW. This message includes LBI of the previous default EPS bearer. The Serving GW deactivates the EPS bearers and releases corresponding EPS bearer context information. The Serving GW sends a DELETE SESSION RESPONSE message to the MME as an acknowledgment.

### 15.1.6 Location Update

A location update procedure is required to be performed in the following cases.

- The UE was registered with a different MME last time.
- The MME does not have valid subscription context for the UE.
- The mobile equipment (i.e., the IMEI) has been changed.
- The UE includes the IMSI of the UE and not the GUTI in the ATTACH REQUEST message.

The location update procedure includes the following steps:

1. The MME sends an UPDATE LOCATION REQUEST message to the HSS. This message includes MME identity, IMSI, ME identity, MME capabilities, and update type. The Update Type field indicates that this is an attach procedure. In the flags, the MME indicates that it requests for subscriber data.
2. If the UE was registered with a different MME last time, then the HSS sends a CANCEL LOCATION message to the old MME. This message includes the IMSI and the cancellation type. The Cancellation Type field indicates that the old MME should release the old serving GW resource.
3. The old MME sends the CANCEL LOCATION ACK message to the HSS as an acknowledgment of its release. This message includes the IMSI of the UE.
4. The HSS sends the UPDATE LOCATION ACK message to the new MME and acknowledges the update location request. This message includes the IMSI and the subscription data of the UE. The subscription data contains one or more APN subscription contexts. For each APN, the subscription data includes the CS/PS mode, the UE-AMBR, and the APN configuration. The APN configuration includes the IP version, the APN-AMBR, the QoS information for the default EPS bearer with the PDN, the PDN GW allocation type, and the APN-NI. The PDN GW allocation type helps the MME to select a PDN GW for the UE from all the PDN GWs of the APN.
5. The new MME constructs an MM context for the UE. The MM context contains mobility management and UE security parameters.

### 15.1.7 PDN Connection Establishment

The establishment of PDN connection or the PDN connectivity procedure includes the steps shown in the following sections.

#### 15.1.7.1 IP Address Allocation

The UE performs IP address allocation procedure in order to obtain its IP address. At least one IP address that can be either IPv4 or IPv6 is assigned to the UE. The version of IP is determined considering the capability of the UE, the subscription of the UE, and the capability of PDN GW. The IP address assigned to the UE is called *PDN address*. The IP address can be allocated by any of the following entities:

- PDN-GW of the HPLMN or EHPLMN;
- PDN-GW of VPLMN;
- External PDN.

The PDN GW may use its internal IPv4 address pool or IPv6 prefix pool for the allocation of the IPv4 address or IPv6 prefix. When an IPv4 address or IPv6 prefix is allocated from an external PDN, the PDN GW obtains it from the external PDN. The allocation of IP address from the PLMN or the external PDN is the same from the perspective of the UE. Also, the allocation of a static IP address or a dynamic IP address is the same from the perspective of the UE.

The IPv4 address or the IPv6 prefix is released when the default EPS bearer is deactivated. The PDN GW should not assign the IPv6 prefix to other user immediately after releasing it.

*PDN Connectivity Request*
The UE initiates the PDN connectivity procedure by sending a PDN CONNECTIVITY REQUEST message piggybacked on the ATTACH REQUEST message to the network as shown in Section 15.1.1.1. If the network accepts this request, it allocates an IP address to the UE and establishes a default EPS bearer context. The PDN CONNECTIVITY REQUEST message includes the following major IEs:

- *Request type:* This IE indicates if a new connectivity to a PDN is requested or an earlier connection made via non-3GPP access should be kept.
- *PDN type:* This IE is set to IPv4 or IPv6 if the UE is only IPv4 capable or only IPv6 capable, respectively. If the UE is both IPv4 and IPv6 capable, then this IE is set to IPv4v6.
- *Access Point Name:* This IE contains an access point name (APN) requested by the UE. The UE may or may not include an APN in this IE.
- *Protocol Configuration Options (PCO):* This IE contains parameters needed to be set up between the UE and the PDN GW. These parameters are sent transparently through the MME and the serving GW. These parameters include, for example, address allocation preference, which may indicate that the UE prefers to obtain an IPv4 address only after the default EPS bearer activation by means of DHCPv4.
- *ESM information transfer flag:* When the UE needs to send PCO or APN or both using ciphering, then this ESM information transfer flag is set.

*IPv4 Address Allocation*
The network may allocate an IPv4 address during activation of the default EPS bearer. To activate the default EPS bearer, the network sends an ACTIVATE DEFAULT EPS BEARER CONTEXT REQUEST message that includes PDN address IE as shown in Section 15.1.7.4. The PDN address IE includes a PDN type field and a PDN address information field. The network sets the PDN type field to IPv4 and includes the IPv4 address in the PDN address information field. Alternatively, instead of allocating IPv4 address while activating the default EPS bearer, the network can optionally allocate IPv4 address using DHCPv4 after the default bearer is established. The network may do so when the UE indicates in the PCO IE of the PDN CONNECTIVITY REQUEST message that it wants to use DHCPv4. In this case, the network sets 0.0.0.0 in the PDN address information field of PDN address IE while sending an ACTIVATE DEFAULT EPS BEARER CONTEXT REQUEST message.

An IPv4 address may be allocated during the establishment of the default EPS bearer context, but then a DNS server address or other IPv4 parameters are not provided. In this case, if the UE needs such additional parameters, the UE can configure them using DHCPv4. Then the UE sends a DHCPDISCOVER message indicating the necessary information and the network replies with that information.

*IPv6 Address Allocation*
The network allocates an IPv6 address by setting the PDN type to IPv6 and including an interface identifier in the PDN address information field of the PDN address IE when it sends an ACTIVATE DEFAULT EPS BEARER CONTEXT REQUEST message as shown in Section 15.1.7.4. The interface identifier is used to configure the link-local address.

The IPv6 stateless address autoconfiguration is mandatory. The PDN-GW sends a Router Advertisement message for a stateless IPv6 address configuration and allocates an IPv6 prefix after the default EPS bearer is established. For this allocation, the UE may wait for the Router Advertisement message from the PDN GW. If the UE does not receive any Router Advertisement message, then it can send a Router Solicitation message to solicit a Router Advertisement message. The Router Advertisement message, whether solicited or unsolicited, allocates the IPv6 prefix. The allocated IPv6 prefix is globally unique. Therefore, the UE does not need to perform duplicate address detection for its IPv6 address. Alternatively, the network can optionally allocate the IPv6 prefix using DHCPv6. The PDN GW keeps a record of the relationship between the IMSI of the UE and the allocated IPv6 prefix.

After the UE has received the IPv6 prefix, it constructs a full IPv6 address via IPv6 stateless address autoconfiguration. When creating a global IPv6 address, the UE can change the interface identifier and there is no restriction on the value of the interface identifier since the prefixes are uniquely allocated to the UE. The UE may actually change the interface identifier as such without involving the network for the purpose of privacy.

If the other configuration flag (O flag) is set in the router advertisement, the UE needs to configure additional IP parameters that are not yet provided. This may be the DNS server address. The UE sends a DHCPv6 Information-Request message indicating the necessary information. The PDN GW acts as the DHCP server and provides the requested parameters. When parameters are needed from the external PDN, the PDN GW obtains the requested parameters from the external PDN and sends them to the UE. When the PDN GW acts as a DHCPv6 server towards the UE, the PDN GW may act as a DHCPv6 client towards the external PDN to request the configuration parameters for the UE.

#### 15.1.7.2 EPC Node Selection

The EPC allocates a default EPS bearer for the UE, which requires a selection of associated nodes in the EPC:

- *Default APN selection:* The MME can be connected to multiple APNs. The MME selects one of these APNs as a default APN for the UE that is used for default EPS bearer activation. The MME checks the subscriber profile downloaded from HSS and uses it for an APN selection. In addition, if the UE has included an APN in the PDN CONNECTIVITY REQUEST message,

the MME considers this APN for selection. These two APNs may actually be the same.
- *PDN GW selection:* There can be multiple PDN GWs for an APN. The MME selects one of these PDN GWs for default EPS bearer activation. For this purpose, the MME uses a PDN GW selection function. The MME checks the subscription context downloaded from HSS if it provides a PDN GW for the default APN. If provided, the MME uses this PDN GW and, if not provided, then the MME selects a PDN GW. The MME uses APN-NI in its selection that the HSS sent earlier using the UPDATE LOCATION ACK message.
- *Serving GW selection:* There can be multiple serving GWs for a PDN GW. The MME selects one of these serving GWs for services. The MME may communicate with the DNS server to find out a serving GW that supports the location of the UE. In addition, the MME may prefer a serving GW that reduces the probability of changing the serving GW. The MME may also prioritize a serving GW higher considering proper load balancing among serving GWs.

### 15.1.7.3 Session Setup

The session setup allows the EPC to establish a default EPS bearer within EPC for the UE and set up necessary parameters for PDN connectivity. It is comprised of the following steps:

1. The MME allocates an EPS Bearer Identity (EBI) to the default EPS bearer to be established. It also generates an MME F-TEID for the control plane. The F-TEID includes the interface type, TEID, and IP address. Then the MME sends a CREATE SESSION REQUEST message to the selected serving GW. This establishes the S11 bearer.

    The CREATE SESSION REQUEST message includes IMSI, MSISDN, MME F-TEID for the control plane, PDN GW address, PDN address, APN, RAT type, default EPS bearer QoS, PDN type, EBI, APN-AMBR, PCO, handover indication, ME identity, ECGI of the serving eNodeB, MS info change reporting support indication, selection mode, charging characteristics, trace reference, trace type, trigger ID, OMC identity, maximum APN restriction, dual address bearer flag, the Protocol Type over S5/S8, and serving network. The PDN GW address is the IP address of the selected PDN GW. The serving GW determines the PDN GW from this address. RAT type should be set to E-UTRAN. The PDN type indicates if it is IPv4, IPv6, or IPv4v6. Default EPS bearer QoS is obtained from a subscription context downloaded from the HSS. PCO is forwarded from the PDN CONNECTIVITY REQUEST message.

2. The serving GW creates a new entry in its EPS bearer table. It generates separate DL F-TEIDs for the control plane and the user plane. Then it sends a CREATE SESSION REQUEST message to the PDN GW. This establishes the S5 bearer.

The CREATE SESSION REQUEST message includes IMSI, MSISDN, APN, serving GW address for the user plane, serving GW F-TEID of the user plane, serving GW F-TEID of the control plane, RAT type, default EPS bearer QoS, PDN type, PDN address, subscribed APN-AMBR, EBI, PCO, handover indication, ME identity, ECGI of the UE, MS info change reporting support indication, selection mode, charging characteristics, trace reference, trace type, trigger Id, OMC identity, maximum APN restriction, dual address bearer flag, and serving network. At this stage, the serving GW starts buffering the downlink packets received from the PDN GW.

3. If dynamic Policy and Charging Control (PCC) is deployed, the PDN GW performs an IP-CAN session establishment procedure and obtains the default PCC rules for the UE. The PDN GW sends a CC-REQUEST message to the PCRF indicating an initial request. The PDN GW provides the PCRF with the IMSI, APN, UE IP address, ECGI as the user location information serving network, RAT type, APN-AMBR, and default EPS bearer QoS. This may lead to the establishment of a number of dedicated bearers in association with the establishment of the default bearer. The PCRF may not have the subscription information in this initial attach procedure, but then it retrieves information from the HSS. The PCRF stores the subscription information and PCC rules. From now on, the PCRF makes PCC-related decisions and provides PCEF with the rules for policy and charging control. PCRF can modify APN-AMBR, QCI, and ARP for the default bearer if requested.

   If dynamic PCC is not deployed, the PDN GW may apply the local QoS policy. This may also lead to the establishment of a number of dedicated bearers in combination with the establishment of the default bearer. The QoS parameter values of the default bearer are assigned based on the subscription data received from the HSS.

4. The PDN GW creates a new entry in its EPS bearer context table and generates a charging ID. Now the PDN GW can route user plane PDUs between the serving GW and the packet data network. The PDN GW also starts charging. Then the PDN GW sends a CREATE SESSION RESPONSE message to the serving GW. This message includes the PDN GW address for the user plane, PDN GW F-TEID of the user plane, PDN GW F-TEID of the control plane, PDN type, PDN address, EBI, EPS bearer QoS, PCO, charging ID, prohibit payload compression, APN restriction, and APN-AMBR. The PDN address is the IP address allocated to the UE. The PCO may contain DNS addresses.

5. The serving GW sends a CREATE SESSION RESPONSE message to the MME. This message includes the PDN type, PDN address, serving GW address for user plane, serving GW F-TEID for user plane, Serving GW TEID for control plane, EBI, EPS bearer QoS, PDN GW addresses and TEIDs (GTP-based S5/S8) or GRE keys (PMIP-based S5/S8) at the PDN GW for uplink traffic, PCO, prohibit payload compression, APN restriction, and APN-AMBR. The eNodeB will use the Serving GW F-TEID for the user plane when it sends the user plane data to the serving GW.

### 15.1.7.4 Initial Context Setup

The MME initiates the initial context setup. This procedure establishes the UE context at the eNodeB and establishes a radio bearer associated with the default EPS bearer. The initial context setup procedure includes the steps shown below.

*Initial Context Setup Request*
The MME sends the INITIAL CONTEXT SETUP REQUEST message to the eNodeB over the S1-MME interface. Upon receipt of this message, the eNodeB establishes the UE context. The INITIAL CONTEXT SETUP REQUEST message includes the following major IEs:

- *E-RAB ID:* This IE gives the E-RAB ID for the E-RAB to be established for the default EPS bearer. The value of E-RAB ID is the same as the EBI value.
- *E-RAB level QoS parameters:* This IE gives QoS parameters for the default EPS bearer. The eNodeB will configure layer 1/layer 2 parameters for the DRB associated with the default EPS bearer based on these QoS parameters.
- *Transport layer address:* This IE gives the IP address that the eNodeB will use to send user data to the serving GW.
- *GTP-TEID:* This IE gives the TEID that the eNodeB will use to send user data to the serving GW.
- *NAS-PDU:* This IE contains the ATTACH ACCEPT message. The eNodeB will forward this message to the UE.
- *UE-AMBR:* This IE contains the UE-AMBR assigned to the UE.
- *UE security capabilities:* This IE contains the security capabilities of the UE.
- *Security key:* This IE contains the security key, $K_{eNB}$, for the UE. The eNodeB creates the AS security context from the security capabilities and $K_{eNB}$ for the UE.
- *Subscriber profile ID for RAT/frequency priority:* This IE contains the RAT/frequency priority list for the UE. The MME received this list in the subscriber profile from the HSS.

*AS Security Activation*
The security function is activated in AS layer as explained in Section 12.4.1.

*UE Capability Transfer*
If the eNodeB requires additional UE capability information, then it performs the UE capability transfer procedure to obtain the information. The eNodeB uses this information to configure layer 1 and layer 2 when it sets up or modifies radio bearers. The UE capability transfer procedure includes the following steps:

1. *UE capabilities enquiry:* The eNodeB instructs the UE to send its radio access capabilities for certain RATs by sending the UECapabilityEnquiry message to the UE. This message includes the UE-CapabilityRequest IE that specifies the RATs for which the radio access capabilities are requested. The UE-CapabilityRequest IE can have any or all of the following values:
    - E-UTRA;

- UTRA;
- GERAN-CS;
- GERAN-PS;
- CDMA2000-1XRTT.

2. *Informing UE capabilities:* The UE sends the UECapabilityInformation message to the eNodeB to inform its radio access capabilities for the RATs that the UE supports and is requested to report. This message contains the UE-CapabilityRAT-ContainerList IE that contains separate UE-CapabilityRAT-Container IEs for each RAT that the UE is reporting. The UE-CapabilityRAT-Container IE contains the following major IEs:
   - *RAT-Type:* This IE specifies the RAT for which the UE capabilities are reported, so it can have any of the following values:
     - E-UTRA;
     - UTRA;
     - GERAN-CS;
     - GERAN-PS;
     - CDMA2000-1XRTT.
   - *UECapabilityRAT-Container:* This IE includes different IEs to specify capabilities with regard to various RATs. In order to specify LTE-related capabilities, it includes the UE-EUTRA-Capability IE, which includes the following IEs or fields:
   - *UE-Category:* This field specifies the UE-Category of the UE. This field can have value 1, 2, 3, 4, or 5 to specify one of the five UE categories.
   - *PDCP-Parameters:* This field includes the SupportedROHC-Profiles IE, which specifies the ROHC profiles supported by the UE.
   - *PhyLayerParameters:* This IE includes the UE-TxAntennaSelectionSupported field that specifies if the UE supports the UE transmit antenna selection. It also includes the UE-SpecificRefSigsSupported field, which specifies if the UE supports UE-specific reference signals.
   - *RF-Parameters:* This IE specifies the radio bands supported by the UE. It includes the SupportedBandListEUTRA IE, which contains operating band numbers for each of the supported radio bands.
   - *MeasParameters:* The MeasParameters includes the BandListEUTRA IE, which contains the InterFreqBandList IE. The InterFreqBandList IE includes InterFreqBandInfo IE for each of the radio bands supported by the UE for E-UTRA as the RAT. The InterFreqBandInfo IE includes the InterFreqNeedForGaps IE, which indicates if measurement gaps are required for the measurement of this band while the UE operates on another band in the list.

   The BandListEUTRA IE also includes the InterRAT-BandList IE, which contains InterRAT-BandInfo for each of the radio bands supported by the UE for a RAT other than E-UTRA. The InterRAT-BandInfo IE includes InterRAT-NeedForGaps IE, which indicates if measurement gaps are required for the measurement of this band while the UE operates on another band.

3. *Updating MME:* The eNodeB sends the UE CAPABILITY INFO INDICATION message to the MME, which contains the UE radio capabilities. The MME stores this information until the UE detaches from the network.

*Radio Bearer Establishment and Attach Accept*
The eNodeB establishes SRB2 as well as a DRB associated with the default EPS bearer. This includes the steps shown below.

*Radio Bearer Configuration*
The eNodeB sends an RRCConnectionReconfiguration message to the UE instructing it to set up SRB2 and a DRB. This message includes the RadioResourceConfigDedicated IE, which contains detailed configuration for layer 1 and layer 2 to be used for the radio bearers. The RRCConnectionReconfiguration message also includes the MeasConfig IE, which configures measurements and reporting.

The RRCConnectionReconfiguration message also includes the DedicatedInfoNAS IE, which contains an ATTACH ACCEPT message. Thus, the MME sends the ATTACH ACCEPT message to the UE in response to the ATTACH REQUEST message, shown in Section 15.1.1.1, indicating the acceptance of the requested attach procedure, so the UE now stops the timer T3410. If T3410 expires before the reception of the ATTACH ACCEPT message, then the UE aborts the attach procedure and releases the NAS signaling connection. The ATTACH ACCEPT message contains the following major IEs:

- *EPS Attach Result:* This IE contains the type of registration that has been accepted. Its value can be "EPS only," indicating attach to EPS only. Alternatively, its value can be "Combined EPS/IMSI Attach," indicating attach to both EPS and 2G/3G (i.e., registration with both CS and PS domain has been accepted, so the UE can fall back to UTRAN/GERAN for a CSFB operation). When the UE requests "Combined EPS/IMSI Attach" in the EPS Attach Type IE of the ATTACH REQUEST message, the MME uses the tracking area of the UE to find out the location area of the UE and also the MSC/VLR serving this location area. Then the MME sends a Location Update Request message to the MSC/VLR. The MSC/VLR sets up an association with the MME over the SGs' interface. The MSC/VLR also performs a location update procedure with the HSS and assigns a TMSI to the UE. Then the MSC/VLR sends a Location Update Accept message to the MME that includes the TMSI. The MME sets the EPS Attach Result IE to "Combined EPS/IMSI Attach."
- *GUTI:* This IE contains the GUTI of the UE assigned by the MME. The UE begins to use this assigned GUTI. The UE might have possessed a GUTI earlier and included it in the ATTACH REQUEST message. In this case, if a new GUTI is assigned to the UE here, then the UE forgets its old GUTI and begins to use the new one. On the other hand, if no GUTI is included here, then the MME is supposed to have retained the old GUTI, so the UE keeps using its old GUTI.
- *TAI list:* This IE contains a TAI list assigned to the UE. The UE considers itself registered to this TAI list and forgets the old TAI list if it had any. Thus, the allocation of GUTI and/or TAI list (i.e., GUTI reallocation) is performed

as a part of the attach procedure. The GUTI reallocation can also occur later independently as shown in Section 15.2.

- *MS identity:* This IE includes the TMSI assigned to the UE when the EPS Attach Result IE is set to "Combined EPS/IMSI Attach."
- *Equivalent PLMNs:* If the MME assigns a new list of equivalent PLMNs to the UE, this IE is included and contains the new list of equivalent PLMNs.
- *Location area identification:* This IE includes the location area identification (LAI) assigned to the UE when the EPS Attach Result IE is set to "Combined EPS/IMSI Attach." The LAI includes MCC, MNC and Location Area Code (LAC).
- *Additional update result:* This IE is included in response to the Additional Update Type IE in the ATTACH REQUEST message. It is used to indicate whether registration in the CN domain has been accepted for only SMS service over the SGs' interface as explained in Section 21.4.2 or the UE can fall back to UTRAN/GERAN for the CSFB operation.
- *T3412 value:* This IE assigns the value of periodic tracking area update timer called T3412. This IE is 2 octets long. The least significant 5 bits of the second octet represents the binary coded value of the timer. The most significant 3 bits of the second octet defines the unit of the timer value. If all these 3 bits are set to 1, then it indicates that the timer is deactivated. The default value of this timer is 54 minutes. The use of this timer is explained in Chapter 22.
- *T3412 extended value:* This IE is introduced in Release 10. This IE assigns a longer value for T3412. When this IE is included in the ATTACH ACCEPT message, the UE applies its value instead of the value of T3412 Value IE for T3412. If this IE is not included, the UE applies the value of T3412 Value IE for T3412. This IE is 3 octets long. The least significant 5 bits of the third octet represents the binary coded value of the timer. The most significant 3 bits of the third octet defines the unit of the timer value. If all these 3 bits are set to 1, then it indicates that the timer is deactivated.
- *ESM message container:* This IE contains the ACTIVATE DEFAULT EPS BEARER CONTEXT REQUEST message. This message instructs for the default EPS bearer context activation and includes the following major IEs:
  - *PDN address:* This IE assigns an IPv4 address or an IPv6 interface identifier or both to the UE.
  - *Access Point Name (APN):* This IE specifies the APN associated with the activated default EPS bearer.
  - *EBI:* This IE specifies the EBI of the activated default EPS bearer.
  - *EPS quality of service (QoS):* This IE specifies the EPS bearer QoS profile for the activated default EPS bearer. It contains the following information for both uplink and downlink:
    - QoS class identifier (QCI);
    - Maximum Bit Rate (MBR);
    - Guaranteed bit rate (GBR).
  - *APN-AMBR:* This IE contains the APN-AMBR for uplink. It specifies the maximum allowed aggregate uplink bit rate that can be expected over all the non-GBR bearers of the UE.

If the network is unable to accept the attach request, then the attach procedure fails and it is terminated. In this case, the network sends an ATTACH REJECT message to the UE instead of an ATTACH ACCEPT message. The ATTACH REJECT message contains the reason for rejection. The ATTACH REJECT message includes the PDN CONNECTIVITY REJECT message in its ESM Message Container IE if the attach procedure fails due to the default EPS bearer setup failure or any ESM procedure failure.

*Radio Bearer Setup Complete*
The UE sends an RRCConnectionReconfigurationComplete message to the eNodeB and thus confirms the successful completion of the establishment of SRB2 and the DRB associated with the default EPS bearer. The RRCConnectionReconfigurationComplete message is sent on SRB1.

*Initial Context Setup Complete*
The eNodeB sends the INITIAL CONTEXT SETUP RESPONSE message to the MME indicating the successful establishment of the UE context and the radio bearer. This message contains the transport address and DL TEID assigned by the eNodeB for the use of downlink traffic at the S1-U interface. The eNodeB sends the INITIAL CONTEXT SETUP FAILURE message to the MME if the UE context setup is not successful.

### 15.1.7.5 Attach Complete and Beginning of Uplink Transmission

The UE sends the ULInformationTransfer message and its DedicatedInfoNAS IE contains an ATTACH COMPLETE message. The eNodeB forwards the ATTACH COMPLETE message to the MME over the S1-MME interface. Thus, the UE sends indication of successful completion of the attach procedure to the MME in response to the ATTACH ACCEPT message. The ESM Message Container IE of the ATTACH COMPLETE message contains the ACTIVATE DEFAULT EPS BEARER CONTEXT ACCEPT message and this message contains the EBI of the activated default EPS bearer. Thus, the UE indicates the successful activation of the default EPS bearer context in response to the ACTIVATE DEFAULT EPS BEARER CONTEXT REQUEST message shown in Section 15.1.7.4. At this stage, the UE can begin transmission of uplink data.

### 15.1.7.6 Bearer Modify and Beginning of Downlink Transmission

The MME sends MODIFY BEARER REQUEST message to the serving GW. This message contains the EBI of the default EPS bearer. This message also contains a transport address and a DL TEID that the eNodeB has provided in the INITIAL CONTEXT SETUP RESPONSE message. The MME forwards this information to the serving GW so that the serving GW can send the downlink data to the eNodeB.

The serving GW sends MODIFY BEARER RESPONSE message to the MME as acknowledgment; this message includes the EBI of the default EPS bearer. At this stage, the default EPS bearer is completely set up between the UE and the network. In case of IPv6, after this default EPS bearer establishment, the UE receives the Router Advertisement message allocating an IPv6 prefix as shown in Section 15.1.7.1. Alternatively, the IPv6 prefix is allocated using DHCPv6.

The serving GW can now begin the transmission of downlink data. The serving GW may first transmit its buffered downlink data.

## 15.2 Independent GUTI Reallocation

The GUTI reallocation is explained in Section 1.6.3. It allocates a new GUTI, a new TAI list, or both to the UE. The GUTI reallocation can be performed as a part of the attach procedure and as a part of the tracking area update procedure as shown in Sections 15.1.7.4 and 22.5. Additionally, the MME can initiate the GUTI reallocation procedure independently to reallocate the GUTI and/or TAI list at any time when there is a NAS signaling connection existing between the UE and the MME. The GUTI reallocation procedure takes place as follows as shown in Figure 15.2:

1. The MME sends a GUTI Reallocation Command message to the eNodeB. The eNodeB transports the GUTI Reallocation Command message to the UE using the DedicatedInfoNAS IE of the DLInformationTransfer message. The GUTI Reallocation Command message contains the GUTI, the TAI list, or both, which are allocated to the UE.
2. The UE sends the GUTI Reallocation Complete message to the MME and thus acknowledges that it has accepted the assigned GUTI, TAI list, or both. The GUTI Reallocation Complete message is carried using the DedicatedInfoNAS IE of the ULInformationTransfer message between the UE and the eNodeB.

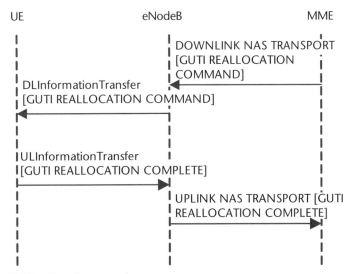

**Figure 15.2** GUTI reallocation procedure.

## 15.3 Detach Procedure

The detach procedure disconnects the UE from the EPS. It causes the UE and the MME to move from the EMM-REGISTERED state to the EMM-DEREGISTERED state. All EPS bearer contexts are deactivated. The UE and the MME delete the EPS Mobility Management (EMM) context. The detach procedure can be classified as follows:

1. *Explicit detach:* The network and the UE explicitly exchange signaling messages for the detach procedure. The explicit detach procedure can be of two categories:
   - *UE initiated detach:* The UE initiates the detach procedure. It may happen, for example, if the UE is switched off, the USIM card is removed from the UE or the EPS capability of the UE is disabled. Also, if the UE receives a paging message including the IMSI instead of the S-TMSI or any change in the E-UTRAN capabilities of the UE takes place, then the UE initiates the detach procedure and after the detachment, the UE reattaches to the EPS.
   
   As shown in Section 15.3.1, the UE initiates the detach procedure by sending a DETACH REQUEST message to the MME. If the UE is switched off, the UE attempts to send the DETACH REQUEST message in 5 seconds before it is actually powered down. Within this 5-second period, as soon as the UE sends the DETACH REQUEST message, the UE may actually be powered down. A field in the DETACH REQUEST message indicates switching off. In this case, the MME typically deletes the current EPS security context. On the other hand, if the UE is not switched off, the MME stores the current EPS security context.
   - *Network-initiated detach:* Either the MME or the HSS initiates the detach procedure. The network initiated detach may occur, for example, to recover an error in the case of network failure. During the detach procedure, the MME may indicate that the UE should reattach after the completion of the detachment. Then the UE deactivates all EPS bearer contexts, detaches from the EPS, moves to the EMM-DEREGISTERED state, and thereafter performs the attach procedure anew.
2. *Implicit detach:* The UE and the EPS get disconnected without any exchange of explicit signaling messages for detach procedure. The MME implicitly detaches the UE as such if no communication is possible with the UE for a long period of time. This may occur, for example, when the UE has moved out of coverage, which is typically indicated by the expiry of the implicit detach timer as explained in Chapter 22.
   
   The implicit detach can also be performed based on indications from the network; such indications are shown next. This may occur, for example, when the MME does not have EMM context data related to the UE subscription, possibly because the MME has been restarted.
   - The MME rejects tracking area update request by sending a TRACKING AREA UPDATE REJECT message to the UE and the MME sets the EMM cause IE of this message to "implicitly detached."

## 15.3 Detach Procedure

- The MME rejects the service request procedure by sending a SERVICE REJECT message to the UE and the MME sets the EMM cause IE of this message to "implicitly detached."
- The network rejects the attach request in the attach procedure.

### 15.3.1 UE-Initiated Detach Procedure

The typical UE initiated detach procedure, shown in Figure 15.3, is comprised of the following steps:

1. *NAS signaling connection establishment and detach request:* If the UE is in an ECM-IDLE state when the detach procedure is invoked, then it first

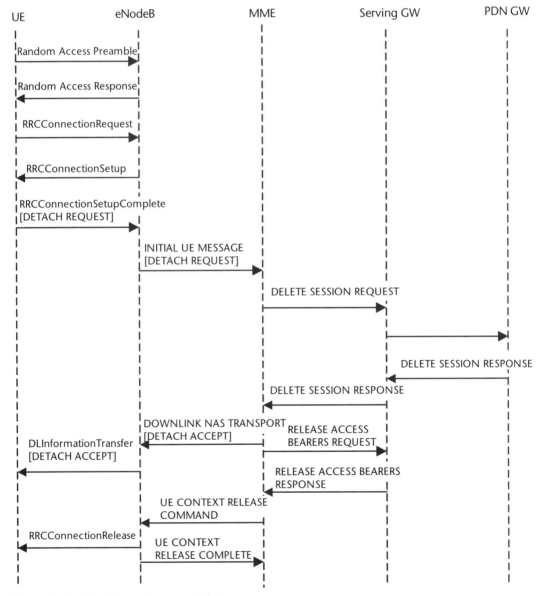

**Figure 15.3** UE initiated detach procedure.

needs to set up a way to communicate with the network in order to send a request for the detachment. Therefore, the UE initiates the establishment of the NAS signaling connection between the UE and the MME. The establishment of the NAS signaling connection is comprised of RRC connection establishment, MME selection and UE-associated logical S1-connection establishment which were explained in Sections 14.1.1, 14.2 and 14.3.2, respectively. During this procedure, when the UE sends an RRCConnectionRequest message to the eNodeB, it uses its S-TMSI in the UE-Identity field and it sets the EstablishmentCause field to Mobile Originated Signaling.

When the UE sends an RRCConnectionSetupComplete message to the eNodeB, the DedicatedInfoNAS IE in this message contains a DETACH REQUEST as the initial NAS message. The eNodeB forwards the DETACH REQUEST message to the MME and thus establishes an UE-associated logical S1-connection. The eNodeB sends an INITIAL UE MESSAGE message to the MME that contains the DETACH REQUEST message. The INITIAL UE MESSAGE also includes the TAI and ECGI of the cell where the UE is located. The activation of security function may also be performed.

On the other hand, if the UE is in the ECM-CONNECTED state, then there is already a NAS signaling connection existing between the UE and the MME, so the UE directly sends DETACH REQUEST message to the eNodeB using the ULInformationTransfer message. The eNodeB forwards the DETACH REQUEST message to the MME using the UPLINK NAS TRANSPORT message.

The DETACH REQUEST message includes the following major IEs:

- *Detach type:* This is a 1-byte-long IE that contains the following fields. Type of Detach is a 3-bit-long field that indicates the type of the requested detach procedure. It can represent any of the following:
- *EPS attach:* This is the normal type. It indicates a request for detachment from EPS services.
- *IMSI detach:* It indicates a request for detachment from non-EPS services.
- *Combined EPS/IMSI detach:* It indicates a request for detachment from both EPS and non-EPS services.
- *Switch off:* This is a 1-bit field that represents any of the following:
  - *Normal detach:* It indicates that the reason for the detach procedure is not switching off the UE.
  - *Switch off:* It indicates that the reason for the detach procedure is switching off the UE.
- *EPS mobile identity:* This IE normally contains the GUTI of the UE. However, if the UE does not possess a valid GUTI, then this IE includes the IMSI of the UE instead.
- *NAS key set identifier, $KSI_{ASME}$:* If the UE has a valid NAS key set identifier, $KSI_{ASME}$, it is included here.

2. *EPS bearer context delete:* The network deletes the EPS bearer contexts as follows:
   - The MME sends a DELETE SESSION REQUEST message to the serving GW for each PDN connection. This message includes a Linked Bearer Identity (LBI) of the default EPS bearer for a particular PDN connection.

The serving GW deactivates all EPS bearers associated with the UE and releases the corresponding EPS bearer context information.

- The serving GW sends DELETE SESSION REQUEST message to the PDN GW for each PDN connection. This message includes the LBI as well as an indication that all EPS bearers associated with the particular PDN connection are released.
- The PDN GW sends the DELETE SESSION RESPONSE message to the serving GW as an acknowledgment.
- The serving GW sends the DELETE SESSION RESPONSE message to the MME as an acknowledgment.

3. *Detach accept:* The MME sends a Detach Accept message to the UE and thus indicates that it accepts the detach request. However, if the Detach Type IE in the DETACH REQUEST message indicates switching off the UE, then the MME does not send the Detach Accept message to the UE assuming that the UE may have already been powered down.
4. *S1 release:* The MME initiates the S1 release procedure. This procedure was explained in Section 14.4. Thus, the NAS signaling connection is released.

### 15.3.2 MME Initiated Detach Procedure

The typical MME initiated detach procedure, shown in Figure 15.4, includes the following steps:

1. *Paging, NAS signaling connection establishment, and service request:* If the UE is in the ECM-IDLE state when the MME initiated detach procedure is invoked, then the MME first needs to set up a way to communicate with the UE in order to send a request for the detachment. Thus, the MME sends a paging message to the particular UE in the way explained in Section 21.1.2. Upon reception of the paging message, the UE initiates the establishment of NAS signaling connection between the UE and the MME. The establishment of the NAS signaling connection is comprised of RRC connection establishment, MME selection and UE-associated logical S1-connection establishment which were explained in Sections 14.1.1, 14.2 and 14.3.2, respectively. During this procedure, when the UE sends an RRCConnectionRequest message to the eNodeB, it uses its S-TMSI in the UE-Identity field and it sets the EstablishmentCause field to the Mobile Terminated Access.

    When the UE sends RRCConnectionSetupComplete message to the eNodeB, the DedicatedInfoNAS IE in this message contains SERVICE REQUEST as the initial NAS message. The eNodeB forwards the SERVICE REQUEST message to the MME using INITIAL UE MESSAGE message and thus establishes the UE-associated logical S1-connection.

    In the case of network failure, S-TMSI may not be available; in this case the paging message includes the IMSI instead. If the UE receives a paging message including the IMSI instead of S-TMSI, then the UE itself initiates the detach procedure by sending a DETACH REQUEST message instead of SERVICE REQUEST message. After the detachment, the UE reattaches to the EPS.

    If the UE is in an ECM-CONNECTED state, then there is already a NAS

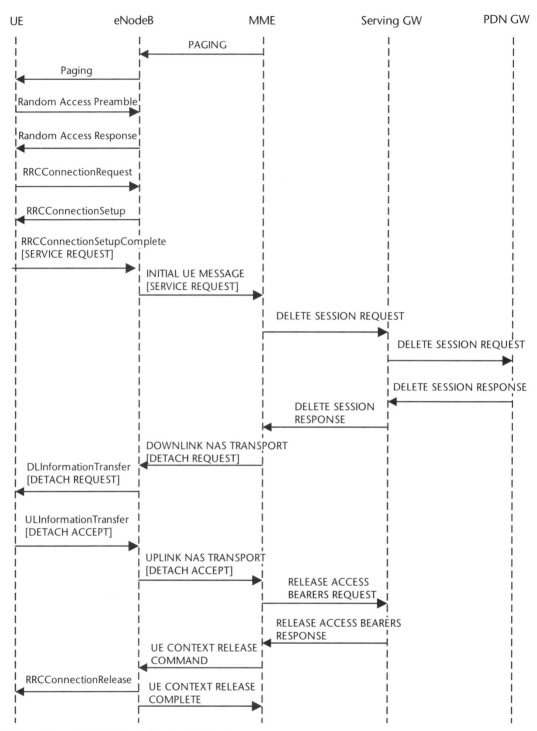

**Figure 15.4** MME initiated detach procedure.

signaling connection existing. Therefore, this step for the NAS signaling connection establishment is not required.

2. *EPS bearer context delete:* The network deletes the EPS bearer contexts the same way as explained in Section 15.3.1.
3. *Detach request:* The MME sends a DETACH REQUEST message to the UE and thus sends a command for the detachment.
   This DETACH REQUEST message used in a network-initiated detach procedure is different from what is used in the case of a UE-initiated detach procedure. Here this message includes the following major IEs:
   - *Detach Type:* The Type of Detach field in this IE represents any of the following:
     - *Reattach Required:* It indicates that the UE needs to reattach at the end of the detach procedure.
     - *Reattach Not Required:* It indicates that the UE does not need to reattach at the end of the detach procedure.
     - *IMSI Detach:* It indicates that the UE detaches from non-EPS services.
   - *EMM Cause:* This IE indicates why the MME is instructing the UE for detachment.
     Upon reception of the DETACH REQUEST message, the UE typically deactivates all EPS bearer contexts.
4. *Detach accept:* In order to indicate that the UE accepts the detach request, the UE sends a DETACH ACCEPT message to the eNodeB and the eNodeB forwards this message to the MME.
5. *S1 release:* The MME initiates the S1 release procedure. This procedure was explained in Section 14.4. Thus, the NAS signaling connection is released. If the Type of Detach field in the DETACH REQUEST message indicates that reattach is required, then the UE performs the attach procedure anew after the RRC connection release is completed.

# CHAPTER 16

# Discontinuous Reception

In order to save battery power and prolong battery life, the UE may periodically switch off the receiver circuitry. This is referred to as Discontinuous Reception (DRX). A periodic cycle is defined that includes a continuous active period and then a continuous inactive period for the UE. This cycle is called *DRX cycle*. The power saving occurs during the inactive period. DRX can be applied in both the RRC_IDLE state and the RRC_CONNECTED state. The length of the DRX cycle in the RRC_CONNECTED state is made, at the maximum, equal to the length of the DRX cycle in the RRC_IDLE state. The maximum possible length of the DRX cycle is 256 radio frames for both the RRC_IDLE state and the RRC_CONNECTED state.

## 16.1 DRX in the RRC_IDLE State

In the RRC_IDLE state, the UE wakes up once in every DRX cycle and looks for a paging message. The active period of the DRX cycle has one Paging Occasion (PO) (i.e., the UE has one PO per DRX cycle). Thus, the DRX cycle is the same as the paging cycle for a UE in the RRC_IDLE state. After the active period of the DRX cycle, the UE exercises DRX (i.e., it turns off its receiver circuitry to save power). The paging cycle and its configuration procedure were explained in Chapter 13.

## 16.2 DRX in the RRC_CONNECTED State

In the RRC_CONNECTED state, the eNodeB can optionally configure the DRX operation. The RRC layer at eNodeB configures DRX by sending the DRX-Config IE. Appendix A.2 shows how the eNodeB sends the DRX-Config IE to the UE while setting up or modifying radio bearers. If DRX is not configured, the UE monitors PDCCH continuously in order to find possible allocations for downlink or uplink transmission. When DRX is configured, the eNodeB specifies a DRX cycle. The DRX cycle consists of an "on duration" at the beginning of the cycle as shown in Figure 16.1. In the rest period of the DRX cycle, the UE exercises RX (i.e., it turns off its receiver circuitry to save power). The UE monitors the PDCCH during on duration and the UE pauses downlink reception during the DRX period. The OnDurationTimer field in the DRX-Config IE specifies the length of the on duration

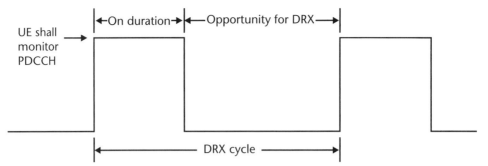

**Figure 16.1** On duration and DRX period in the DRX cycle.

in the number of subframes. It can specify 1, 2, 3, 4, 5, 6, 8, 10, 20, 30, 40, 50, 60, 80, 100, or 200 subframes.

### 16.2.1 Configuration of the DRX Cycle

Two types of DRX cycles are defined:

1. *Long DRX cycle:* The long DRX cycle allows increased power saving, although it reduces the frequency of opportunities for scheduling via a PDCCH.
2. *Short DRX cycle:* The short DRX cycle is optional. It is used only when there are significant chances of scheduling and the long DRX cycle is used otherwise. The short DRX cycle is suitable when there are transmissions of small data at short but regular intervals, for example, VoIP. If the short DRX cycle is configured, the UE initiates DRX operation with short DRX cycles, but transitions to long DRX cycles at the expiry of a timer. On the other hand, if the short DRX cycle is not configured, the UE initiates a DRX operation with long DRX cycles.

If the short DRX cycle is configured, the DRX-Config IE includes the Short-DRX IE. The ShortDRX IE includes the ShortDRX-Cycle field, which specifies the length of the short DRX cycle in the number of subframes. It can be 2, 5, 8, 10, 16, 20, 32, 40, 64, 80, 128, 160, 256, 320, 512, or 640 subframes. When the UE initiates DRX with short DRX cycles, it starts a timer called the DRXShortCycleTimer and keeps on going through short DRX cycles one after another until the DRX-ShortCycleTimer expires. Once the DRXShortCycleTimer expires, the UE initiates long DRX cycles in order to save more battery power. The ShortDRX IE includes the DRXShortCycleTimer field, which specifies the length of the DRXShortCycle-Timer in terms of multiples of the ShortDRX-Cycle. The value of the DRXShortCycleTimer can be any integer up to 16. The DRX-Config IE also includes the LongDRX-CycleStartOffset IE, which specifies the values for the following fields:

1. *LongDRX-Cycle:* This field specifies the length of the long DRX cycle in the number of subframes. It can be 10, 20, 32, 40, 64, 80, 128, 160, 256, 320, 512, 640, 1,024, 1,280, 2,048, or 2,560 subframes. When the short

DRX cycle is configured, the value of LongDRX-Cycle is chosen such that it becomes a multiple of the ShortDRX-Cycle.
2. *DRXStartOffset*: This field specifies the particular radio frame and subframe in which the long or short DRX cycle begins. The radio frame and subframe that initiate the long or short DRX cycle are determined by satisfying the following relations:
   - Long DRX Cycle:DRXStartOffset = [(SFN × 10) + (subframe number initiating long DRX cycle)] mod LongDRX-Cycle;
   - Short DRX Cycle:DRXStartOffset mod ShortDRX-Cycle = [(SFN × 10) + (subframe number initiating short DRX cycle)] mod ShortDRX-Cycle.

### 16.2.2 Continuous Reception and Resumption of DRX

If the UE detects any scheduling while it is monitoring the PDCCH during an on duration, then the UE stops the DRX operation, begins continuous reception, and starts a timer called the DRX-InactivityTimer. The UE maintains continuous reception with no DRX Cycle until the DRX-InactivityTimer expires. If the UE receives any scheduling again on the PDCCH while the DRX-InactivityTimer is running, the UE restarts the DRX-InactivityTimer. Once the DRX-InactivityTimer expires, the UE immediately initiates a short DRX cycle provided that the short DRX cycle is configured. If the short DRX cycle is not configured, then the UE initiates a long DRX cycle instead. The DRX-Config IE includes the DRX-InactivityTimer field, which specifies the length of the DRX-InactivityTimer in the number of subframes. It can be 1, 2, 3, 4, 5, 6, 8, 10, 20, 30, 40, 50, 60, 80, 100, 200, 300, 500, 750, 1,280, 1,920, or 2,560 subframes.

Alternatively, the eNodeB can send an explicit command by including the DRX command MAC control element in the MAC PDU during continuous reception. Then the UE initiates a short DRX cycle provided that the short DRX cycle is configured. Again, if the short DRX cycle is not configured, the UE initiates a long DRX cycle instead. The DRX command MAC control element was explained in Section 6.2.1.2.

### 16.2.3 HARQ During DRX

HARQ operation is independent of DRX operation. The UE wakes up from a sleeping state to monitor the PDCCH for possible HARQ retransmissions and ACK/NAK signaling on PHICH and to transmit HARQ feedbacks when they are due. As will be shown in Section 20.4.1, each HARQ process expects HARQ round trip time (HARQ RTT) between the transmission and retransmission of a transport block when the decoding fails. The HARQ RTT is estimated as 8 ms. Therefore, in the case of downlink, the HARQ process at the UE maintains a HARQ RTT timer to allow the UE to sleep during the HARQ RTT. The HARQ RTT timer is set to 8 ms.

If the DRX operation is configured, then the UE pauses the monitoring of the PDCCH for a certain period. If decoding a downlink transport block fails, the UE still pauses the monitoring of the PDCCH but for the HARQ RTT period assuming that the retransmission of the transport block might take place after this HARQ RTT. Of course, the UE can pause the monitoring of the PDCCH as such only if any other schedule does not come up for monitoring the PDCCH. When the HARQ RTT timer indicates that the HARQ RTT is over, the UE starts monitoring the PDCCH for the retransmission. At this moment, the UE monitors the PDCCH for a maximum period specified by the DRX-RetransmissionTimer. The DRX-Config IE includes the DRX-RetransmissionTimer field, which gives the length of the DRX-RetransmissionTimer in the number of subframes. It can specify 1, 2, 4, 6, 8, 16, 24, or 33 subframes. Also, the UE does not enter DRX when uplink HARQ retransmission is due.

### 16.2.4 Active Time

The active time refers to the duration in which the UE is awake and it monitors PDCCH. Thus, it includes the following durations:

1. On duration in the beginning of the DRX cycle;
2. During the data transfer and the DRX-InactivityTimer period after the last detection of scheduling on the PDCCH;
3. When the downlink HARQ retransmission is expected and when the uplink HARQ retransmission is due;
4. If the UE has sent a scheduling request;
5. A random access process is in progress and the contention resolution timer is running.

While the UE is not in active time, it does not perform any uplink transmissions including the transmission of the PUCCH and the SRS.

# CHAPTER 17
# Uplink Power Control

The uplink power control is performed to enable the UE to determine the appropriate average power to be used over an SC-FDMA symbol for uplink transmission.. The power control procedure is specified for uplink transmissions on the PUSCH, PUCCH, and SRS, while the downlink power control mechanism is very much dependent on eNodeB implementation. The major objectives of uplink power control are as follows:

1. The uplink power control may be performed in an attempt to maintain a desired level of SINR. This can help maintain a sufficient transmitted energy per bit and a certain error rate under changing channel conditions. Thus, it attempts to guarantee the required data rate and quality of service (QoS).
2. The uplink power control helps in the improvement of system coverage and capacity.
3. The uplink power control helps in the reduction of battery power consumption.
4. The uplink power control helps in interference management.

## 17.1 Uplink Power Control Considerations

The UEs in the cell use separate time-frequency resources in LTE, and thus, intracell interference and management of Rise Over Thermal (RoT) are not critical. Nevertheless, a number of points need to be considered in uplink power control:

1. *Radio link characteristics:* The uplink power level adapts to the changing characteristics of the radio propagation channel. This includes adaptation with path loss, shadowing, fast fading, interference, and noise.

   For the estimation of uplink channel characteristics, the eNodeB primarily uses a Sounding Reference Signal (SRS). It may also use a Demodulation Reference Signal (DM RS) and the error rate in the current data transfer on the PUSCH. The eNodeB can ask for transmission of SRS when no PUCCH or PUSCH is scheduled in the uplink to estimate the uplink channel characteristics. Also, the UE may transmit the SRS in different parts of the bandwidth where no uplink data transmission is available.

   The eNodeB can configure the bandwidth of the SRS. A wider

bandwidth for the SRS makes the available power lower per resource block especially for cell edge users. On the other hand, limiting the SRS to a small bandwidth allows precise information for only the limited bandwidth. The eNodeB can also configure the frequency of transmission of SRS in time as shown in Section 5.2.1.2. The eNodeB can configure either an individual SRS transmission or periodic transmission of SRS. When periodic SRS transmissions are configured, the eNodeB can configure the periodicity. A more frequent transmission of SRS requires more resources but it allows more precise power adjustments.

In order to compensate the path loss in uplink, the uplink power control adds an estimated value while setting the power level. The UE estimates the path loss in downlink from cell-specific reference signals. Then the UE uses this value as the estimated path loss in uplink.

2. *Other link adaptation techniques:* The uplink power control is not a unique technique to adapt to the changing radio link conditions. Channel-Dependent Scheduling (CDS), Adaptive Modulation and Coding (AMC), and multiple antenna schemes may also be simultaneously applied for link adaptation. Thus the uplink power control needs to consider the effects of other link adaptation techniques. AMC may be found as a more efficient link adaptation technique than power control.

3. *Open-loop and closed-loop control:* The uplink power control uses a closed-loop control around open-loop set points. The closed-loop feedback is used to compensate for the deficit in the open-loop estimate; for example, closed-loop control can be used to compensate the power amplifier error, the path loss estimation error, and the change in the intercell interference level. Since closed-loop control is combined with open-loop control, the necessary feedback is less than what would be required for a fully closed-loop control. The eNodeB primarily uses the SRS as the feedback. It may also use DM RS and the error rate in the current data transfer on the PUSCH.

4. *Cell-specific and UE-specific control:* The uplink power control considers the overall situation in the cell and applies cell-specific control. The eNodeB applies cell-specific parameters in the uplink power control, which is common for all UEs in the cell. Additionally, the uplink power control considers the situation for a particular UE and applies UE-specific control. The eNodeB sends UE-specific parameters to make adjustments necessary for the UE. This may be used to compensate an inappropriate estimation of path loss or any other power settings for the particular UE.

5. *Semistatic and dynamic control:* The uplink power control adapts both semistatically and dynamically with time-varying radio link characteristics. The dynamic adaptation requires greater overhead, but it offers quick adjustments. On the contrary, semistatic adaptation requires reduced overhead, but it offers infrequent adjustments. Therefore, some parameters in uplink power control are configured semistatically or once in a while via layer 3 messages and they are used for major adjustments. Additionally, some parameters are configured dynamically, in every allocated subframe, as the data transfer goes on and they are used for minor adjustments.

6. *Dynamic scheduling and semipersistent scheduling:* The uplink power control may allow different power levels for dynamic scheduling and semipersistent scheduling. The semipersistent scheduling is typically used to provide reduced signaling for scheduling and thus used for applications such as VoIP. However, the HARQ retransmissions for semipersistent transmission are scheduled dynamically, which requires extra signaling for scheduling. Therefore, a higher power level may be used for semipersistent scheduling than for dynamic scheduling to achieve a lower probability of error and thus to achieve fewer retransmissions.
7. *Periodic and aperiodic control:* Although there is no specific mechanism for sending periodic power control commands, there is a provision for periodic update. As shown in Section 17.2.2.3, the instructions for closed-loop power adjustments can be sent via DCI format 0 as well as via DCI format 3 or 3A. In order to achieve adjustments at a regular interval, the commands on DCI format 3 or 3A may be sent periodically. However, the command on DCI format 0 is sent along with scheduling resource grants, so it is aperiodic.
8. *Power capability:* The uplink power control considers the power transmit capability of the UE. It sets the uplink power level within its maximum uplink transmission power. The UE sends power headroom reports to notify the eNodeB how much more or less power it is capable of transmitting. Power headroom reporting was explained in Section 6.2.1.2. The UE may be found to have less power capability than required, especially if the UE is far away from the eNodeB.
9. *Uplink bandwidth:* The uplink power control considers the bandwidth of uplink transmission. The transmit power needs to be higher for transmission over the wider bandwidth.
10. *Intercell Interference (ICI):* The transmit power control needs to account for ICI. If a cell edge UE uses high uplink transmit power, it can cause substantial interference to neighboring cells.
11. *Control and data transmit power:* A small difference in the transmit power for control and data helps reduce PAPR at the UE.
12. *Battery life:* The uplink power control attempts to reduce the uplink transmit power level as much as possible in order to minimize the power consumption and thus maximize the battery life of the UE.

## 17.2 Power Control on the PUSCH

The UE computes the uplink transmit power on the PUSCH in a particular subframe as $P_{PUSCH}$:

$$P_{PUSCH} = \text{MIN}\,[P_{CMAX}, P_{O\_PUSCH} + \alpha \times PL + 10\log_{10} M_{PUSCH} + \Delta_{TF} + f_{TPC}]$$

The different terms in the expression of $P_{PUSCH}$ are configured for appropriate power control as explained in the following sections.

### 17.2.1 Semistatically Configurable Terms

#### 17.2.1.1 $P_{CMAX}$

$P_{CMAX}$ provides the maximum allowed uplink transmission power of the UE. Thus, if the long expression, ($10\log_{10} M_{PUSCH} + P_{O\_PUSCH} + \alpha \times \text{PL} + \Delta_{TF} + f_{TPC}$) generates a power level greater than $P_{CMAX}$, then this power level is not used and $P_{PUSCH}$ is rather made equal to $P_{CMAX}$.

#### 17.2.1.2 $P_{O\_PUSCH}$

$P_{O\_PUSCH}$ is the basic parameter for open-loop set points in power control used to achieve proper SINR at the eNodeB. It is composed as the sum of a cell-specific parameter and a UE-specific parameter. The eNodeB sends their values separately for dynamic scheduling and semipersistent scheduling as shown next.

*Cell-Specific Parameter Value*
- *Dynamic scheduling:* The parameter P0-NominalPUSCH gives the value. Appendix A.1 shows how the eNodeB broadcasts this parameter on SIB type 2. Appendix A.3 shows how the source eNodeB sends this parameter to the UE for use at the target eNodeB when handover takes place. The value of P0-NominalPUSCH can range between −126 dBm and 24 dBm.
- *Semipersistent scheduling:* The parameter P0-NominalPUSCH-Persistent gives the value. Appendix A.2 shows how the eNodeB sends this parameter to the UE while setting up or modifying radio bearers. The value of P0-NominalPUSCH-Persistent can range between −126 dBm and 24 dBM. If the P0-NominalPUSCH-Persistent field is not provided, then the UE uses the value of P0-NominalPUSCH instead.

*UE-Specific Parameter Value*
- *Dynamic scheduling:* The parameter P0-UE-PUSCH gives the value. Appendix A.2 shows how the eNodeB sends this parameter to the UE while setting up or modifying radio bearers. The value of P0-UE-PUSCH can range between −8 dB and 7 dB.
- *Semipersistent scheduling:* The parameter P0-UE-PUSCH-Persistent gives the value. Appendix A.2 shows how the eNodeB sends this parameter to the UE while setting up or modifying radio bearers. The value of P0-UE-PUSCH-Persistent can range between −8 dB and 7 dB. If the P0-UE-PUSCH-Persistent field is not provided, then the UE uses the value of P0-UE-PUSCH instead.

#### 17.2.1.3 Path Loss (PL) and $\alpha$ (Alpha)

The PL term is used to compensate the path loss in uplink. The eNodeB sends the parameter ReferenceSignalPower to assist the UE in the estimation of path loss. ReferenceSignalPower provides the linear average of transmit power in resource elements that carry downlink cell-specific reference signals. Appendix A.1 shows how the eNodeB broadcasts ReferenceSignalPower on SIB type 2. Appendix A.3

shows how the source eNodeB sends this parameter to the UE for use at the target eNodeB when handover takes place. The value of ReferenceSignalPower can range between –60 dBm and 50 dBm.

The UE estimates the path loss in the downlink from cell-specific reference signals. The UE measures RSRP as the linear average over the power contributions of the resource elements on cell-specific reference signals within the considered measurement frequency bandwidth. Then the UE performs filtering of the measured results to calculate the estimated path loss in downlink and then considers it equal to the estimated path loss in uplink.

Estimated path loss in downlink = ReferenceSignalPower − Filtered RSRP = Estimated path loss in uplink, PL

The parameter $\alpha$ (alpha) is used to reduce the effect of PL. Appendix A.1 shows how the eNodeB broadcasts $\alpha$ on SIB type 2. Appendix A.3 shows how the source eNodeB sends $\alpha$ to the UE for use at the target eNodeB when handover takes place. The value of $\alpha$ can be 0, 0.4, 0.5, 0.6, 0.7, 0.8, 0.9, or 1. Thus, if $\alpha$ is set to 0, then there is no compensation of path loss, and if is set to 1, then there is full compensation of path loss. All other values of $\alpha$ allow fractional compensation of path loss. $\alpha$ can serve the following purposes:

1. $\alpha$ allows the eNodeB to configure the degree to which the UE's estimate of the path loss would be taken into account. The eNodeB can use a high value of $P_{O\_PUSCH}$ incorporating the path loss partially and rely only a little on the UE's estimate by setting a low value of $\alpha$. Conversely, the eNodeB can use a low value of $P_{O\_PUSCH}$ and rely greatly on the UE's estimate by setting a high value of $\alpha$.
2. As shown earlier, the UE considers the estimated path loss in downlink equal to the estimated path loss in uplink when it determines PL. However, in the case of FDD, the path loss can actually be slightly less in uplink because of the use of lower frequencies. $\alpha$ can be used to reduce estimated path loss in uplink considering lower frequencies or any other effects.
3. $\alpha$ is used to control the amount of ICI caused to neighboring cells. A lower compensation of path loss using lower value of $\alpha$ can allow the UEs at the cell edge to use relatively less transmit power and thus generate relatively less interference to neighboring cells. On the other hand, the high value of $\alpha$ can allow the UEs at the cell edge to compensate the path loss better and avoid significant degradation of data rate. Thus, there is a trade-off in the choice of $\alpha$ and it depends on ICI management schemes in the system.

### 17.2.2 Dynamically Configurable Terms

#### 17.2.2.1 $M_{PUSCH}$

$M_{PUSCH}$ indicates the number of resource blocks allocated on the subframe for which the uplink power is calculated. The function of $M_{PUSCH}$ allows the use of more power as a wider uplink bandwidth is used and it attempts to keep transmit Power Spectral Density (PSD) in a resource block constant.

### 17.2.2.2 $\Delta_{TF}$ ($\Delta_{Transport\ Format}$)

The $\Delta_{TF}$ term allows the transmitted power per resource block to be adapted according to the transmitted information data rate. If the UE uses a higher level of Modulation and Coding Scheme (MCS), then the number of information bits will be higher per resource element. This requires a higher transmit power, which is facilitated by the $\Delta_{TF}$ term. $\Delta_{TF}$ is formulated based on the determination of SNR as $2^C - 1$ from Shannon's channel capacity theorem. Representing in decibels, $\Delta_{TF}$ is given by the following expression for the user data transfer:

$$\Delta_{TF} = 10\log_{10}(2^{BPRE \times K} - 1)$$

Here BPRE depends on the number of information bits per resource element. The value of $K$ is given by DeltaMCS-Enabled IE and Appendix A.2 shows how the eNodeB sends the DeltaMCSEnabled IE to the UE while setting up or modifying radio bearers. The eNodeB sets the value of $K$ to either 1.25 or 0 semistatically using DeltaMCS-Enabled field. The value for $K$ is made larger than 1 and set to 1.25 considering that the whole PUSCH resources are not used for data transfer and the PUSCH resources may also contain DeModulation Reference Signal (DM RS) and Uplink Control Information (UCI). The 1.25 value for $K$ makes $1/K = 0.8$ and this assumes that 80% of the uplink resource elements are used in calculating $\Delta_{TF}$. If this compensation for MCS needs to be deactivated, then $K$ is set to 0. The deactivation may be adopted, for example, if fast AMC is used. The value of the DeltaMCS-Enabled IE can be en0 or en1. Here, en1 is interpreted as activation of the compensation for the MCS setting $K = 1.25$, and en0 is interpreted as the deactivation of the compensation for the MCS setting $K = 0$.

The eNodeB may use $\Delta_{TF}$ for changing the power level indirectly. The eNodeB changes the MCS level in the uplink using DCI format 0. Thus, the eNodeB, if necessary, can command a change of uplink MCS level in order to change the uplink transmit power.

### 17.2.2.3 $f_{TPC}$

$f_{TPC}$ is used for closed-loop power adjustments with time-varying radio link characteristics. The eNodeB sends the instruction for necessary power adjustments known as the Transmit Power Control (TPC) command. The TPC command is used to calculate $f_{TPC}$. The TPC command applies in two ways:

- *Accumulation enabled:* When accumulation is enabled, the TPC command asks for a change of the power level relative to the previous power level. Enabling accumulation can be a better choice when the uplink resource scheduling has a continuous and regular pattern.
- *Accumulation disabled:* When accumulation is disabled, the TPC command asks for an absolute change of power level without relating to the previous TPC command. Disabling accumulation can be a better choice when the uplink resource scheduling has an intermittent pattern.

The eNodeB sends the AccumulationEnabled field to specify which of these two types the UE would use. The eNodeB can vary this command semistatically. Appendix A.2 shows how the eNodeB sends the AccumulationEnabled field to the UE while setting up or modifying radio bearers.

When the eNodeB sends DCI format 0 on the PDCCH to convey resource grants on the PUSCH, the DCI format 0 also includes the TPC command for the allocated resources. This TPC command field is 2 bits long and can have values 0, 1, 2, or 3.

*Accumulation Enabled*
Apart from DCI format 0, the TPC command can be also be sent via DCI format 3 or 3A. The TPC command on DCI format 3 or 3A may address multiple UEs. This may be used for regular adjustments of power on the PUSCH. In this case, the CRC of the PDCCH instance is scrambled by TPC-PUSCH-RNTI. All the UEs that are assigned the same TPC-PUSCH-RNTI will decode the PDCCH instance and apply the TPC command. In order to assign the TPC-PUSCH-RNTI to the UE, the eNodeB sends the TPC-PDCCH-ConfigPUSCH IE to the UE as shown in Appendix A.2 while setting up or modifying radio bearers. The value of the TPC-PDCCH-ConfigPUSCH IE is given by the TPC-PDCCH-Config IE and the TPC-PDCCH-Config IE includes the assigned TPC-PUSCH-RNTI.

The calculation of $f_{TPC}$ based on the TPC command is explained below. If the UE detects both DCI format 0 and DCI format 3/3A in the same subframe, then the UE applies the TPC command on DCI format 0.

*TPC Command on DCI Format 0 and DCI Format 3*
The TPC command field on DCI format 3 is 2 bits long as it is in DCI format 0. The value of the TPC command field on DCI format 0 and DCI format 3 corresponds to the change of power level, $\delta_{PUSCH}$, as shown in Table 17.1.

DCI format 3 transports a number of TPC commands such as TPC command number 1, TPC command number 2, and so forth. A particular UE is assigned an index in order to select its own TPC command. For this purpose, the aforementioned TPC-PDCCH-Config IE also includes TPC-Index IE. The TPC-Index IE contains an IndexOfFormat3 field, which provides the index for a particular

**Table 17.1** PUSCH Power Adjustment Based on the TPC Command on DCI Format 0 and DCI Format 3 for Accumulation Enabled

| TPC Command Field | Change of Power Level, $\delta_{PUSCH}$ [dB] |
|---|---|
| 0 | −1 |
| 1 | 0 |
| 2 | 1 |
| 3 | 3 |

UE. This index can range between 1 and 15 as DCI format 3 transports 15 TPC commands at the maximum.

*TPC Command on DCI Format 3A*
The TPC command field on DCI format 3A is 1 bit long. The value of this TPC command field corresponds to the change of power level, $\delta_{PUSCH}$, as shown in Table 17.2.

DCI format 3A also transports a number of TPC commands such as TPC command number 1, TPC command number 2, and so forth, and they are assigned to particular UEs similar with DCI format 3. However, in this case, the TPC-Index contains the IndexOfFormat3A field, which provides the index for a particular UE. This index can range between 1 and 31 as DCI format 3A transports 31 TPC commands at the maximum.

The TPC command on DCI format 3A provides only single-bit power adjustments, whereas the TPC command on DCI format 3 provides power adjustments in a greater extent using 2 bits. However, DCI format 3A can accommodate almost twice as many TPC commands as DCI format 3.

*Calculation of $f_{TPC}$*
Once the UE determines $\delta_{PUSCH}$ as shown above, it calculates the value of $f_{TPC}$ for subframe $i$, using the following equation

$$f_{TPC}(i) = f_{TPC}(i-1) + \delta_{PUSCH}(i-4)$$

As the equation shows, $\delta_{PUSCH}$ increases or decreases $f_{TPC}$ of the current subframe compared to the $f_{TPC}$ value in the previous subframe. The eNodeB sends the PDCCH signaling four subframes earlier, and therefore, $f_{TPC}(i)$ is calculated using $\delta_{PUSCH}(i-4)$.

With accumulation enabled, it is possible for $f_{TPC}$ to attain very high positive or negative values. For example, if the TPC command field keeps indicating positive values for $\delta_{PUSCH}$ consecutively, the value of $f_{TPC}$ will also keep increasing.

*Accumulation Disabled*
When accumulation is disabled, only DCI format 0 of the PDCCH can convey the TPC command. In addition, in the random access procedure, the random access response (RAR) message carries the TPC command with accumulation disabled. The calculation of $f_{TPC}$ based on the TPC command is explained below.

Table 17.2 PUSCH Power Adjustment Based on the TPC Command on DCI Format 3A

| TPC Command Field | Change of Power Level, $\delta_{PUSCH}$ [dB] |
|---|---|
| 0 | −1 |
| 1 | 1 |

*TPC Command on DCI Format 0*
The TPC command field on DCI format 0 corresponds to the change of power level, $\delta_{PUSCH}$ as shown in Table 17.3.

*TPC Command on RAR*
In the case of random access, the UE first transmits random access preamble without any good estimate of required uplink power level. In response, the eNodeB sends a RAR message as was shown in Section 6.2.2.2. The RAR message includes the TPC command field for necessary adjustments in the subsequent uplink transmission on PUSCH. The value of TPC command field corresponds to the required change of power level, $\delta_{PUSCH}$, as shown in Table 17.4. Since the UE might have a poor estimate of the required uplink power, there should be provision for a wider range of correction in power level. Therefore, the TPC command field is 3 bits long here allowing correction between −6 dB and 8 dB.

*Calculation of $f_{TPC}$*
Once the UE determines $\delta_{PUSCH}$ as shown above, the value of $f_{TPC}$ is made equal to $\delta_{PUSCH}$ (i.e., the value of $f_{TPC}$ is made equal to the last instructed change of the power level)

**Table 17.3** PUSCH Power Adjustment Based on the TPC Command on DCI Format 0 for Accumulation Disabled

| TPC Command Field | Change of Power Level, $\delta_{PUSCH}$ [dB] |
|---|---|
| 0 | −4 |
| 1 | −1 |
| 2 | 1 |
| 3 | 4 |

**Table 17.4** PUSCH Power Adjustment Based on the TPC Command on the RAR Message

| TPC Command Field | Change of Power Level, $\delta_{PUSCH}$ [dB] |
|---|---|
| 0 | −6 |
| 1 | −4 |
| 2 | −2 |
| 3 | 0 |
| 4 | 2 |
| 5 | 4 |
| 6 | 6 |
| 7 | 8 |

Thus, the value of $f_{TPC}$ in the current subframe is independent of its value in the previous subframe. In case of DCI format 0, the PDCCH signals uplink scheduling information four subframes earlier, and therefore, $f_{TPC}(i)$ is equal to $\delta_{PUSCH}(i-4)$. The possible value of $f_{TPC}$ is limited within ±4 dB here.

## 17.3 Power Control on PUCCH

The UE computes uplink transmit power on the PUCCH in a particular subframe as $P_{PUCCH}$:

$$P_{PUCCH} = \text{MIN}\,[P_{CMAX}, P_{O\_PUCCH} + PL + h(n_{CQI}, n_{HARQ}) + \Delta_{F\_PUCCH} + g_{TPC}]$$

The different terms in the expression of $P_{PUCCH}$ are configured for appropriate power control as explained in the following sections.

### 17.3.1 Semistatically Configurable Terms

#### 17.3.1.1 $P_{CMAX}$

$P_{CMAX}$ was explained in Section 17.2.1.1.

#### 17.3.1.2 $P_{O\_PUCCH}$

$P_{O\_PUCCH}$ is the basic parameter for open-loop set points in power control used to achieve proper SINR at the eNodeB. It is composed as the sum of a cell-specific parameter and a UE specific parameter. The eNodeB sends their values as shown here:

- *Cell-specific parameter value:* The parameter P0-NominalPUCCH gives the value. Appendix A.1 shows how the eNodeB broadcasts this parameter on SIB type 2. Appendix A.3 shows how the source eNodeB sends this parameter to the UE for use at the target eNodeB when handover takes place. The value of P0-NominalPUCCH can range between −127 dBm and −96 dBm.
- *UE-specific parameter value:* The parameter P0-UE-PUCCH gives the value. Appendix A.2 shows how the eNodeB sends this parameter to the UE while setting up or modifying radio bearers. The value of P0-UE-PUCCH can range between −8 dB and 7 dB.

#### 17.3.1.3 Path Loss (PL)

The PL term is used to compensate the path loss in uplink. Unlike the power control on PUSCH, PL is not multiplied by any parameter like $\alpha$ (alpha) and full path loss compensation is rather used. The full path loss compensation helps in limiting interference between different UEs and thus maximizes the number of UEs multiplexed on the same resource block.

### 17.3.2 Dynamically Configurable Terms

#### 17.3.2.1 $h(n_{CQI}, n_{HARQ})$

The UE sends periodic CQI/PMI/RI report on PUCCH if there is no uplink data transfer on PUSCH. The function $h(n_{CQI}, n_{HARQ})$ allows the use of more power as wider uplink bandwidth is used on the PUCCH. $n_{CQI}$ and $n_{HARQ}$ represent the number information bits for the CQI/PMI/RI report and HARQ ACK/NACK, respectively, in the function $h(n_{CQI}, n_{HARQ})$. The function $h(n_{CQI}, n_{HARQ})$ takes on zero value for the PUCCH formats 1, 1a, and 1b, which carry only a few bits. The function $h(n_{CQI}, n_{HARQ})$ adds value for the PUCCH formats 2, 2a, and 2b, depending on the number of information bits for the CQI/PMI/RI report and HARQ ACK/NACK.

#### 17.3.2.2 $\Delta_{F\_PUCCH}$

The parameter $\Delta_{F\_PUCCH}$ applies an offset depending on the format of the PUCCH. The formats of the PUCCH include different number of bits and thus require different levels of power. Also, $\Delta_{F\_PUCCH}$ allows setting a proper relative probability of error to different formats of the PUCCH. The DeltaFList-PUCCH IE provides the offset values separately for each of the PUCCH format types. Appendix A.1 shows how the eNodeB broadcasts the DeltaFList-PUCCH IE on SIB type 2. Appendix A.3 shows how the source eNodeB sends the DeltaFList-PUCCH IE to the UE for use at the target eNodeB when handover takes place.

#### 17.3.2.3 $g_{TPC}$

$g_{TPC}$ is used for closed-loop power adjustments with time-varying radio link characteristics. The eNodeB sends the UE-specific TPC command, and the value of the TPC command field corresponds to a change of the power level $\delta_{PUCCH}$, which is used to calculate $g_{TPC}$. The calculation of $g_{TPC}$ always applies accumulation.

The DCI formats 1, 1A, 1B, 1D, 2, and 2A of the PDCCH can include the TPC commands for the PUCCH. It may be noted that these DCI formats are basically used for downlink resource allocations on the PDSCH. The UE may send HARQ ACK/NACK and CQI/PMI/RI on the PUCCH during the downlink data transfer. Therefore, along with the downlink resource allocations, the DCI formats can also carry TPC commands for associated signaling on the PUCCH.

Additionally, DCI formats 3 and 3A can include the TPC command and then the PDCCH instance can address multiple UEs. This may be used for regular adjustments of power on PUCCH. The CRC of the PDCCH instance for DCI formats 3 or 3A is scrambled by the TPC-PUCCH-RNTI. All the UEs that are assigned the same TPC-PUCCH-RNTI will decode the PDCCH instance and apply the TPC command. In order to assign the TPC-PUCCH-RNTI to the UE, the eNodeB sends the TPC-PDCCH-ConfigPUCCH IE to the UE as shown in Appendix A.2 while setting up or modifying radio bearers. The value of TPC-PDCCH-ConfigPUCCH IE is given by the TPC-PDCCH-Config IE, and the TPC-PDCCH-Config IE includes the assigned TPC-PUCCH-RNTI.

The calculation of gTPC based on the TPC command is explained below.

*TPC Command on DCI Formats 1, 1A, 1B, 1D, 2, 2A, and 3*

The TPC command field on DCI formats 1, 1A, 1B, 1D, 2, 2A, and 3 is 2 bits long. The value of the TPC command field corresponds to the change of power level, $\delta_{PUCCH}$ as shown in Table 17.5.

DCI format 3 transports a number of TPC commands such as TPC command number 1, TPC command number 2, and so forth. A particular UE is assigned an index in order to select its own TPC command. For this purpose, the aforementioned TPC-PDCCH-Config IE also includes the TPC-Index IE. The TPC-Index IE contains an IndexOfFormat3 field that provides the index for a particular UE. This index can range between 1 and 15 as DCI format 3 transports 15 TPC commands at the maximum.

*TPC Command on DCI Format 3A*

The TPC command field on DCI format 3A is 1 bit long. The value of TPC command field corresponds to the change of power level, $\delta_{PUCCH}$, as shown in Table 17.6.

DCI format 3A also transports a number of TPC commands such as TPC command number 1, TPC command number 2, and so forth, and they are assigned to particular UEs similar with DCI format 3. However, in this case, the TPC-Index contains an IndexOfFormat3A field that provides the index for a particular UE. This index can range between 1 and 31 as DCI format 3A transports 31 TPC commands at the maximum.

**Table 17.5** PUCCH Power Adjustment Based on the TPC Command on DCI Formats 1, 1A, 1B, 1D, 2, 2A, and 3

| TPC Command Field | Change of Power Level, $\delta_{PUCCH}$ [dB] |
|---|---|
| 0 | −1 |
| 1 | 0 |
| 2 | 1 |
| 3 | 3 |

**Table 17.6** PUCCH Power Adjustment Based on TPC Command on DCI Format 3A

| TPC Command Field | Change of Power Level, $\delta_{PUCCH}$ [dB] |
|---|---|
| 0 | −1 |
| 1 | 1 |

The TPC command on DCI format 3A provides only single-bit power adjustments, whereas the TPC command on DCI format 3 provides power adjustments in greater extent using 2 bits. However, DCI format 3A can accommodate almost twice as many TPC commands as DCI format 3.

*Calculation of $g_{TPC}$*

Once the UE determines $\delta_{PUCCH}$ as shown above, it calculates the value of $g_{TPC}$ for subframe $i$, using the following equation.

$$g_{TPC}(i) = g_{TPC}(i-1) + \delta_{PUCCH}(i-4)$$

As the equation shows, $\delta_{PUSCH}$ increases or decreases $g_{TPC}$ of the current subframe compared to the $g_{TPC}$ value in the previous subframe. Such accumulation allows $g_{TPC}$ to attain very high positive or negative values. The eNodeB sends the PDCCH signaling four subframes earlier, and therefore, $g_{TPC}(i)$ is calculated using $\delta_{PUSCH}(i-4)$.

# Multiple Antenna Techniques in the Downlink

The use of multiple transmit antennas is primarily of interest for the downlink. There are various multiple antenna schemes supported in the downlink:

1. *Multiple input single output (MISO):* This uses two or more transmitters and one as shown in Figure 18.1 (a). MISO is commonly referred to as transmit diversity. The same data is sent on both transmitting antennas, but coded so that the receiver can identify each transmitter. LTE downlink can use open-loop transmit diversity, which is also supported in UMTS Release 99. However, closed-loop transmit diversity techniques have not been adopted in the LTE from the UMTS.

2. *Single input multiple output (SIMO):* This uses one transmitter and two or more receivers as shown in Figure 18.1 (b). SIMO is often referred to as receive diversity. Similar to transmit diversity, it is particularly well suited for low SNR conditions in which a theoretical gain of 3 dB is possible when two receivers are used. As with transmit diversity, there is no change in the data rate since only one data stream is transmitted, but coverage at the cell edge is improved due to the lowering of the usable SNR.

3. *Multiple input multiple output (MIMO):* MIMO is used to achieve array gain, multiplexing gain, and diversity gain. It may also help reduce cochannel interference. MIMO uses two or more transmitters and two or more receivers as shown in Figure 18.1 (c). It must have at least as many receivers as there are transmitted streams. The number of transmit data streams must be equal to or less than the number of transmit antennas. The signals in transmit antennas vary in amplitude, phase, and waveform. The transmitted signals from each antenna undergo different channel fading on their way to the different receiving antennas. Thus, by spatially separating $N$ data streams, the use of at least $N$ receiving antennas will allow to fully reconstruct the original data streams provided that the crosstalk and noise in the radio channel are low enough. MIMO exploits space diversity and time diversity of the multipath propagation channel to generate multiple orthogonal data streams. The data streams vary in their data rate capabilities. A feedback from the receiver about the propagation channel status can be optionally used at the transmitter to enhance performance in MIMO technique. Thus, MIMO can be applied in either open-loop mode or closed-loop mode.

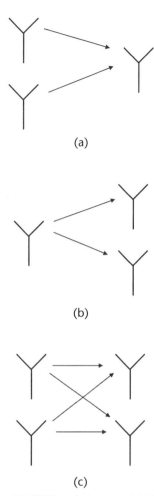

**Figure 18.1** (a) MISO technique, (b) SIMO technique, and (c) MIMO technique.

MIMO can be used to attain various multiple antenna solutions. The MIMO techniques supported in the downlink are: spatial multiplexing (SM), transmit diversity (TxD), and beamforming.

The multiple antenna techniques used in different downlink channels are as follows:

1. PBCH, PDCCH, PCFICH, and PHICH, as well as random access response (RAR) message, paging message, and system information (SI) message on PDSCH, use the same set of antenna ports. If they use more than one antenna port, then transmit diversity is applied.
2. Data on PDSCH may use any of the three techniques: spatial multiplexing, transmit diversity, and beamforming. PDSCH is configured with one of the following transmission modes:
   - *Transmission mode 1:* Using a single antenna at eNodeB.
   - *Transmission mode 2:* Transmit diversity, explained in Section 18.1.
   - *Transmission mode 3:* SU-MIMO spatial multiplexing: open-loop, explained in Section 18.2.1.2.

- *Transmission mode 4:* SU-MIMO spatial multiplexing: closed-loop. This is explained in Section 18.2.1.1.
- *Transmission mode 5:* MU-MIMO spatial multiplexing, explained in Section 18.2.2.
- *Transmission mode 6:* Beamforming using closed-loop rank-1 precoding. It can also be seen as a special case of SU-MIMO spatial multiplexing, explained in Sections 18.2.1.1 and 18.3.2.
- *Transmission mode 7:* Beamforming using UE-specific reference signals, explained in Section 18.3.1.
- *Transmission mode 8:* This is introduced in Release 9. It employs dual-layer SU-MIMO and single-layer MU-MIMO with UE-specific RS. It allows dynamic switching between SU-MIMO and MU-MIMO.
- *Transmission mode 9:* This is introduced in Release 10. This facilitates the application of SU-MIMO and MU-MIMO while RRC signaling is not required for switching between them. It supports up to eight transmit antennas using a dual codebook.

The choice of transmission mode for PDSCH depends on the instantaneous radio channel conditions. The eNodeB uses CQI/PMI/RI reports from the UE to select a suitable transmission mode. The selection of transmission mode can be adapted semi-statically. Appendix A.2 shows how the eNodeB sends AntennaInfoDedicated IE to the UE while setting up or modifying radio bearers. The AntennaInfoDedicated IE includes a TransmissionMode field that specifies the transmission mode. It thus indicates a mode between 1 and 9.

Release 8 supports up to four transmit antennas for the eNodeB and the UE. The common MIMO configurations are $2 \times 2$ and $4 \times 2$. It can also be $4 \times 4$ if the UE uses four antennas. Release 10 supports eight transmit antennas for the eNodeB and the UE and thus supports up to an $8 \times 8$ antenna configuration. MIMO involves the following terms:

- *Layer:* The *layers* refer to different data streams in MIMO implementation. For various layers, data is mapped onto different transmit antenna ports. Each layer generates a radiation pattern from the antenna. Thus, the number of layers can be 1, 2, 3, or 4 in Release 8. The transmit power is divided among layers.
- *Rank:* The *rank* of the transmission indicates the number of layers transmitted. The highest possible rank depends on the highest number of antennas existing on both the transmitting and the receiving sides (i.e., for a $4 \times 2$ configuration, the highest possible rank is 2). The rank can be 1, 2, 3, or 4. Adaptive control of the rank is made to control number of transmitted layers according to the spatial channel condition (e.g., received SINR and fading correlation between antennas). As the LOS component overpowers multipath components, the rank gets lower. If the rank is lower than the number of transmit antennas, then the additional antennas can be used for spatial diversity gain. In practice, most users need to use rank 1 or 2 and only a few users can use the higher rank.

The UE sends the following feedback to the eNodeB to assist MIMO operation:

- *Rank Indication (RI):* The UE sends the RI to the eNodeB, indicating its preferred rank to be used on PDSCH for certain transmission modes. A common rank is indicated for all resource blocks.
- *Precoding Matrix Indicator (PMI):* The UE sends the PMI to help the eNodeB select the appropriate precoding matrix from the codebook. The PMI can be optionally subband based because the optimum precoding matrix can vary among resource blocks.

The RI and PMI reporting depends on transmission modes, as shown in Table 18.1.

## 18.1 Transmit Diversity: Transmission Mode 2

Transmit diversity is configured by selecting transmission mode 2. Transmit diversity increases the signal-to-noise ratio at the receiver instead of directly increasing the data rate. Each transmit antenna transmits essentially the same stream of data and so the receiver gets replicas of the same signal. A suitable signal combining technique reduces fading variation and increases the signal-to-noise ratio at the receiver side. Thus, the robustness of data transmission is achieved especially in fading scenarios. It also improves the cell edge user data rate and coverage range. An additional antenna-specific coding is applied to the signals before the transmission to increase the diversity effect. The transmit diversity is an open-loop scheme and feedback from the UE is not required.

Transmit diversity is only defined for two and four transmit antennas and one data stream. The number of layers is equal to the number of antenna ports. Transmit diversity uses only one code word. The code word is mapped to two or four layers when there are two or four transmit antennas, respectively. To maximize the diversity gain, the antennas typically need to be uncorrelated. Thus, the antennas need to be well separated relative to the wavelength or they need to have a different polarization.

Transmit diversity is applied using space-frequency block coding (SFBC) and frequency-switched transmit diversity (FSTD). Cyclic delay diversity (CDD) is commonly associated with OFDM as a transmit diversity technique. However, in case of LTE, CDD is not used as a transmit diversity scheme but rather as a precoding scheme for spatial multiplexing.

SFBC is a frequency-domain version of the well-known space-time block codes (STBCs), also known as the Alamouti codes. This family of codes is designed so that the transmitted diversity streams are orthogonal and achieve the optimal SNR with a linear receiver. Such orthogonal codes only exist in the case of two transmit antennas. The multiple subcarriers of OFDM allow a straightforward application of SFBC.

- *Two transmit antennas:* Pure SFBC is used when there are two transmit antennas. SFBC uses modulation symbols in pairs and modifies them before

## 18.1 Transmit Diversity: Transmission Mode 2

**Table 18.1** PMI/RI Reporting

| Transmission Mode | RI | PMI |
|---|---|---|
| 1 | Not Reported | Not reported |
| 2 | Not Reported | Not reported |
| 3 | Reported | Not reported |
| | Two antennas: RI can indicate rank 1 or 2 | |
| | Four antennas: RI can indicate rank 1, 2, 3, or 4 | |
| 4 | Reported | Reported |
| | Two antennas: RI can indicate rank 1 or 2 | Two antennas: For rank 1, PMI can indicate a codebook index from 0 through 3. |
| | Four antennas: RI can indicate rank 1, 2, 3, or 4 | For rank 2, PMI can indicate a codebook index 1 or 2. |
| | | Four antennas: PMI can indicate a codebook index from 0 through 15 for any of the four ranks. |
| 5 | Not reported | Reported |
| | (rank is always 1) | Two antennas: PMI can indicate a codebook index from 0 through 3. |
| | | Four antennas: PMI can indicate a codebook index from 0 through 15. |
| 6 | Not reported | Reported |
| | (rank is always 1) | Two antennas: PMI can indicate a codebook index from 0 through 3. |
| | | Four antennas: PMI can indicate a codebook index from 0 through 15. |
| 7 | Not reported | Not reported |
| 8 | Reported only if CQI-ReportConfigIE configures PMI-RI-Report IE | Reported only if CQI-ReportConfigIE configures PMI-RI-Report IE |
| 9 | Reported only if CQI-ReportConfigIE configures PMI-RI-Report IE | Reported only if CQI-ReportConfigIE configures PMI-RI-Report IE |

feeding to antennas. Alamouti's rate-1 block coding is applied in SFBC. The symbols transmitted from the two antenna ports on each pair of adjacent subcarriers are defined as follows.

$$\begin{bmatrix} y^{(0)}(1) & y^{(0)}(2) \\ y^{(1)}(1) & y^{(1)}(2) \end{bmatrix} = \begin{bmatrix} S_1 & S_2 \\ -S_2^* & S_1^* \end{bmatrix}$$

As shown in Figure 18.2(a), SFBC encodes a pair of symbols, $S_1$ and $S_2$, into four variants, $S_1$, $S_2$, $-S_2^*$, and $S_1^*$, and transmits $S_1$ and $-S_2^*$ over a certain subcarrier from the two antennas. However, the other two variants, $S_2$ and $S_1^*$, are transmitted from the subsequent contiguous or noncontiguous subcarriers.

- *Four transmit antennas:* SFBC cannot be directly used beyond 2 × 2 since no orthogonal codes exist. Thus, in case of four transmit antennas, SFBC is combined with FSTD. It is, in fact, a combination of two 2 × 2 SFBC schemes

| Subcarrier | Tx Antenna 1 | Tx Antenna 2 |
|---|---|---|
| ... | | |
| ... | | |
| k | $S_1$ | $S_2$ |
| k+1 | $-S_2^*$ | $S_1^*$ |
| k+2 | $S_3$ | $S_4$ |
| k+3 | $-S_4^*$ | $S_3^*$ |
| ... | | |
| ... | | |

(a)

| Subcarrier | Tx Antenna1 | Tx Antenna2 | Tx Antenna3 | Tx Antenna4 |
|---|---|---|---|---|
| ... | | | | |
| ... | | | | |
| k | $S_1$ | $S_2$ | × | × |
| k+1 | $-S_2^*$ | $S_1^*$ | × | × |
| k+2 | × | × | $S_3$ | $S_4$ |
| k+3 | × | × | $-S_4^*$ | $S_3^*$ |
| ... | | | | |
| ... | | | | |

(b)

**Figure 18.2** (a) Transmit diversity with two transmit antennas. (b) Transmit diversity with four transmit antennas.

mapped to independent subcarriers where Alamouti's rate-1 block coding is applied in SFBC and the antenna mapping is (1, 3) and (2, 4) as shown here:

$$\begin{bmatrix} y^{(0)}(1) & y^{(0)}(2) & y^{(0)}(3) & y^{(0)}(4) \\ y^{(1)}(1) & y^{(1)}(2) & y^{(1)}(3) & y^{(1)}(4) \\ y^{(2)}(1) & y^{(2)}(2) & y^{(2)}(3) & y^{(2)}(4) \\ y^{(3)}(1) & y^{(3)}(2) & y^{(3)}(3) & y^{(3)}(4) \end{bmatrix} = \begin{bmatrix} S_1 & S_2 & 0 & 0 \\ 0 & 0 & S_3 & S_4 \\ -S_2^* & S_1^* & 0 & 0 \\ 0 & 0 & -S_4^* & S_3^* \end{bmatrix}$$

The mapping of symbols to antenna ports is different in the four transmit-antenna case compared to the two transmit-antenna SFBC scheme. This is because the cell-specific reference signal symbol density on the third and fourth antenna ports is half of that on the first and second antenna ports, as shown in Section 5.1.5.1. Therefore, the channel estimation accuracy may be lower on the third and fourth antenna ports. Thus, the difference in the mapping avoids concentrating the channel estimation losses in one of the SFBC codes and results in a slight coding gain.

## 18.2 Spatial Multiplexing

Spatial multiplexing allows multiple antennas to transmit multiple independent streams, and it is thus sometimes referred to as the true MIMO technique. If the receiver also has multiple antennas, the streams can be separated out using space-time processing. Instead of increasing diversity, multiple antennas are used here to increase the data rate or the capacity of the system. Assuming a rich multipath environment, the capacity of the system can be increased linearly with the number of antennas when performing spatial multiplexing. A $2 \times 2$ MIMO system therefore doubles the peak throughput capability. However, in practice, the multiplexing gain is not that high because the SINR of the two parallel streams are not high enough to support the same modulation and coding scheme. When the radio link is good, it is more efficient to share the transmit power among multiple data streams. Thus, unlike transmit diversity and beamforming, spatial multiplexing works efficiently only under good SINR conditions. When SINR is poor, the multiplexing gain becomes very low, and because of interlayer interference the gain can even be negative.

Each receive antenna may receive the data streams from all transmit antennas. The number of layers represents the number of data streams. The number of layers is equal to or less than the number of antenna ports. The channel is described by a channel matrix $H$ for a specific delay. If $N_t$ is the number of transmit antennas and $N_r$ is the number of receive antennas, then the number of data streams that can be transmitted in parallel is given by $MIN(N_t, N_r)$ and is limited by the rank of the matrix $H$. The transmission quality degrades significantly if the singular values of the matrix $H$ are not sufficiently strong. This can happen if the two antennas are not sufficiently decorrelated, for example, in an environment with little scattering or when the antennas are too closely spaced. The rank of the channel matrix $H$ is an important criterion to determine whether spatial multiplexing can be done with good performance. In practice, the data streams are often weighted and added so that each antenna actually transmits a combination of the streams.

Spatial multiplexing can be used in the following modes.

1. *Single-user MIMO (SU-MIMO)*: The SU-MIMO is expected to be used more commonly. In this case, each UE uses the allocated resource exclusively (i.e., a particular resource block is allocated to only one UE). It increases the data rate for the particular UE. The increase of the individual data rate also corresponds to an increase in the capacity of the cell. Multiple data streams are allocated to one UE. The MIMO technique allows the receiver to separate the original data streams and thus increase the data rate. It uses precoding in LTE.
2. *Multiple-user MIMO (MU-MIMO)*: Multiple UEs are served by one common eNodeB by sharing the same radio resource simultaneously. The eNodeB can transmit to multiple UEs using the same resource blocks

simultaneously using space-division multiple access (SDMA). This allows the UEs to be differentiated not only in time and frequency, but also in space. Instead of using multiple layers for one UE, all antenna elements concentrate on the strongest layer for the particular UE. This results in a beam shaping towards the UE and higher SNR. The improvement of the SNR is referred to as array gain. To get the most gain out of MU-MIMO, the UE must be well aligned in time and power. The users need to use the same precoding matrix as they are using the same resource block.

The comparison between SU-MIMO and MU-MIMO techniques is shown here:

- In the case of MU-MIMO, the antennas belong to different UEs and thus are much farther apart than in the SU-MIMO case. The extra spatial separation makes the paths more uncorrelated and maximizes the potential capacity gain. It reduces the cochannel interference.
- MU-MIMO does not increase an individual user's data rate, but it does offer cell capacity gains that are similar to, or better than, those provided by SU-MIMO.
- In the case of MU-MIMO, the transmitter antennas belong to different UEs, and thus, there is no physical connection between the transmitter antennas and no way to optimize the coding to the channel eigenmodes by mixing the two data streams.
- In the case of MU-MIMO, the UE does not require the expense and power drain of two transmitters. In fact, SU-MIMO necessitates that the UE has at least $N$ antennas in order to multiplex $N$ streams, but MU-MIMO allows multiplexing up to $N$ streams with $N$ number of UEs all of which are equipped with single antennas.
- MU-MIMO is associated with a large number of users, but each user experiences low control overhead. Each user has sparse reporting granularity, single layer transmission, and so forth.
- MU-MIMO is favorable for heavily loaded systems where the maximization of the overall system throughput is the primary concern. It requires a closely spaced antenna array at eNodeB.
- SU-MIMO is appropriate for high peak user throughput, bursty traffic, and lightly loaded cell.

### 18.2.1 Single-User MIMO (SU-MIMO)

The SU-MIMO technique employs the adaptive control of the rank. A higher rank can increase spatial multiplexing gain, but it also increases interlayer interference. Only one code word or two simultaneous code words can be sent and this also depends on the rank as shown here:

1. *Rank 1:* One code word can be sent for rank 1 transmission.

2. *Rank 2/3/4:* Rank 2/3/4 transmissions are performed using one code word or two simultaneous code words as shown here:
   - *One code word:* One code word is sent that is mapped on all the available layers. This requires less control signaling as follows:
     - For CQI reporting, a single value is needed for all layers.
     - For HARQ ACK/NACK feedback, one ACK/NACK is needed per subframe per UE. The use of one code word allows a low complexity MMSE receiver, but it is not capacity achieving and it results in lower throughput in the low rank channels.
   - *Two simultaneous code words:* Two code words are simultaneously sent that are mapped on different layers as follows:
     - Rank 2: CW1 → Layer1, CW2 → Layer2
     - Rank 3: CW1 → Layer1, CW2 → Layer2+Layer3
     - Rank 4: CW1 → Layer1+Layer2, CW2 → Layer3+Layer4

   All resource blocks belonging to the same code word use the same MCS, even if a code word is mapped onto multiple layers. However, the MCS of two code words are independent.

   For two simultaneous code words, significant gains are possible by using successive interference cancellation (SIC) at receiver. The SIC receiver is used to decouple the transmission layers, allowing better throughput and more tolerance to spatial correlation. Thus, it is capacity achieving and results in optimal performance. The use of two code words also allows per antenna rate control (PARC). However, the two code words require more signaling, receiver complexity, and memory than for one code word.

   If two code words are simultaneously sent, then the eNodeB uses the transport block to code word swap flag of DCI format 2 or 2A on PDCCH in order to indicate how transport block to code word mapping has been performed while allocating resources. The transport block to code word swap flag has 1 bit. The value 0 of the flag dictates the use of transport block 1 as code word 0 and transport block 2 as code word 1. Value 1 of the flag dictates the use of transport block 2 as code word 0 and transport block 1 as code word 1.

   If only one code word transmission is used, then this flag is reserved. In this case, the transport block that is enabled is mapped onto code word 0 and code word 1 is disabled.

The SU-MIMO technique uses precoding in LTE in order to achieve the optimal method of transmission over the MIMO channel. The channel-dependent precoder provides for transmit beamforming, power allocation across the transmitted streams, and a matching receive beamforming structure. The optimal precoder is the concatenation of optimal beamforming and optimal power allocation. The optimal spatial multiplexing scheme uses transmit-and-receive beamforming to decompose the MIMO channel into a number of parallel noninterfering subchannels known as eigenchannels.

Precoding from a defined codebook is used to form the transmitted layers in LTE. This reduces the control signaling overhead. Each codebook consists of a set of predefined precoding matrices. The size of the set of the matrices is set considering a trade-off between the number of signaling bits required to indicate a particular matrix in the codebook and the suitability of the resulting transmitted beam direction. The following PDSCH transmission modes use precoding from the defined codebook to form the transmitted layers.

- *Transmission mode 3:* Open-loop spatial multiplexing;
- *Transmission mode 4:* Closed-loop spatial multiplexing;
- *Transmission mode 6:* Closed-loop rank-1 precoding.

The eNodeB can optionally limit the use of the codebook within a part of it instead of using the whole codebook using CodebookSubsetRestriction field. The AntennaInfoDedicated IE includes the CodebookSubsetRestriction field. Appendix A.2 shows how the eNodeB sends AntennaInfoDedicated IE to the UE while setting up or modifying radio bearers. The CodebookSubsetRestriction field can define subsets of the codebooks separately for different transmission modes for both two and four transmit antennas. Then the UE must restrict itself to report PMI/RI within the specified codebook subset.

### 18.2.1.1 Closed-Loop Spatial Multiplexing: Transmission Mode 4 and 6

The UE estimates the radio channel and selects the most desirable entry from a predefined codebook. Then the UE sends a feedback to the eNodeB as shown in Section 20.1 and hence is called a closed loop. The feedback can be optionally subband based because the optimum precoding matrix can vary among resource blocks. The preferred precoder is the matrix that would maximize the capacity based on the receiver capabilities. In an interference-free environment, the UE will typically indicate the precoder that would result in a transmission with an effective SNR by choosing most closely the largest singular values of its estimated channel matrix. The codebook has equal amplitude for each antenna component in precoding weight matrices. The precoders comprise pure phase corrections with no amplitude changes. This allows the power amplifier connected to each antenna to be loaded equally. The codebook has its lower rank part as subset of the higher rank codebook vectors. This nested property reduces the number of calculations required for the UE to generate the feedback.

1. *Two-antenna codebook:* The codebook used is shown in Table 18.2. For rank 1, the codebook index used can be one of six, 0, 1, 2, 3, 4, or 5. For rank 2, the two-antenna codebook has identity, discrete Fourier transform (DFT), π/2-rotated DFT matrices. The codebook has three entries, but the codebook index 0 is not used, so the codebook index can be either 1 or 2.
2. *Four-antenna codebook:* The four-transmit antenna codebook uses a Householder generating function, $W_H = I - 2uu^H/u^Hu$, which generates unitary matrices from input vectors $u$. The Householder codebook has 16 precoding matrices for ranks 1, 2, 3, and 4. The Householder function simplifies the CQI calculation by reducing the number of matrix inversions.

The closed-loop spatial multiplexing is used for transmission modes 4 and 6.

## 18.2 Spatial Multiplexing

**Table 18.2** Two-Antenna Codebook

| Codebook Index | Number of Layers $v$ | |
|---|---|---|
| | 1 | 2 |
| 0 | $\frac{1}{\sqrt{2}}\begin{bmatrix}1\\1\end{bmatrix}$ | $\frac{1}{\sqrt{2}}\begin{bmatrix}1 & 0\\0 & 1\end{bmatrix}$ |
| 1 | $\frac{1}{\sqrt{2}}\begin{bmatrix}1\\-1\end{bmatrix}$ | $\frac{1}{2}\begin{bmatrix}1 & 1\\1 & -1\end{bmatrix}$ |
| 2 | $\frac{1}{\sqrt{2}}\begin{bmatrix}1\\j\end{bmatrix}$ | $\frac{1}{2}\begin{bmatrix}1 & 1\\j & -j\end{bmatrix}$ |
| 3 | $\frac{1}{\sqrt{2}}\begin{bmatrix}1\\-j\end{bmatrix}$ | — |

*Transmission Mode 4*

The transmission mode 4 can offer the peak user throughput. The UE sends RI to the eNodeB indicating its preferred rank to be used. Also, the UE sends PMI as a feedback to help the eNodeB select the precoding matrix from the codebook.

When the eNodeB allocates resources, it uses Precoding Information field in DCI format 2 on PDCCH in order to indicate how precoding has been used. The Precoding Information field is 3 bits long for two antenna ports at the eNodeB and 6 bits long for four antenna ports at the eNodeB. The value of this field corresponds to a value of TRI which indicates the rank for spatial multiplexing. It also corresponds to a value of TPMI which indicates the codebook index used from the two-antenna codebook or the four-antenna codebook. It may also indicate precoding according to the latest PMI report on PUSCH. The same value of the Precoding Information field indicates different information between the use of one code word and two code words. If the value is 0 with one code word enabled, it indicates the use of transmit diversity as opposed to spatial multiplexing.

*Transmission Mode 6*

When rank 1 is used from the predefined codebook, it becomes transmission mode 6. Since there is only one spatial layer, spatial multiplexing is not possible. However, there are still user dependent optimum beams to increase user throughput and tracking of instantaneous channel variations. The receive beamforming works using two receive antennas. The UE combines the signals from its two antennas through the use of weights. The UE can set the beamforming vector to its optimal value based on the UE's channel to maximize the received SNR. The UE sends the PMI as a feedback to help the eNodeB select the precoding matrix from the codebook. Reporting RI is not required since rank always 1.

When the eNodeB allocates resources, it uses the following fields in DCI format 1B on PDCCH in order to indicate how precoding has been used.

- *TPMI information for precoding*: The number of bits of this field is 2 or 4 for two or four antenna ports at the eNodeB, respectively. This field indicates which codebook index is used from the two-antenna codebook or the four-antenna codebook for the single layer transmission.
- *PMI confirmation for precoding*: This field has 1 bit. If this field is 0, then precoding takes place according to the aforementioned TPMI Information for Precoding field. If this field is 1, then precoding takes place according to the latest PMI report on PUSCH.

### 18.2.1.2 Open-Loop Spatial Multiplexing: Transmission Mode 3

Open-loop spatial multiplexing is configured by selecting transmission mode 3. It makes use of the spatial dimension of the propagation channel and transmits multiple data streams on the same resource blocks. The feedback from the UE indicates only the rank of the channel using RI and not a preferred precoding matrix, and hence it is called open-loop. A channel-independent fixed precoding with large delay cyclic delay diversity (CDD) is used. For high-speed users, the PMI fails to indicate optimum use of precoding matrix and then the fixed precoding may achieve better performance.

CDD provides for another type of diversity. It employs transmitting the same set of OFDM symbols on the same set of OFDM subcarriers from multiple transmit antennas, with a different delay on each antenna. The delay is applied before the Cyclic Prefix (CP) is added, which guarantees that the delay is cyclic over the Fast Fourier Transform (FFT) size. Since the delay is added before the addition of the CP, any delay value can be used without increasing the overall delay spread of the channel. Adding a time delay is identical to applying a phase shift in the frequency domain. Therefore, in general, it is up to the designer if he or she implements CDD in the time domain or in the frequency domain.

As the same time delay is applied to all subcarriers, the phase shift increases linearly across the subcarriers with increasing subcarrier frequency. Each subcarrier will therefore experience a different beamforming pattern. Thus, the different subcarriers will pick out different spatial paths in the propagation channel and increase the frequency selectivity of the channel. The channel coding, which is applied to a whole transport block across the subcarriers, ensures that the whole transport block benefits from the diversity of spatial paths.

The eNodeB transmitter combines CDD delay-based phase shifts with additional precoding using fixed unitary DFT-based precoding matrices. This precoding is helpful because then virtual antennas formed by channel-independent precoding will typically be uncorrelated and antenna correlation will degrade.

CDD does not optimally exploit the channel similar with an ideal precoding, which attempts to match with the eigenvectors of the channel, but CDD helps to keep any destructive fading constrained to individual subcarriers rather than

## 18.2 Spatial Multiplexing

affecting a whole transport block. This can be particularly beneficial if the channel information at the transmitter is unreliable, for example, if there is limited feedback from the UE or the velocity of the UE is very high. This justifies the use of CDD for transmission mode 3, which uses open-loop spatial multiplexing with no feedback.

For a multilayer CDD operation, the mapping of the layers to antenna ports is carried out using precoding matrices selected from the same codebooks used for closed-loop spatial multiplexing. However, the UE does not indicate a preferred precoding matrix because CDD is used with open loop spatial multiplexing. Thus, in this case, the particular matrices selected from the codebooks are predetermined. Precoding for spatial multiplexing is accomplished as follows.

$$\begin{bmatrix} y^{(0)}(i) \\ \vdots \\ y^{(u-1)}(i) \end{bmatrix} = W(i)D(i)U \begin{bmatrix} s^{(0)}(i) \\ \vdots \\ s^{(u-1)}(i) \end{bmatrix}$$

Here $W(i)$ is the precoding matrix and $D(i)$ and $U$ are supporting CDD. $D(i)$ and $U$ are shown in Table 18.3.

For two antenna ports, the precoder, $W(i)$ is the precoding matrix corresponding to precoder index 0 in the two-antenna codebook used for closed-loop spatial multiplexing.

For four antenna ports, the eNodeB cyclically assigns different precoders, $W(i)$, to the input of different antenna ports. Thus, a different precoder is used for every $u$ vector, where $u$ denotes the number of transmission layers. The precoder is selected as $W(i) = C_k$, where $k$ is the precoder index given by $k = \left(\left\lfloor \dfrac{i}{u} \right\rfloor \mod 4\right) + 1$ from the four-antenna codebook used for closed-loop spatial multiplexing.

When the eNodeB allocates resources, it uses the Precoding Information field in DCI format 2A on PDCCH in order to convey the Transmitted Rank Indication (TRI) which indicates the rank used. The Precoding Information field is 0 bits long

**Table 18.3** $D(i)$ and $U$ Precoding for Open-Loop Spatial Multiplexing

| Number of Layers, $u$ | $U$ | $D(i)$ |
|---|---|---|
| 2 | $\dfrac{1}{\sqrt{2}}\begin{bmatrix} 1 & 1 \\ 1 & e^{-j2\pi/2} \end{bmatrix}$ | $\begin{bmatrix} 1 & 0 \\ 0 & e^{-j2\pi/2} \end{bmatrix}$ |
| 3 | $\dfrac{1}{\sqrt{3}}\begin{bmatrix} 1 & 1 & 1 \\ 1 & e^{-j2\pi/3} & e^{-j4\pi/3} \\ 1 & e^{-j4\pi/3} & e^{-j8\pi/4} \end{bmatrix}$ | $\begin{bmatrix} 1 & 0 & 0 \\ 0 & e^{-j2\pi/3} & 0 \\ 0 & 0 & e^{-j4\pi/3} \end{bmatrix}$ |
| 4 | $\dfrac{1}{2}\begin{bmatrix} 1 & 1 & 1 & 1 \\ 1 & e^{-j2\pi/4} & e^{-j4\pi/4} & e^{-j6\pi/4} \\ 1 & e^{-j4\pi/4} & e^{-j8\pi/4} & e^{-j12\pi/4} \\ 1 & e^{-j6\pi/4} & e^{-j12\pi/4} & e^{-j18\pi/4} \end{bmatrix}$ | $\begin{bmatrix} 1 & 0 & 0 & 0 \\ 0 & e^{-j2\pi/4} & 0 & 0 \\ 0 & 0 & e^{-j4\pi/4} & 0 \\ 0 & 0 & 0 & e^{-j6\pi/4} \end{bmatrix}$ |

for two antenna ports at the eNodeB and 2 bits long for four antenna ports at the eNodeB.

If the eNodeB has four antenna ports, then the 2-bit field corresponds to a value of TRI that indicates the rank. For TRI > 1, spatial multiplexing with large delay CDD is used. However, if TRI = 1, then transmit diversity is used.

If the eNodeB has two antenna ports, then there is no need of precoding information, so the number of bits is 0. If both code words are enabled, TRI = 2 and spatial multiplexing with large delay CDD is used. If one code word is enabled (i.e., code word 0 is enabled and code word 1 is disabled), then TRI = 1 and transmit diversity is used.

### 18.2.2 Multiple Users MIMO (MU-MIMO): Transmission Mode 5

Multiple users MIMO (MU-MIMO) is configured by selecting transmission mode 5. The transmission mode 5 can offer the peak cell throughput. The eNodeB transmits to multiple UEs using the same resource blocks simultaneously making use of the spatial dimension of the propagation channel. The number of UEs supported is up to two in Release 8. It supports one transmission layer per UE.

The MU-MIMO scheme focuses on highly correlated antenna elements. It supports only rank 1 transmission (i.e., only one layer is allowed for each of the UEs). Thus, reporting RI is not required. MU-MIMO is supported only with precoding. It uses the same SU-MIMO precoding codebooks defined for two transmit antennas and four transmit antennas but only for single layer transmission. The CQI calculation and precoding feedback are the same as the rank-1 SU-MIMO. The UE sends the PMI as a feedback to help the eNodeB select the precoding matrix from the codebook. The eNodeB can optionally limit the use of the codebook within a part of it instead of using the whole codebook using the CodebookSubsetRestriction field, as explained in Section 18.2.1.

When the eNodeB allocates resources, it uses the following fields in DCI format 1D on PDCCH in order to indicate how precoding has been used.

- *TPMI information for precoding:* The number of bits of this field is 2 or 4 for two or four antenna ports at the eNodeB, respectively. This field indicates which codebook index is used from the two-antenna codebook or the four-antenna codebook for the single layer transmission.
- *Downlink power offset:* This field has 1 bit. This indicates whether a power offset is applied to the data symbols. If this field is 0, then a power offset, $\delta_{power\text{-}offset} = -10\log_{10} 2$ dB is applied. If this field is 1, then no power offset is applied assuming $\delta_{power\text{-}offset} = 0$ dB. The power offset allows the transmission power to be shared between two UEs.

## 18.3 Beamforming: Transmission Mode 6 and 7

Beamforming allows the focusing of the transmitted beam in the direction of the UE. All antenna elements concentrate on the strongest layer and create a beam towards the UE. This provides a high SNR known as array gain. Moreover, because of the directional radiation, beamforming helps reduce intercell interference. Beamforming can provide significant improvement in the cell edge user data rate, coverage range, capacity, and reliability. The UE does not find the beamformed signal special except that the signal level is expectedly better.

The beamforming may be chosen when the radio link is poor with high interference. Beamforming then concentrates the whole transmit power on the strongest layer. Beamforming can help especially in rural deployments, which have larger cells. However, beamforming is only used for PDSCH and not for control channels. Thus, the overall cell range may still be limited by the range of the control channels unless other measures are taken.

The beamforming is archived using precoding. The eNodeB determines appropriate precoding based on feedback from the UE. Thus, the improvement may be better in a beamforming technique than in a transmit diversity for slow-moving UEs. However, the beamforming technique may be worse for fast-moving UEs because then the UE feedback cannot track the channel variation quickly enough.

Beamforming is used for transmission modes 6 and 7.

### 18.3.1 Transmission Mode 7

Transmission Mode 7 is a pure beamforming technique. The eNodeB transmits UE-specific RS in addition to the cell-specific RSs. The UE-specific RS enable the UE to demodulate the beamformed data coherently, which is needed because the narrow beam in the direction of the UE will experience a different channel response. The UE-specific RS uses resources on PDSCH that are dedicated for the particular UE. The UE-specific RS uses a distinct antenna port that has its own downlink channel response.

The eNodeB attempts to determine the relative position of the UE and creates antenna beam in that direction. The eNodeB may use Direction of Arrival (DOA) estimations from the uplink signal measurements, but the UE does not feed back any precoding-related information or indicate any preference for the direction of the beam.

Phased array can be used for beamforming with closely-spaced antenna elements (i.e., highly correlated antenna arrays in order to transmit in the direction of the UE). This is in contrast to the transmit diversity case where at least a few wavelengths of antenna spacing are required. Thus, for beamforming, the channels between the different antenna elements and the receiver are essentially the same and undergo the same fading. The signals from the different antenna elements are phased appropriately so that they all add up constructively in the desired direction. The overall transmission beam is steered in the direction of the UE by applying appropriate phase shifts to the signals at the different antenna elements. The adjustments may be based on DOA estimations from the uplink signal measurements.

A probable implementation of beamforming may use one of the common antenna ports as one of the elements of the phased array and apply beamforming to UEs near the edge of the cell. All the other antenna ports may be used to apply SU-MIMO spatial multiplexing to UEs closer to the eNodeB to deliver high data rates.

The CQI feedback is derived using the cell-specific RSs. The eNodeB may, over time, establish a suitable offset to apply to the CQI reports received from the UE in order to adapt the CQI reports to the actual quality of the beamformed signal. Such an offset may, for example, be derived from the proportion of transport blocks positively acknowledged by the UE.

### 18.3.2 Transmission Mode 6

Apart from Transmission Mode 7, Transmission Mode 6, known as *closed-loop rank 1 precoding*, is also considered as beamforming technique and used to improve SNR. In this case, rank 1 is used from the codebook considering it predefined for spatial multiplexing. But since, there is only one spatial layer, spatial multiplexing is actually not effective. However, it provides for beamforming using two receive antennas of the UE. The UE combines the signals from its two antennas through the use of weights. The UE sends channel information as feedback to the eNodeB to help determine the optimal beamforming vector and maximize the received SNR.

# CHAPTER 19
# Scheduling and Allocation of Resources

The scheduler in the MAC layer of the eNodeB allocates the available radio resources among different UEs and among the radio bearers of each UE for both uplink and downlink transmission in a cell through the proper handling of priority. The various DCI formats on the PDCCH assign resource blocks on the PDSCH on the same subframe for downlink transmission (i.e., if the PDCCH signaling is sent on subframe $n$, then it allocates resources on the PDSCH on the same subframe $n$). On the other hand, the uplink resource blocks on the PUSCH are allocated 4 subframes after the transmission of the resource grant on DCI format 0 (i.e., if PDCCH signaling is sent on subframe $n$, then it allocates resources for uplink transmission on the subframe $n + 4$). This time interval allows the UE to de-queue its data and prepare its transport block according to the specified attributes. Figures 20.1 and 20.2 demonstrate the time intervals for the downlink and the uplink, respectively.

## 19.1 Scheduling Decision

The MAC layer of eNodeB is responsible for scheduling both uplink and downlink resources. The scheduler attempts to make appropriate apportionment of the resources with certain objectives as follows:

- Required QoS for applications;
- Optimized spectral efficiency ensuring high cell throughput under existing channel conditions;
- Fairness among UEs and applications;
- Limiting the impact of interference through special handling of cell edge users;
- Load balancing among cells.

The scheduling decision can be modified every Transmission Time Interval (TTI), which is a subframe. The resource block is the basic scheduling unit for both downlink and uplink.

### 19.1.1 Information for the Scheduling Decision

The following information is made available in order to aid the scheduling decision for the apportionment of downlink resources:

1. Channel Quality Indicator (CQI) reports from the UEs to estimate the downlink channel quality;
2. QoS description of the EPS bearers for each UE, available in the eNodeB from the downlink data flow;
3. Relative Narrowband Tx Power (RNTP) from a neighboring eNodeB indicating the level of transmit power it will use in different parts of the bandwidth;
4. The status of the usage of resource blocks, transport network load in the S1 interface, and hardware load from a neighboring eNodeB to indicate the level of loading.

The following information is made available in order to aid scheduling decision for the apportionment of uplink resources:

1. The estimation of channel characteristics primarily based on the SRS (in addition, DM RS and the error rate in the current data transfer on the PUSCH may also be used for estimation);
2. The Scheduling Request (SR) from the UEs requesting initiation of uplink data transfer;
3. The Buffer Status Report (BSR) from the UEs reporting pending data in uplink buffers;
4. The power headroom reporting from the UEs indicating the difference between the maximum transmit power capability and the estimated power for uplink transmission;
5. The QoS description of the EPS bearers for each UE;
6. The UL interference overload indication from a neighboring eNodeB indicating the level of interference it is experiencing in different parts of the bandwidth;
7. The UL High Interference Indicator (HII) from a neighboring eNodeB indicating the level of interference it will create in different parts of the bandwidth;
8. The status of the usage of resource blocks, transport network load in the S1 interface, and hardware load from a neighboring eNodeB to indicate the level of loading.

### 19.1.2 Considerations in Scheduling Decision

The actual scheduling algorithm depends on the implementation at eNodeB. The scheduler may consider at least the following factors while making decisions in an attempt to make appropriate apportionment of the resources:

1. *Channel dependent scheduling (CDS):* The CDS can be made in both the time and frequency domains. The scheduling adapts to channel variations

and attains link adaptation. A part of the time-frequency resource grid is allocated to a user who has the best channel quality in that part (i.e. every set of resource blocks are allocated a user with the best channel quality in both time and frequency domain for those resource blocks). In other words, the transmission occurs at fading peaks. However, this may deprive users with constant poor channel quality. In practice, each user is attempted to be allocated the best possible part from the time-frequency resource grid. CDS allows very good use of the resources leading to a higher cell throughput. The CDS requires the availability of sufficient information on uplink and downlink channel conditions. This information needs to be available for each part of the bandwidth to allow the CDS in frequency domain. The eNodeB can configure the availability of information with more or less granularity in frequency domain where a higher granularity consumes more resources for this information. Moreover, the eNodeB can configure frequency of the availability of the information in time domain. A more frequent availability represents the variation of radio channel better, but again at the cost of more resources for this information. The eNodeB uses the following information for channel estimation:

- *Information for downlink CDS:* The eNodeB uses CQI reporting as will be explained in Section 20.1. The CQI reporting can be configured to be subband or wideband based. Also, the frequency of CQI reporting in time can be configured.
- *Information for uplink CDS:* The eNodeB primarily uses SRS. It may also use DM RS and the error rate in the current data transfer on the PUSCH. The UE may send the SRS in different parts of the bandwidth where no uplink data transmission is available. Also, the eNodeB can ask the UE for transmission of the SRS when no PUCCH or PUSCH is scheduled. The eNodeB can configure the bandwidth of the SRS. The wider the bandwidth for the SRS, the lower the available power per resource block. On the other hand, limiting the SRS to a small bandwidth makes it difficult for the scheduler to find an optimal scheduling solution for all users. An appropriate frequency-domain scheduling can be more difficult for cell edge users because they need higher uplink transmit power. The eNodeB can also configure the frequency of transmission of the SRS in time as was shown in Section 5.2.1.2. A more frequent transmission of SRS requires more resources but it allows better adjustments of resource scheduling.

2. *Apportionment among UEs:* The channel-dependent scheduling leads to higher cell throughput and, on the other hand, the scheduling should maintain some fairness among the users in their resource allocations. Thus, there is a trade-off between fairness and cell throughput. The scheduler can exercise various methods as shown here in order to address this trade-off:

- *Round robin (RR):* The scheduler assigns resources cyclically to the users without taking channel conditions into account. This is a simple procedure giving the best fairness. However, it would offer poor performance in terms of cell throughput.

- *Maximum C/I:* The scheduler assigns resources to the user with the best channel quality. This offers excellent cell throughput but it is not fair. However, if the users require resource allocations intermittently, then the fairness is not much violated. In this case, a user with low C/I may be allocated enough resources when the users with high C/I do not need allocations.
- *Proportional fair (PF):* The scheduler can exercise PF scheduling by allocating more resources to a user with relatively better channel quality. This offers high cell throughput as well as satisfactory fairness. Thus, PF scheduling may be the best option.
- *Scheduling for delay-limited capacity:* Some applications have very strict latency constraints, so their QoS requires a certain guaranteed data rate independent of the fading states. This guaranteed data rate is called the delay-limited capacity. The scheduler can allocate resources considering such special requirements.

3. *Queue and priority aware scheduling:* The scheduler performs queue-aware scheduling and adapts to instantaneous traffic situation for dynamic sharing of the common resources among UEs as well as among logical channels. This is assisted by the QoS profile of each logical channel. The eNodeB needs to be aware of the queue or data available for transmission. In the case of the downlink, the eNodeB already has the information of the buffered data waiting for transmission. In the case of the uplink, the UE notifies when it has data pending in its buffer as will be explained in Section 19.2.2.1. The scheduler allocates appropriate resources for all buffered data among UEs and among logical channels depending on the QoS profile of the logical channels. The apportionment of resources among logical channels will be explained in Section 19.3. The scheduler also considers services with specific priorities, for example, emergency calls.

4. *Dynamic or semipersistent scheduling:* The scheduler employs either dynamic or semipersistent resource scheduling in both uplink and downlink as will be explained in Section 19.2.

5. *Modulation and Coding Scheme (MCS) and Transport Block Size (TBS):* The scheduling decision depends on the MCS used on the allocated resources to achieve a certain data rate. Adaptive Modulation and Coding (AMC) may be used for link adaptation. A particular MCS on a particular number of allocated resource blocks in the subframe corresponds to a particular transport block size. If the number of allocated resource blocks increases for a fixed MCS and a fixed transmit power is available, the required code rate decreases and the coding gain increases.

6. *Localized or distributed scheduling in downlink:* The scheduler employs either localized or distributed scheduling in the downlink:
    - *Localized scheduling:* The localized scheduling uses contiguous blocks of subcarriers. It attempts to make use of the best transmission band. Thus, it achieves Frequency Selective Scheduling (FSS). However, the localized scheduling may not provide good frequency diversity. Also, in this case, a low rate user may block the required channel allocation for a high rate user, especially if channel-dependent scheduling is used. The FSS

is performed typically based on subband CQI reporting. The localized scheduling is suitable for large payloads and when channel dependent scheduling is intended. It is typically suitable for low-speed UEs.

- *Distributed scheduling:* The distributed scheduling spreads the physical resource blocks across the subcarriers in the frequency domain. It attempts to maximize the frequency diversity. Thus, it achieves Frequency Diverse Scheduling (FDS). In this case, a low rate user does not block required channel allocation for a high rate user if channel-dependent scheduling is used. However, in this case, the channel estimation is degraded, especially for very large repetition factors.

  The FDS is typically performed based on wideband CQI reporting. It can be used when the proper channel information for different frequencies is not available. Thus, the resource allocation is typically initiated using the FDS, and later the FSS or FDS is chosen. Also, the FDS is suitable for very high-speed users because the high Doppler causes the channel information for different frequencies to vary before it can be used. The FDS can be preferred to the FSS when wideband MCS becomes close to the subband MCS or there are resource blocks available at other frequencies.

7. *Frequency hopping in uplink:* In the case of the uplink, the single carrier properties of the SC-FDMA must be maintained. Thus, all the resource blocks allocated to a single user are required to be contiguous in frequency. Therefore, except SRS, all uplink data transmissions use localized transmission on contiguous subcarriers. In order to achieve the FDS and the average interference level for the PUSCH, frequency hopping may be applied in the uplink.

8. *Power or bandwidth limitation:* The scheduling decision accounts for a power or bandwidth limitation for a particular scenario. The cell edge users may be power limited, whereas the users close to the eNodeB may be bandwidth limited. The cell edge users suffer from greater path loss in the signal and there also may be higher intercell interference around the cell edge. Thus, the cell edge users experience low SINR and may not be able to operate at a high data rate. The available transmission power may be insufficient to improve SINR. Thus, the improvement of their data rate can be power limited. This may even limit the coverage area. The uplink transmission is more likely to be power limited than the downlink transmission because of the lower power availability at the UE compared to the eNodeB. In the case of power limitation, a wider bandwidth allocation may further aggravate the SINR and there may not be significant improvement in the data rate. This is because, assuming that fixed power is available, the transmit power per resource block decreases as the bandwidth increases. Moreover, noise bandwidth increases for wider bandwidth allocation.

   The UEs close to the eNodeB may experience sufficient power capabilities in both downlink and uplink. Thus, an increased bandwidth allocation can result in an increased data rate. The improvement of the data rate can be assumed to be bandwidth limited here.

   The UE sends power headroom reports to notify the eNodeB how much more or less power it is capable of transmitting as explained in Section

6.2.1.2. These reports assist the eNodeB to allocate uplink resources among different UEs appropriately. The eNodeB may not allocate more uplink resources to a UE than what the UE's transmit power capability supports.

9. *TTI bundling in uplink:* The uplink scheduling depends on whether TTI bundling is used or not. The TTI bundling will be explained in Section 19.2.2.4.
10. *UE capabilities:* The scheduling depends on the capabilities of the UE. For example, the scheduler must consider that a category 4 UE cannot support 64-QAM in the uplink.
11. *Measurement gaps:* The scheduling avoids the measurement gaps configured for interfrequency and inter-RAT measurements of the UE.
12. *Limitation of scheduling in a TTI:* The number of UEs that can be scheduled in a TTI has a maximum limit, as described next. However, this does not provide the limit of the total number of UEs scheduled because the UEs can also be time multiplexed, scheduling different UEs on various TTIs:
    - *Number of resource blocks:* Since a pair of resource blocks located in the two slots of a subframe is the basic unit for scheduling, the number of resource blocks in the frequency can limit the number of UEs scheduled in the subframe. For example, there are 25 resource blocks available in a 5-MHz channel bandwidth. Thus, 25 UEs can be scheduled in a TTI at the maximum. It may happen that the implementation uses more than one pair of resource blocks as the minimum scheduling unit, for example, with Resource Block Group (RBG) size more than 1. In this case, the maximum number of UEs scheduled in a TTI will be even fewer. In addition, in uplink, the PUCCH region takes up a part of the bandwidth, leaving the rest for the PUSCH.
    - *PDCCH signaling:* The limitation in scheduling can limit the capacity. The dynamic scheduling requires the PDCCH signaling for downlink or uplink allocation in every TTI. Therefore, the available PDCCH resources can limit the maximum possible scheduling in the TTI. Such limitation is exemplified here.

      The example assumes a 5-MHz channel bandwidth using three OFDM symbols in the control region. It also assumes absence of the MBSFN subframes. There are 300 subcarriers on 25 resource blocks, so there are $300 \times 3 = 900$ resource elements in the control region. The control region can be used as follows:
        - *Cell-specific reference signals:* Assuming the use of two antenna ports, the cell-specific reference uses one out of every three resource elements on the first symbol. Thus, the cell-specific reference signal takes up $300/3 = 100$ resource elements.
        - *PCFICH:* The PCFICH instance takes up 16 resource elements on the first symbol.
        - *PHICH:* Considering a value of 1 for PHICH-Resource, $N_g$, and normal cyclic prefix, the number of the PHICH groups is derived from the relation shown in Section 5.1.4.

$$\text{CEIL}\left[\frac{1}{8} \times N_g \times N_{RB}^{DL}\right]$$
$$= \text{CEIL}\left[\frac{1}{8} \times 1 \times 25\right] = 4$$

- Four PHICH groups require 4 × 3 × 4 = 48 resource elements.
- *PDCCH:* The PDCCH may use the resource elements left after these allocations in the control region. This shows the availability of 900 − 100 − 16 − 48 = 736 resource elements. This allows accommodation of 20 CCEs since each CCE consists of 36 resource elements. Assuming the use of only PDCCH format 1, there can be 10 PDCCH instances. Now, considering both downlink and uplink allocations in the 10 PDCCH instances, five UEs can be simultaneously scheduled in the subframe at the maximum. The number of UEs scheduled in the example would be further smaller if PDCCH allocations were used with common search spaces for purposes such as TPC command and indications for system information, paging, and random access response.

13. *Antenna configuration:* The scheduling decision can depend on the use of antenna configuration and transmission mode. For example, when beamforming is used intermittently, the CQI reporting of the UEs may become less accurate because of intermittent interference in the adjacent cells. This is known as the flashlight effect and can reduce throughput in the neighboring cells. This problem can be effectively minimized through careful scheduling in the frequency domain.
14. *Intercell interference coordination (ICIC):* The scheduling depends on the implementation of ICIC among neighboring cells, as will be shown in Section 25.1.
15. *Load balancing among neighboring cells:* The scheduling depends on load balancing among neighboring cells, as will be shown in Section 25.2.
16. *Backhaul support:* In some cases, the backhaul available at the eNodeB may be quite limited and needs to be considered in scheduling decision.

## 19.2 Resource Allocation Procedure

The procedure for resource allocation in uplink and downlink is explained in this section. Both uplink and downlink resource scheduling can be either dynamic or semipersistent:

*Dynamic scheduling*
The dynamic scheduling allocates uplink and downlink resources subframe by subframe. It requires separate signaling on the PDCCH for allocation in each subframe. The dynamic scheduling is suitable for large and bursty payloads, so it conforms well to TCP, which is used in most applications. The CRC of the PDCCH is scrambled by the C-RNTI of the UE, allowing the UE to detect the PDCCH instance intended for it.

*Semipersistent scheduling*

The semipersistent scheduling allocates downlink or uplink resources for a number of subframes. It sets up a regular allocation with fixed MCS and the allocation persists until it is changed or deactivated. Thus, allocation in each subframe is not individually signaled and the control signaling overhead on PDCCH is sufficiently reduced. The semipersistent scheduling is suitable for services with small, predictable, and periodic payloads, for example, VoIP. In such cases, the control signaling required for dynamic scheduling would be too large compared to the amount of user data transmitted. The semipersistent scheduling is required as an alternative solution of circuit switching (CS), although semipersistent scheduling is not completely free from control signaling such as CS. The CRC of the PDCCH is scrambled by semipersistent scheduling C-RNTI (SPS C-RNTI) when semipersistent scheduling is used. The UE differentiates between dynamic scheduling and semipersistent scheduling checking if the CRC is scrambled by C-RNTI or SPS C-RNTI.

The RRC layer of the eNodeB first configures different parameters for semipersistent scheduling using the SPS-Config IE. Appendix A.2 shows how the eNodeB sends the SPS-Config IE to the UE while setting up or modifying radio bearers. The SPS-Config IE includes SemiPersistSchedC-RNTI field, which provides the SPS C-RNTI assigned to the particular UE.

The SPS-Config IE includes SPS-ConfigDL IE and SPS-ConfigUL IE for the configuration of the downlink and uplink semipersistent scheduling, respectively. The SPS-ConfigDL IE and SPS-ConfigUL IE include fields to activate or deactivate downlink and uplink semipersistent scheduling, respectively. The SPS-ConfigDL IE and SPS-ConfigUL IE include SemiPersistSchedIntervalDL and SemiPersistSchedIntervalUL fields that give the interval between successive semipersistent allocations in number of subframes for downlink and uplink, respectively. The interval can be 10, 20, 32, 40, 64, 80, 128, 160, 320, or 640 subframes.

The SPS-ConfigDL IE includes the NumberOfConfSPS-Processes field, which gives the maximum number of parallel HARQ processes allowed for downlink semipersistent scheduling. Its value can be up to 8, indicating 8 HARQ processes at the maximum. The SPS-ConfigDL IE also includes the N1Pucch-AN-Persistent field, which gives the resource index, n(1)PUCCH, indicating the PUCCH region for transmission of HARQ ACK/NACK. The UE sends ACK/NACK on the PUCCH on subframe $n + 4$ for downlink transmission on subframe $n$.

After this RRC configuration, signaling on the PDCCH notifies the initiation of semipersistent scheduling. Then the downlink transmission starts and keeps taking place at regular intervals without any further signaling. A dynamic scheduling can also be made while semipersistent scheduling is running. Thus, there can be a dynamically scheduled resource in between two semipersistently scheduled resources. Also, if resources are scheduled dynamically where a semipersistent transmission is due, then the dynamic scheduling overrides the semipersistent scheduling.

### 19.2.1 Downlink Allocation

The eNodeB uses DCI format 1, 1A, 1B, 1C, 1D, 2, 2A, or 2B on the PDCCH to convey the resource allocations on the PDSCH for the downlink transmission. The DCI formats use one of the following resource allocation methods:

1. Resource Allocation Type 0;
2. Resource Allocation Type 1;
3. Resource Allocation Type 2.

The DCI formats include the Resource Block Assignment field to indicate which resource blocks are allocated to the UE. The DCI formats include a few other fields to configure the resource allocation as shown in Section 5.1.3.2.

#### 19.2.1.1 Using DCI Formats 1, 2, 2A, and 2B

DCI formats 1, 2, 2A, and 2B on the PDCCH use either resource allocation type 0 or resource allocation type 1. These DCI formats include the Resource Allocation Header field, which indicates the resource allocation type. Resource allocation type 0 is indicated by value 0 and resource allocation type 1 is indicated by value 1. If the downlink bandwidth is less than or equal to 10 PRBs, then this resource allocation header does not exist and resource allocation type 0 is assumed.

In order to avoid the PDCCH overhead to schedule every resource block, a set of Virtual Resource Blocks (VRBs) of localized type is defined as RBG. The resource blocks are allocated to the UE in terms of the RBGs. Since localized virtual resource blocks are directly mapped onto Physical Resource Blocks (PRBs), an RBG, in fact, is nothing but a set of consecutive PRBs. The number of VRBs in the RBG, known as the RBG size and denoted as $P$, is made larger when the number of available resource blocks in the total system bandwidth, $N_{RB}^{DL}$, is higher, as shown in Table 19.1. This limits the increase in the PDCCH overhead for scheduling with larger system bandwidth.

*Resource Allocation Type 0*
The resource blocks are allocated to the UE from the whole frequency resource for resource allocation type 0. It can be used for FSS. The bitmap of the Resource Block Assignment field on PDCCH indicates which RBGs are allocated to the UE. The RBGs are numbered from 0 to ($N_{RBG}$ – 1) in the order of increasing frequency where $N_{RBG}$ is the total number of RBGs available. Then the bits from MSB to LSB in the bitmap correspond to RBGs from RBG0 to RBG ($N_{RBG}$ – 1). Each bit in the bitmap represents whether the corresponding RBG is allocated or not. Value 1 of the bit represents the allocation of the corresponding RBG. Thus, the bitmap of the Resource Block Assignment field is $N_{RBG}$ long.

Table 19.1  RBG Size

| Number of Resource Blocks in the System Bandwidth, $N_{RB}^{DL}$ | RBG Size (P) |
|---|---|
| ≤10 | 1 |
| 11–26 | 2 |
| 27–63 | 3 |
| 64–110 | 4 |

- If $N_{RB}^{DL}$ mod $P = 0$, then $N_{RBG} = N_{RB}^{DL}/P$, where each of the RBGs has size $P$.
- If $N_{RB}^{DL}$ mod $P > 0$, then $N_{RBG} = ceil(N_{RB}^{DL}/P)$ where $floor(N_{RB}^{DL}/P)$ number of RBGs has size $P$ and the remaining one RBGs has size $[N_{RB}^{DL} - P \times floor(N_{RB}^{DL}/P)]$ (i.e., the last RBG is smaller than $P$).

*Resource Allocation Type 1*

The resource blocks are allocated to the UE by first splitting the whole frequency resource for resource allocation type 1. It attempts for FDS. The total number of available virtual resource blocks, $N_{VRB}^{DL}$ ($= N_{RB}^{DL}$) is divided into $P$ number of subsets where $P$ is the same RBG size. One RBG from every subset is allocated one after another in the frequency resources, which creates one set. Then such a whole set from all the subsets is repeated again and again as shown in Figure 19.1. Therefore, within one repetition, $P^2$ numbers of resource blocks exist. Allocation to the UE is made from one of these subsets. From the selected subset, RBs are chosen for the UE. The Resource Block Assignment field on the PDCCH includes the following parts for this allocation:

- *Header:* It is $ceil[\log_2(P)]$ bits long field. It indicates the selected RBG subset from the $P$ number of RBG subsets.
- *Resource allocation:* It is $[ceil(N_{RB}^{DL}/P) - ceil[\log_2(P)] - 1]$ bits long field. The bitmap of this field indicates which RBs from the selected RBG subset are allocated to the UE. Then the bits from the MSB to the LSB in the bitmap correspond to VRBs in the increasing frequency. Each bit in the bitmap represents whether the corresponding RB is allocated or not. Value 1 of the bit represents the allocation of the corresponding RB.
- *Shift:* This 1-bit field indicates whether an offset is used to shift the resource allocation span or not. If the offset is applied, then the VRB with the smallest number in the selected subset is shifted by a particular value. The offset value depends on $N_{RB}^{DL}$.

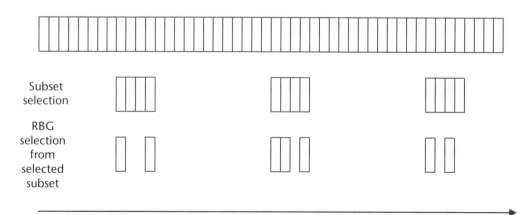

**Figure 19.1** Resource Allocation Type 1

### 19.2.1.2 Using DCI Formats 1A, 1B, 1C, and 1D: Resource Allocation Type 2

The DCI formats 1A, 1B, 1C, and 1D use only resource allocation type 2. It requires fewer overhead bits compared to resource allocation types 0 and 1. The DCI formats do not include any resource allocation header field. DCI formats 1A, 1B, and 1D can allocate either localized or distributed VRBs. Therefore, DCI format 1A, 1B, or 1D includes a 1-bit flag that indicates whether localized or distributed virtual resource blocks are assigned. Value 0 indicates a localized VRB assignment and value 1 indicates a distributed VRB assignment. However, in the case of DCI format 1C, only the distributed VRBs are assigned, so it does not include the differentiating flag.

In the case of DCI format 1A, 1B, or 1D, when localized VRBs are assigned, the Resource Block Assignment field consists of a resource indication value (RIV) that represents both the starting resource block number, $RB_{START}$, and the number of contiguously allocated resource blocks, $L_{CRBs}$, from the starting resource block. RIV is related to $RB_{START}$ and $L_{CRBs}$ as shown here:

$$RIV = N_{RB}^{DL}(L_{CRBs} - 1) + RB_{START} \text{ if } L_{CRBs} - 1 \leq \text{floor}(N_{RB}^{DL}/2)$$
$$RIV = N_{RB}^{DL}(N_{RB}^{DL} - L_{CRBs} + 1) + (N_{RB}^{DL} - 1 - RB_{START}) \text{ if } L_{CRBs} - 1 > \text{floor}(N_{RB}^{DL}/2)$$

This distributed virtual resource blocks were explained in Section 3.2.2.2. The virtual resource blocks are mapped on physical resource blocks in a distributed manner in the frequency and provide good frequency diversity. For uniform distribution, interleaving is performed between the VRB numbers and the physical resource block numbers. The procedure involves the use of a specific gap value which depends on the system bandwidth. If $N_{RB}^{DL} \geq 50$, then two gap values are specified and a 1-bit part of the Resource Block Assignment field indicates which one between the two is used.

### 19.2.2 Uplink Allocation

The eNodeB uses DCI format 0 on the PDCCH to convey the resource grants on the PUSCH for the uplink transmission of the UE, and it is explained in this section. The eNodeB also allocates resources on the PUSCH for the UE using random access response grant during random access. This was explained in Section 6.2.2.2.

The number of allocated PRBs is made from multiples of 2, 3, or 5 for a given user in a given subframe in order to allow DFT implementation. The resource blocks on the PUSCH need to be contiguous in frequency in order to maintain the single carrier properties. Nevertheless, frequency diversity can be optionally achieved using frequency hopping. The allocated resources hop to different frequencies with time and thus achieves frequency diversity maintaining the single carrier properties. The DCI format 0 includes a 1-bit hopping flag to indicate whether frequency hopping is enabled or not. The UE performs frequency hopping if this flag is set to 1 and it does not perform frequency hopping if the flag is set to 0.

The DCI format 0 also includes the Resource Block Assignment and Hopping Resource Allocation field, which contains the allocated resource information and hopping information for the PUSCH. It uses ceil[(log$_2$ ($N_{RB}^{UL}$($N_{RB}^{UL}$ + 1)/2)] number of bits. When there is uplink transmission due on the PUSCH but there is no PDCCH indication for the same transport block, for example, in the case of subsequent transmissions of persistent scheduling and HARQ retransmissions, the UE uses the frequency hopping information of the initial resource grant.

The other fields in DCI format 0 used to configure the resource allocation were explained in Section 5.1.3.2. The eNodeB semistatically configures some parameters related to the uplink resource allocation using the PUSCH-ConfigCommon IE. Appendix A.1 shows how the eNodeB broadcasts the PUSCH-ConfigCommon IE on SIB type 2. Appendix A.3 shows how the source eNodeB sends the PUSCH-ConfigCommon IE to the UE for use at the target eNodeB when handover takes place. The PUSCH-ConfigCommon IE includes the Enable64QAM field, which indicates whether 64-QAM is allowed for PUSCH or not.

All UEs in the cell use the same length for the cyclic prefix. The eNodeB configures if the UE would use a normal or extended cyclic prefix in uplink using UL-CyclicPrefixLength field. Appendix A.1 shows how the eNodeB broadcasts the UL-CyclicPrefixLength field on SIB type 2. Appendix A.3 shows how the source eNodeB sends the UL-CyclicPrefixLength field to the UE for use at the target eNodeB when handover takes place. The UL-CyclicPrefixLength field can have values len1 or len2 where len1 and len2 indicate the use of normal and extended cyclic prefixes, respectively.

#### 19.2.2.1 Seeking Resources

When a UE has data available for transmission in the RLC or PDCP layer and it needs uplink resources allocation, it may notify the eNodeB seeking the allocation in the following ways:

1. *Buffer Status Report (BSR):* If there is already PUSCH resources allocated for uplink transmission, the UE sends BSRs on PUSCH in order to report pending data status in its uplink buffers. For this purpose, the UE includes BSR MAC control element in the MAC PDU as was explained in Section 6.2.1.2.
2. *Scheduling Request (SR):* When there is no ongoing uplink data transfer on the PUSCH, the UE sends an SR on the PUCCH seeking resources provided it has the PUCCH resource allocation. An SR may also be sent after the transmission of a regular BSR when the PUSCH is allocated. The SR is not sent when there is measurement gap. The SR has 1-bit information 1 or 0 indicating whether resources have been sought or not. The UE sends the SR using the PUCCH Format 1 explained in Section 5.2.2.1. If resources on the PUSCH are not allocated, the UE keeps sending SR in every TTI until the transmission reaches a maximum limit. The SchedulingRequestConfig IE includes the DSR-TransMax field, which provides the maximum number of transmission and it can be 4, 8, 16, 32, or 64. Appendix A.2 shows how the eNodeB sends the SchedulingRequestConfig IE to the UE while setting up or modifying radio bearers.

3. *Random access procedure:* There is no ongoing uplink data transfer and there are no PUSCH resources allocated. Additionally, if the UE does not have the PUCCH resources for sending an SR or if the UE does not have its uplink synchronized, then the UE goes through contention-based random access procedure and sends a request for uplink resources allocation on PRACH as explained in Section 11.2. As shown in Section 10.2.2, the UE loses uplink synchronization when the TimeAlignmentTimer expires.

### 19.2.2.2 Resource Allocation Without Frequency Hopping

The DCI format 0 simply provides the allocation of contiguous resource blocks without frequency hopping. The $ceil[(\log_2 (N_{RB}^{UL}(N_{RB}^{UL} + 1)/2)]$ bits of the Resource Block Assignment and Hopping Resource Allocation fields consist of the resource indication value (RIV), which represents both the starting resource block number, $RB_{START}$, and the number of contiguously allocated resource blocks, $L_{CRBs}$, from the starting resource block. RIV is related to $RB_{START}$ and $L_{CRBs}$ as follows.

$$RIV = N_{RB}^{UL}(L_{CRBs} - 1) + RB_{START} \text{ if } L_{CRBs} - 1 \leq \text{floor}\left(N_{RB}^{UL}/2\right)$$
$$RIV = N_{RB}^{UL}(N_{RB}^{UL} - L_{CRBs} + 1) + (N_{RB}^{UL} - 1 - RB_{START}) \text{ if } L_{CRBs} - 1 > \text{floor}\left(N_{RB}^{UL}/2\right)$$

### 19.2.2.3 Resource Allocation with Frequency Hopping

The frequency hopping is performed at slot boundaries or at subframe boundaries. Transmission in one slot or subframe uses one set of contiguous subcarriers and the next slot or subframe uses a different set of contiguous subcarriers. Thus, it still maintains the single carrier properties. Depending on slot or subframe boundary, it can be intersubframe hopping or intrasubframe hopping. The PUSCH-Config-Common IE includes the HoppingMode field that indicates if only intersubframe hopping is supported or both intrasubframe hopping and intersubframe hopping are supported.

To indicate the offset between hopped allocations, the PUSCH-ConfigCommon IE includes the PUSCH-HoppingOffset field. This field gives an offset, $N_{RB}^{HO}$, in number of resource blocks. Its value can be any integer up to 98.

The frequency hopping can be classified as follows:

- Type 1: It explicitly indicates the hopping pattern.
- Type 2: It uses a predefined hopping pattern.

The UE first picks the $N_{UL\_hop}$ number of MSB bits from the $ceil[(\log_2 (N_{RB}^{UL}(N_{RB}^{UL} + 1)/2)]$ bits of the Resource Block Assignment and Hopping Resource Allocation fields. As shown in Table 19.2, the value of $N_{UL\_hop}$ can be 1 or 2 depending on the number of resource blocks in the system bandwidth. This $N_{UL\_hop}$ number of bits indicates if Type 1 or Type 2 is configured.

**Table 19.2** Hopping Parameters

| $N_{RB}^{UL}$ | $N_{UL\_hop}$ | $N_{UL\_hop}$ bits | Hopping Type | $n$ |
|---|---|---|---|---|
| Less than 50 | 1 | 0 | 1 | $[floor(N_{RB}^{PUSCH}/2) + RB_{START} - N_{RB}^{HO\_NEW}/2] \mod N_{RB}^{PUSCH}$ |
|  |  | 1 | 2 | N/A |
| 50 or more | 2 | 00 | 1 | $[floor(N_{RB}^{PUSCH}/4) + RB_{START} - N_{RB}^{HO\_NEW}/2] \mod N_{RB}^{PUSCH}$ |
|  |  | 01 | 1 | $[-floor(N_{RB}^{PUSCH}/4) + RB_{START} - N_{RB}^{HO\_NEW}/2] \mod N_{RB}^{PUSCH}$ |
|  |  | 10 | 1 | $[floor(N_{RB}^{PUSCH}/2) + RB_{START} - N_{RB}^{HO\_NEW}/2] \mod N_{RB}^{PUSCH}$ |
|  |  | 11 | 2 | N/A |

*Type 1 Hopping*

If both intrasubframe hopping and intersubframe hopping are supported, then the two slots of a subframe belong to two separate resource blocks with an offset between them. After picking $N_{UL\_hop}$ number of MSB bits, the UE uses the rest $[ceil[(\log_2(N_{RB}^{UL}(N_{RB}^{UL}+1)/2)] - N_{UL\_hop}]$ number of bits for the information of the first slot of the allocated resource subframe. These bits consist of a Resource Indication Value (RIV) which represents the starting resource block, $RB_{START}$, and the length of contiguously allocated resource blocks, $L_{CRBs}$. The $RB_{START}$ is used to locate the lowest index resource block for the first slot and $L_{CRBs}$ gives the length of contiguously allocated resource blocks for the first slot. Then the lowest index resource block for the second slot is given by $n + N_{RB}^{HO\_NEW}/2$ where the value of $n$ is shown in Table 19.2. $L_{CRBs}$ gives the length of contiguously allocated resource blocks for the second slot.

Here,

$$N_{RB}^{HO\_NEW} = N_{RB}^{HO} \quad \text{if } N_{RB}^{HO} \text{ is an even number}$$
$$N_{RB}^{HO\_NEW} = N_{RB}^{HO} + 1 \quad \text{if } N_{RB}^{HO} \text{ is an odd number}$$
$$N_{RB}^{PUSCH} = N_{RB}^{UL} - N_{RB}^{HO\_NEW} - (N_{RB}^{UL} \mod 2)$$

Alternatively, if the HoppingMode indicates that only intersubframe hopping is supported, then both slots in the subframe belong to the same resource block. In this case, the resource allocations explained above for the first slot and the second slot are applied to consecutive transport blocks instead.

*Type 2 Hopping*

A particular predefined pseudo-random frequency hopping pattern is applied. The bandwidth available for the PUSCH is split into $N_{sb}$ number of subbands of equal size. The PUSCH-ConfigCommon IE includes the N-SB field, which gives the value of $N_{sb}$. The value can be 1, 2, 3, or 4. Then the hopping pattern is determined based on the number of subbands $N_{sb}$, $N_{RB}^{HO}$ and whether only intersubframe hopping is supported or both intrasubframe hopping and intersubframe hopping are supported. The UE hops at slot boundaries or at subframe boundaries among the subbands. The resource allocation must be less than or equal to the size of the subband.

Mirroring is applied in alternate turns. When mirroring is applied, the allocation of the resource block in the subband is taken from the other direction.

### 19.2.2.4 TTI Bundling

The UEs far away from the eNodeB suffer from greater path loss in the signal, and also there may be higher intercell interference around the cell edge. The weak signal and strong cochannel interference result in low SINR for the cell edge users and they may not be able to operate at a high data rate. The improvement of the data rate can be especially power limited in uplink because of the lower power availability at the UE compared to the eNodeB. Thus, the coverage area can turn out to be limited. Therefore, TTI bundling is introduced in the uplink to improve the coverage. In this case, the same transport block is transmitted repeatedly in four consecutive TTIs. These four TTIs are called a *TTI bundle* and are treated as a single resource. The TTI bundling is applied when it appears that the single transmission of a transport block would not have acceptable error rate. The retransmissions of the transport block are made in consecutive subframes within the TTI bundle without waiting for any ACK/NACK feedback. A single uplink resource grant on DCI format 0 allocates the whole TTI bundle. The resource allocation is limited to three resource blocks for TTI bundling. The TTI bundling always uses QPSK for modulation.

The RRC layer of the eNodeB activates and deactivates TTI bundling using the TTIBundling field. Appendix A.2 shows how the eNodeB sends the TTIBundling field to the UE while setting up or modifying radio bearers. The TTIBundling field can be true or false indicating the activation or deactivation of TTI bundling, respectively. If periodic CQI/PMI/RI reporting is due in a subframe that has TTI bundling, the UE drops the periodic CQI/PMI/RI report. The HARQ operation procedure with TTI bundling will be explained in Section 20.3.6.

## 19.3 Apportionment of Resources Among Logical Channels

The MAC layer is responsible for apportionment of the available resources among the radio bearers or logical channels for a particular UE in both downlink and uplink. There are individual RLC buffers for different logical channels. In the case of a downlink, the MAC layer of the eNodeB performs the apportionment of the resources with priority among logical channels depending on the QoS profile of different radio bearers and the inclusion of MAC control elements if there are any. The QoS profile is explained in Section 2.4.2. The actual algorithm for this prioritization depends on implementation at the eNodeB. In the case of uplink, the RLC buffers for different logical channels are located at the UE, and the eNodeB does not directly control the apportionment of the resources as it would require high overhead. The eNodeB instead indicates relative priority among the logical channels depending on their QoS profile and the MAC layer of the UE attempts to apportion the available resources among the logical channels in the optimum way. In addition, the UE may need to add MAC control elements to MAC PDUs that require the allocation of resources. The UE includes data from a logical channel and MAC control elements in the MAC PDU with the following relative priority:

1. MAC control element for the C-RNTI or data from the UL CCCH;
2. MAC control element for regular BSR or periodic BSR;
3. MAC control element for PHR;
4. Data from the logical channel with relative priority, except data from the UL CCCH, explained later;
5. MAC control element for padding BSR.

The RRC layer at eNodeB configures the prioritization of the logical channels by sending the LogicalChannelConfig IE for each of the logical channels. Appendix A.2 shows how the eNodeB sends the LogicalChannelConfig IE. The LogicalChannelConfig IE includes the Priority field, which assigns the priority level to a logical channel. The Priority field can be any integer between 1 and 16 and a higher value of the Priority field indicates a lower priority level. The UE allows the logical channel of the highest priority to be first included into the MAC PDU, and, similarly, it includes other logical channels one after another according to their priority. The priority-based allocation may keep depriving a logical channel with low priority if insufficient resource allocation continues. Therefore, the eNodeB can ensure a particular minimum amount of allocation for a logical channel configuring the prioritized Bit Rate (PBR). The LogicalChannelConfig IE includes the PrioritizedBitRate field, which assigns the value of the PBR. Its value can be 0, 8, 16, 32, 64, 128, or 256 kbps or infinity. The UE allocates resources for a logical channel in a subframe in two steps:

1. The UE allocates resources for logical channels in decreasing order of priority, but it allocates only an amount of resources that corresponds to the PBR. This allocated amount, $B$, for a particular logical channel is set to zero when the logical channel is established. Thereafter, $B$ is incremented by $PBR \times TTI = PBR \times 1$ ms in each subframe and the amount $B$ is used for initial allocation in the subframe. If the PBR is infinity for a logical channel, the UE allocates resources for all the data available for the logical channel before it allocates resources according to the PBR of a lower priority logical channel.
2. After including all logical channels in the MAC PDU meeting their prioritized bit rate, the leftover space of the MAC PDU, if there is any, is allocated again in a decreasing order of priority. Once the minimum requirement has been fulfilled for all logical channels, a logical channel with higher priority is allowed to put all of its data in the leftover space if there is any. Thus, the logical channel with next lower priority will get this chance only if there is still room in the MAC PDU. If there are logical channels with equal priority, the UE attempts to serve them equally.

After complete allocation for the logical channel in the TTI, $B$ is decremented by the total allocation in the TTI and this $B$ is forwarded for calculation in the next TTI. Thus, it may happen sometimes that the allocation in a step makes $B$ negative.

The eNodeB sets a maximum limit of $B$ by configuring the Bucket Size Duration (BSD). The LogicalChannelConfig IE includes the BucketSizeDuration field, which assigns the value of BSD. Its value can be 50, 100, 150, 300, 500, or 1,000 ms. The maximum limit of $B$ is determined as $PBR \times BSD$. If the aforementioned calculation leads to an even higher value of $B$, then $B$ is set to this maximum limit.

# CHAPTER 20
# Basic Activities During Data Transfer

The data transfer in downlink involves a few basic steps:

1. *Feedback from the UE:* The UE sends the Channel Quality Indicator (CQI), Precoding Matrix Indicator (PMI), and Rank Indication (RI) to the eNodeB on the PUCCH if there is no uplink data transfer and on the PUSCH if there is ongoing uplink data transfer. This is explained in Section 20.1. The CQI, PMI, and RI are sent on the High-Speed Dedicated Physical Control Channel (HS-DPCCH) in the case of the HSPA.
2. *Allocation of resources:* Based on the feedback from the UE, the eNodeB allocates radio resources on the PDSCH for downlink transmission. The eNodeB uses PDCCH signaling to inform the UE about the allocated resource blocks and configuration used in data transmission. The configuration information includes the Modulation and Coding Scheme (MCS) used, the Transmitted Rank Indication (TRI) indicating the rank used, and the Transmitted Precoding Matrix Indicator (TPMI) indicating the codebook index used from the codebook. The CRC of the PDCCH instance is normally scrambled by the C-RNTI or SPS C-RNTI so that the UE can identify its PDCCH instance for dynamic or semipersistent scheduling, respectively. The RNTI is an RA-RNTI when the eNodeB is going to send a Random Access Response (RAR) message. It is a temporary C-RNTI during random access. The resource allocation information is sent on the High-Speed Shared Control Channel (HS-SCCH) in the case of HSPA.
3. *Data transfer:* The eNodeB sends user data on the PDSCH. The data is sent on the High-Speed Physical Downlink Shared Channel (HS-PDSCH) in the case of HSPA.
4. *Link adaptation:* Signal transmission parameters are changed with time and frequency in order to better adjust to the changing radio link conditions. The set of techniques used for this purpose is known as the link adaptation techniques, explained in Section 20.2.
5. *HARQ ACK/NACK:* The UE attempts to decode the received packets. The UE sends HARQ ACK or NACK based on its success or failure to decode the packets, respectively. The HARQ ACK/NACK is sent on the PUCCH if there is no uplink data transfer and on the PUSCH if there is ongoing uplink data transfer. The eNodeB also changes the New Data Indicator (NDI)

on the PDCCH accordingly, explained in Section 20.3. The HARQ ACK/NACK is sent on the HS-DPCCH in the case of HSPA.

The data transfer in uplink involves a few basic steps:

1. *Seeking resource allocation:* When the UE has data available for transmission and it needs uplink resources allocation, it notifies the eNodeB seeking resources as explained in Section 19.2.2.1. The UE sends buffer status reports (BSRs) on the PUSCH if there is no ongoing data transfer on the PUSCH and it sends a scheduling request (SR) on the PUCCH if there is no ongoing data transfer over the PUSCH. In the case of HSPA, scheduling information (SI) is sent on the E-DCH dedicated physical data channel (E-DPDCH).
2. *Allocation of resources:* The eNodeB informs the UE about the allocated resource blocks on the PUSCH and the configuration for data transmission using DCI format 0 on the PDCCH. The configuration includes the MCS indicating transport format. The CRC of the PDCCH instance is scrambled by the C-RNTI or the SPS C-RNTI so that the UE can identify its PDCCH instance for dynamic or semipersistent scheduling, respectively. The RNTI is temporary C-RNTI during random access. Since the eNodeB fully commands the transport format to be used by the UE in uplink, unlike UMTS, the UE does not need to send any indication about the transport format used. The resource allocation information is sent on the E-DCH absolute grant channel (E-AGCH) and the E-DCH relative grant channel (E-RGCH) in the case of HSPA.
3. *Data transfer:* The UE sends user data over the allocated resources on the PUSCH. The data is sent on the E-DCH dedicated physical data channel (E-DPDCH) in the case of HSPA.
4. *Link adaptation:* Link adaptation techniques are applied to adapt the signal transmission parameters to the changing radio link conditions, explained in Section 20.2.
5. *HARQ ACK/NACK:* The eNodeB attempts to decode the received packets and sends HARQ ACK/NACK on the PHICH indicating success or failure in decoding. The eNodeB also changes the NDI on PDCCH accordingly. This is explained in Section 20.3. The HARQ ACK/NACK is sent on the E-DCH hybrid ARQ indicator channel (E-HICH) in the case of HSPA.

## 20.1 Feedback from the UE

The UE sends four types of feedback to the eNodeB in order to support downlink transmission of data:

1. *CQI:* The UE sends CQI feedback as an indication of the data rate that can be supported by the downlink channel. This helps the eNodeB to select the modulation scheme and the code rate for downlink transmission. The eNodeB may also use the CQI from UEs in determination of the downlink power level. The UE determines the CQI to be reported based on measurements of the downlink reference signals. A number of CQI indexes are

defined as shown in Table 20.1. The UE selects a CQI based on the estimation that the MCS corresponding to the CQI index shown in Table 20.1 is the highest possible MCS that will allow the UE to decode transport blocks with an error rate probability not exceeding 10%.

The reported CQI is not actually a direct indication of the downlink channel quality. A UE with a receiver of better quality can report a better CQI for the same downlink channel quality and can receive downlink data with a higher MCS. Thus, the CQI report indicates the downlink channel quality while taking the capabilities of the UE's receiver into account.

2. *PMI:* The UE sends the PMI to help the eNodeB select the appropriate precoding matrix from the codebook as explained in Chapter 18.
3. *RI:* The UE sends the RI to the eNodeB indicating its preferred rank to be used in the downlink as explained in Chapter 18. A common rank is indicated for all resource blocks.
4. *HARQ ACK/NACK:* This is explained in Section 20.3.

The UE always sends CQI reports for all transmission modes of the PDSCH. The requirement of PMI or RI reporting depends on the transmission mode of the PDSCH as shown in Table 18.1.

For low-speed users, the CQI/PMI/RI reports remain valid for longer durations, and thus, the downlink performance becomes better. The eNodeB configures a set of sizes and formats of the CQI/PMI/RI reports depending on the type of the report using CQI-ReportConfig IE. Appendix A.2 shows how the eNodeB sends CQI-ReportConfig IE to the UE while setting up or modifying radio bearers. The UE sends CQI/PMI/RI reports depending on how the reporting is configured.

The CQI/PMI/RI reporting can represent the whole downlink bandwidth or a part of it. When the reporting is used for partial bandwidth, separate feedback

Table 20.1  CQI Index

| CQI Index | Modulation | Code Rate × 1,024 | Efficiency |
|---|---|---|---|
| 0 | No transmission | | |
| 1 | QPSK | 78 | 0.1523 |
| 2 | QPSK | 120 | 0.2344 |
| 3 | QPSK | 193 | 0.3770 |
| 4 | QPSK | 308 | 0.6016 |
| 5 | QPSK | 449 | 0.8770 |
| 6 | QPSK | 602 | 1.1758 |
| 7 | 16QAM | 378 | 1.4766 |
| 8 | 16QAM | 490 | 1.9141 |
| 9 | 16QAM | 616 | 2.4063 |
| 10 | 64QAM | 466 | 2.7305 |
| 11 | 64QAM | 567 | 3.3223 |
| 12 | 64QAM | 666 | 3.9023 |
| 13 | 64QAM | 772 | 4.5234 |
| 14 | 64QAM | 873 | 5.1152 |
| 15 | 64QAM | 948 | 5.5547 |

values are needed to represent different parts of the whole bandwidth, so the reporting requires more uplink resources. Each part of the bandwidth for which reporting is made is called a subband. If a subband consists of *k* number of resource blocks, then the number of subbands required to cover the whole bandwidth is $N = N_{RB}^{DL}/k$. The eNodeB can configure how the CQI/PMI/RI reporting would be associated with the frequency range and it can occur in the following ways:

1. *Wideband reporting:* The reported CQI represents to the whole downlink bandwidth;
2. eNodeB configured subband reporting;
3. UE selected subband reporting.

If the UE reports CQI/PMI/RI more frequently in time, then it can represent the variation of downlink radio channel better, but at the cost of the use of more uplink resources. The eNodeB can configure the frequency of the CQI/PMI/RI reporting in time. The CQI/PMI/RI reporting can be either periodic or aperiodic. If both periodic and aperiodic reportings are scheduled in the same subframe, the UE transmits only the aperiodic report in that subframe.

### 20.1.1 Aperiodic CQI/PMI/RI Reporting

The aperiodic CQI/PMI/RI reports are always sent on the PUSCH, and the CQI/PMI/RI report is multiplexed with the uplink data on PUSCH. The RI reporting is made only if the configured CQI/PMI/RI feedback type supports RI reporting. The minimum CQI/PMI/RI reporting interval is 1 subframe. The aperiodic reporting takes place only when the number of resource blocks in the system bandwidth is more than 7.

The eNodeB instructs the UE to send an individual CQI/PMI/RI report. The eNodeB uses the 1-bit CQI Request field of DCI format 0 on PDCCH for the instruction. If the CQI Request field is set to 1 in subframe *n*, then the UE sends an aperiodic CQI/PMI/RI report in the subframe *n* + 4 on the PUSCH. Also, the UE performs aperiodic CQI/PMI/RI reporting when the eNodeB sends a random access response with its CQI request field set to 1 in case of a noncontention-based random access procedure.

There are different reporting modes of aperiodic CQI/PMI/RI reports as shown in Table 20.2. The CQI-ReportConfig IE contains the CQI-ReportModeAperiodic field, which specifies the reporting mode. The value of this field is set to rm12 to indicate Mode 1-2, rm20 to indicate Mode 2-0, rm22 to indicate Mode 2-2, and so forth.

### 20.1.2 Periodic CQI/PMI/RI Reporting

The UE sends periodic CQI/PMI/RI report together with uplink data on the PUSCH if there is ongoing uplink data transfer and it sends periodic CQI/PMI/RI report on the PUCCH if there is no uplink data transfer. The UE uses the same CQI/PMI/RI reporting format on the PUCCH and the PUSCH. If periodic CQI/PMI/RI reporting is due in a subframe that is scheduled for transmission from the TTI bundle, the

**Table 20.2** Reporting Modes for Aperiodic CQI/PMI/RI Reporting

| Reporting Mode | Feedback Type | Transmission Mode for PDSCH | Feedback Procedure |
|---|---|---|---|
| Mode 1-2 | Wideband CQI with multiple PMIs | 4 and 6 | The UE reports the PMI for each subband assuming transmission only in the subband. The subband size is given in Table 20.3. Then the UE reports a wideband CQI value per code word assuming the use of the precoding matrix in each subband corresponding to the PMIs reported. |
| | | | For transmission mode 4, the PMI and CQI values are calculated conditioned on the reported RI. For transmission mode 6, the PMI and CQI values are calculated conditioned on rank 1. |
| Mode 2-2 | UE-selected subbands with multiple PMIs | 4 and 6 | The UE reports a wideband CQI value and a single PMI value assuming transmission on all subbands. |
| | | | Additionally, the UE selects a particular number of preferred subbands each of particular size according to Table 20.4. The UE reports a single PMI value assuming transmission on all the selected subbands. The UE also reports a single CQI value per code word assuming transmission on all the selected subbands using the single precoding matrix indicated by the reported PMI in each of the subbands. |
| | | | The UE uses a differential value for the CQI value for the selected subbands as shown in Table 20.5. |
| | | | For transmission mode 4, the PMI and CQI values are calculated conditioned on the reported RI. For transmission mode 6, the PMI and CQI values are calculated conditioned on rank 1. |
| Mode 2-0 | UE-selected subbands with no PMI | 1, 2, 3, and 7 | The UE reports a wideband CQI value assuming transmission on all subbands. |
| | | | Additionally, the UE selects a particular number each of preferred subbands of particular size according to Table 20.4. The UE also reports a single CQI value assuming transmission on all the selected subbands. |
| | | | The UE uses a differential value for the CQI value for the selected subbands as shown in Table 20.5. |
| | | | For transmission mode 3, the CQI values are calculated conditioned on the reported RI. For transmission modes 1, 2, and 7, they are calculated conditioned on rank 1. |
| Mode 3-0 | eNodeB configured subbands with no PMI | 1, 2, 3, and 7 | The UE reports a wideband CQI value assuming transmission on all subbands. The UE also reports CQI values for each subband assuming transmission on only in the particular subband. All the CQI values represent channel quality for the first code word if there are two code words. |
| | | | For transmission mode 3, the CQI values are calculated conditioned on the reported RI. For transmission modes 1, 2, and 7, they are calculated conditioned on rank 1. |
| Mode 3-1 | eNodeB configured subbands with single PMI | 4, 5, and 6 | The UE reports a wideband CQI value and a single PMI value assuming transmission on all subbands. |
| | | | The UE reports CQI values for each subband assuming transmission on only in the particular subband using the precoding matrix corresponding to the PMI reported. |
| | | | The UE uses a differential value for the CQI value for the subbands as shown in Table 20.6. |
| | | | For transmission mode 4, the CQI values are calculated conditioned on the reported RI. For transmission modes 5 and 6, they are calculated conditioned on rank 1. |

**Table 20.3** Subband Size for Aperiodic CQI/PMI/RI Reports

| Number of Resource Blocks in the System Bandwidth | Number of Resource Blocks in Each Subband |
|---|---|
| 6–7 | N/A |
| 8–10 | 4 |
| 11–26 | 4 |
| 27–63 | 6 |
| 64–110 | 8 |

**Table 20.4** Subband Size and Number of Subbands for Aperiodic CQI/PMI/RI Reports

| Number of Resource Blocks in the System Bandwidth | Number of Resource Blocks in Each Subband | Number of Subbands |
|---|---|---|
| 6–7 | N/A | N/A |
| 8–10 | 2 | 1 |
| 11–26 | 2 | 3 |
| 27–63 | 3 | 5 |
| 64–110 | 4 | 6 |

**Table 20.5** Mapping differential CQI value to offset level

| Differential CQI Value | CQI Index for the Selected Subbands: CQI Index for the Wideband |
|---|---|
| 0 | ≤1 |
| 1 | 2 |
| 2 | 3 |
| 3 | ≥4 |

**Table 20.6** Mapping subband differential CQI value to offset level

| Differential CQI Value | CQI Index for the Selected Subbands: CQI Index for the Wideband |
|---|---|
| 0 | 0 |
| 1 | 1 |
| 2 | ≥2 |
| 3 | −1 |

UE drops the periodic CQI/PMI/RI report. Also, if there is conflicting schedule of CQI/PMI/RI and a positive SR in the same subframe, then CQI/PMI/RI is dropped.

The periodic CQI/PMI/RI reporting is semistatically configured by the RRC layer. The CQI-ReportConfig IE contains the CQI-ReportPeriodic field, which sets up or releases periodic CQI/PMI/RI reporting. The CQI-ReportConfig IE also includes the SimultaneousAckNackAndCQI field, which determines the action when both CQI/PMI/RI and ACK/NACK are scheduled in the same subframe. If the SimultaneousAckNackAndCQI field is set to false, then CQI/PMI/RI is dropped. Conversely, if this field is set to true, then CQI/PMI/RI is multiplexed with ACK/NACK.

The UE-selected subband reporting on the PUCCH takes place only when the number of resource blocks in the system bandwidth is more than 7. In the case of subband reporting, the whole bandwidth is split into $N$ subbands where each subband consists of $k$ resource blocks. The $N$ number of subbands is divided into $J$ number of parts called bandwidth parts (BPs). The number of bandwidth parts, $J$, depends on the system bandwidth as shown in Table 20.7. The number of subbands in a bandwidth part is given by $N_{RB}^{DL}/k/J$ or $(N_{RB}^{DL}/k/J) - 1$ depending on the values of $N_{RB}^{DL}$, $k$, and $J$.

There are different reporting modes for periodic CQI/PMI/RI reporting as shown in Table 20.8. The CQI-ReportConfig IE includes the CQI-FormatIndicatorPeriodic IE, which indicates if the reporting is wideband or subband. In addition, the RRC layer configures the transmission mode as shown in Chapter 18. Based on this information, the UE can determine the reporting mode according to Table 20.8.

There are four PUCCH report types defined for periodic CQI/PMI/RI reporting as shown in Table 20.9 in order to allow the payload size of PUCCH instances only as much as necessary.

## 20.2 Link Adaptation

The multipath channel between the UE and the eNodeB keeps varying because of the variation in the environment and the velocity of the UE. Thus, the quality of the radio link between the UE and the eNodeB also keeps varying because of continuous variation in the signal received from the serving cell, in the level of

Table 20.7 Subband Size and Number of Bandwidth Parts for Periodic CQI/PMI/RI Reports

| Number of Resource Blocks in the System Bandwidth, $N_{RB}^{DL}$ | Number of Resource Blocks in Each Subband (k) | Number of Bandwidth Parts (J) |
| --- | --- | --- |
| 6–7 | N/A | N/A |
| 8–10 | 4 | 1 |
| 11–26 | 4 | 2 |
| 27–63 | 6 | 3 |
| 64–110 | 8 | 4 |

**Table 20.8** Reporting Modes for Periodic CQI/PMI/RI Reporting

| Reporting Modes | Feedback Type | Transmission Mode | Feedback Procedure |
| --- | --- | --- | --- |
| Mode 1-0 | Wideband CQI with no PMI | Transmission modes 1, 2, 3, and 7 | For transmission mode 3, the UE sends a PUCCH report type 3 consisting of a single RI and a PUCCH report type 4 consisting of a single CQI value assuming transmission on all subbands. The CQI is calculated conditioned on the last reported periodic RI. |
| | | | For transmission modes 1, 2, and 7, the UE sends a PUCCH report type 4 consisting of a single CQI value assuming transmission on all subbands. The CQI is calculated conditioned on rank 1. |
| Mode 1-1 | Wideband CQI with single PMI | Transmission modes 4, 5, and 6 | For transmission mode 4, the UE sends a PUCCH report type 3 consisting of a single RI and a PUCCH report type 2 consisting of single CQI and PMI values assuming transmission on all subbands. The PMI is calculated conditioned on the last reported periodic RI. The CQI is calculated conditioned on the selected PMI and the last reported periodic RI. |
| | | | For transmission modes 5 and 6, the UE sends a PUCCH report type 2 consisting of single CQI and PMI values assuming transmission on all subbands. The PMI is calculated conditioned on rank 1. The CQI is calculated conditioned on the selected PMI and rank 1. |
| Mode 2-0 | UE-selected subbands with no PMI | Transmission modes 1, 2, 3, and 7 | For transmission mode 3, the UE sends a PUCCH report type 3 consisting of a single RI and a PUCCH report type 4 consisting of a single CQI values assuming transmission on all subbands. |
| | | | For transmission modes 1, 2, and 7, the UE sends a PUCCH report type 4 consisting of a single CQI value assuming transmission on all subbands. The CQI is calculated conditioned on rank 1. |
| | | | Additionally, for all transmission modes 1, 2, 3, and 7, the UE selects a preferred subband within each of the bandwidth parts shown in Table 20.7. Then the UE sends a PUCCH report type 1 consisting of single CQI values for each of the selected subbands assuming transmission only on the respective subbands. |
| | | | For transmission mode 3, the preferred subband selection and CQI calculation are conditioned on the last reported RI. For transmission modes 1, 2, and 7, they are calculated conditioned on rank 1. |
| Mode 2-1 | UE-selected subbands with single PMI | Transmission modes 4, 5, and 6 | For transmission mode 4, the UE sends a PUCCH report type 3 consisting of a single RI and a PUCCH report type 2 consisting of single CQI and PMI values assuming transmission on all subbands. The PMI is calculated conditioned on the last reported periodic RI. |
| | | | For transmission modes 5 and 6, the UE sends a PUCCH report type 2 consisting of single CQI and PMI values assuming transmission on all subbands. |
| | | | Additionally, for all transmission modes 4, 5, and 6, the UE selects a preferred subband within each of the bandwidth parts shown in Table 20.7. Then the UE sends a PUCCH report type 1 consisting of single CQI values for each of the selected subbands assuming transmission only on the respective subbands. |
| | | | For transmission mode 4, the preferred subband selection and CQI calculation are conditioned on the last reported wideband PMI and RI. For transmission modes 5 and 6, they are calculated conditioned on the last reported PMI and rank 1. |

Table 20.9  PUCCH Report Types

| PUCCH Report Type | Reported Parameters |
|---|---|
| 1 | CQI feedback for the UE-selected subbands |
| 2 | Wideband CQI and PMI feedback |
| 3 | RI feedback |
| 4 | Wideband CQI |

interference from other cells, and in the noise level. Link adaptation refers to a set of techniques where signal transmission parameters are changed on the fly to better adjust to the changing radio link conditions. The various link adaptation techniques are explained next. A combination of these link adaptation techniques may be applied while taking the individual impact of each of the techniques into account. Channel-dependent scheduling (CDS) and adaptive modulation and coding (AMC) are generally more efficient link adaptation techniques than others.

1. *CDS:* The CDS can be applied in both time and frequency domains, explained in Section 19.1.2.
2. *AMC:* The AMC can be used for link adaptation by varying MCS in the time domain. It attempts to achieve higher data rate when the radio link is favorable. It may consider the same transmit power available. All resource blocks allocated to a particular user in one subframe use the same MCS. Thus, frequency-dependent MCS is not used, although the eNodeB may have channel information for different subbands (i.e., parts of the bandwidth). This is because the maximum data size in a resource block is small. Thus, if MCS is defined per resource block or per a small group of resource blocks, then it would give only a little improvement but at the cost turbo coding gain. Moreover, it would increase the control signaling overhead. The eNodeB selects MCS to be used for downlink and uplink transmission as follows:
   - *MCS selection for downlink transmission:* The UE sends CQI/PMI/RI reports as explained in Section 20.1. The eNodeB checks on the amount of data available for downlink transmission, CQI/PMI/RI reports, and the available resource blocks. The eNodeB makes a decision about the MCS, the number of allocated resource blocks, the rank, and the codebook index from the codebook. The actual algorithm for making this decision depends on the implementation at the eNodeB and it is not mandatory for the eNodeB to always respect the CQI/PMI/RI reports from the UE. The eNodeB may select MCS such that the modulation order and the transport block size correspond to a code rate as closest as possible to what is indicated by the CQI index. If more than one combination of modulation order and the transport block size correspond to a code rate equally close to the one indicated by the CQI index, then the combination with the smallest of such transport block sizes may be selected.
   - *MCS selection for uplink transmission:* The eNodeB can estimate uplink channel quality based on the sounding reference signal (SRS), the demodulation reference signal (DM RS), and the error rate in the current

uplink data transfer. The eNodeB checks on the amount of data available for the uplink transmission from BSRs and the available resource blocks. Then the eNodeB makes a decision about the MCS. The actual algorithm for making this decision depends on the implementation at eNodeB.

The eNodeB notifies the UE about the MCS to be used using DCI formats on PDCCH while allocating resources for both downlink and uplink transmission as explained in Section 5.1.3.2.

3. *Transmission power control:* The transmit power can be adjusted in an attempt to maintain a desired level of SINR. This can help maintain sufficient transmitted energy per bit and a certain error rate under changing channel conditions. Thus, it attempts to guarantee the required data rate and quality of service (QoS). The uplink power control was explained in Chapter 17.

4. *Multiple antenna schemes:* The multiple antenna schemes are adapted depending on the channel conditions. The multiple antenna schemes in downlink were explained in Chapter 18.

## 20.3  Hybrid Automatic Repeat Request (HARQ)

Automatic repeat request (ARQ) is an error control method for data transmission to achieve reliable data transmission. The receiver sends an ACK/NACK to indicate whether it has correctly received the packet or not. When the transmitter receives the NACK or a certain time elapses without any feedback, the transmitter retransmits the packet until it receives an acknowledgment or exceeds a predefined number of retransmissions.

HARQ is a technique combining forward error correction (FEC) and ARQ methods that save information from previous failed decoding attempts for use in future decoding. When the receiver fails to decode the data packet, it sends the NACK to the transmitter, but it keeps bits from the failed attempt for future use. When the transmitter receives the NACK or a certain time elapses without any feedback, the transmitter retransmits the transport block. The receiver HARQ process then soft-combines bits from the previous failed decodes with the currently received retransmission. This helps minimize the number of retransmissions.

The RLC acknowledged mode and the TCP {**AU: Spell out acronym?**} can also offer retransmission for error correction, but this additional error correction by HARQ is warranted because the retransmissions at lower layer require less bandwidth, overhead, and delay.

### 20.3.1  HARQ Process

The MAC layer is responsible for the HARQ process. The HARQ process uses the stop-and-wait (SAW) operation. In this case, the transmitter sends a transport block and waits for an ACK from the receiver before transmitting the next transport block. If the transmitter receives a NACK or a certain time elapses without any feedback, the transmitter retransmits the transport block. The HARQ process is fast and consumes small radio resources for feedback.

## 20.3 Hybrid Automatic Repeat Request (HARQ)

There are two types of the HARQ process. LTE uses the second type, incremental redundancy (IR).

1. *Chase combining (CC):* CC uses the transmission of a number of repeats of coded data so every retransmission contains the same information. Through a soft combination of identical code words, the SINR gets better, but the code rate remains the same. This scheme achieves a gain with a small buffer size in a receiver. The buffer size becomes the number of coded symbols of one coded packet.
2. *IR:* IR uses the transmission of additional redundant information in each retransmission so every retransmission contains different information than the previous one. The retransmissions are called redundancy versions (RVs) of the same code word. Through a soft combination of the redundancy versions, the code rate gets better, but the SINR remains the same. IR requires a larger size of buffer in a receiver than CC. The buffer size becomes the number of coded bits of total transmitted coded packets. As shown in Section 3.3, the rate matching generates four redundancy versions for the initial transmission and retransmissions.

Multiple HARQ processes are required to run in parallel in order to keep up the transmission of transport blocks while the receiver is decoding already received transport blocks. This allows the continuous use of the whole transmission resources available. Each HARQ process runs its own SAW operation.

Each HARQ process expects a HARQ round trip time (HARQ RTT) between the transmission and retransmission of a transport block when the decoding fails. This expected HARQ RTT includes the following parts as shown in Figures 20.1 and 20.2 for downlink and uplink, respectively:

1. Propagation delay for the original transport block (i.e., the time the transport block takes to reach the receiver after it is transmitted);
2. The time for processing and generation of ACK/NACK at the receiver (this duration is about 3 ms);
3. Propagation delay for the ACK/NACK;
4. The time for processing before the transmitter retransmits the transport block (this duration is about 3 ms).

It is shown that the HARQ RTT is around 8 ms, taking the aforementioned durations into account. Thus, the retransmission of the transport block may take place 8 subframes after the transmission. Therefore, in order to use the available transmission resources continuously, the maximum number of parallel HARQ processes is set to 8.

The HARQ entity in the MAC layer maintains a number of parallel HARQ processes for both uplink and downlink transmission. The HARQ entity in the MAC layer has two parts of operation in both the eNodeB and the UE:

1. Transmit HARQ operation:
   - Transmission and retransmission of transport blocks;
   - Processing of the ACK/NACK.

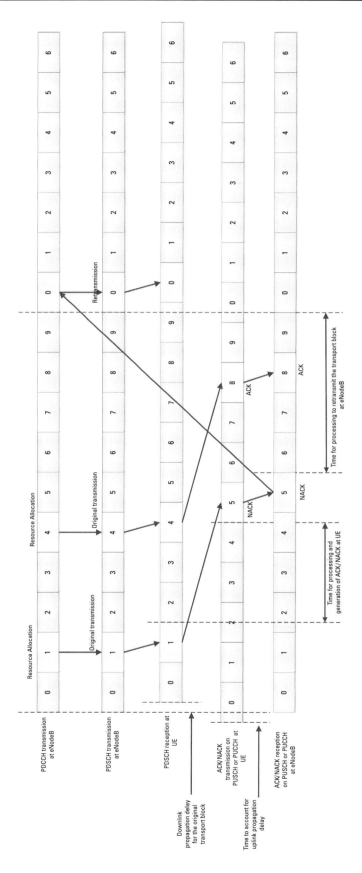

**Figure 20.1** HARQ for downlink transmission.

## 20.3 Hybrid Automatic Repeat Request (HARQ)

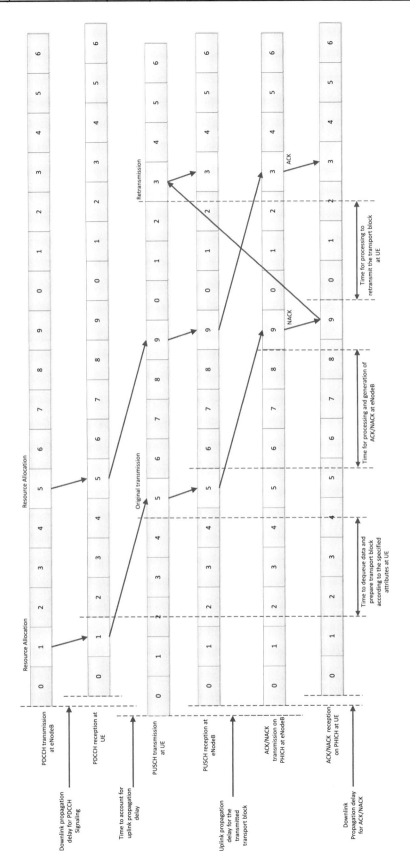

**Figure 20.2** HARQ for uplink transmission.

2. Receive HARQ operation:
   - Reception of transport blocks. The receiver buffers the transport blocks which could not be decoded in order to soft combine them with future retransmissions;
   - Soft combining the received transport blocks until the data packet is successfully decoded or aborted;
   - Generation of the ACK/NACK.

The eNodeB may configure discontinuous reception (DRX) for saving battery power in the RRC_CONNECTED state. The HARQ operation along with the DRX was explained in Section 16.2.3.

### 20.3.2 Interrelationship Between HARQ and ARQ

The RLC layer performs a sliding window–based ARQ procedure only for the acknowledged mode (AM). The ARQ and HARQ protocols run together and complement each other as described here:

1. The HARQ procedure is complemented by the ARQ procedure. The data that cannot be corrected at a lower layer through HARQ retransmissions is retransmitted by the ARQ. The feedback in HARQ has very low overhead requiring only 1 bit, so the HARQ uses fast retransmissions. Conversely, the ARQ has significant overhead for feedback as it requires a number of bits, so the ARQ uses less frequent feedback. Thus, the HARQ attempts for necessary correction of data and, if it fails, then the correction is left for the ARQ.
2. Since the HARQ uses a single uncoded bit as feedback, it may experience an incorrect decoding of its feedback. It can be misinterpreting an ACK as a NACK or misinterpreting a NACK as an ACK. The ARQ procedure at a higher layer then provides for the correction of data. In the case of the ARQ, the receiving side sends a status report indicating the missing RLC PDUs or the missing portions of RLC PDUs. The status report is a multiple bit feedback sent just like data, so it is reliable. Moreover, the MAC layer treats the status report as any other data and applies the HARQ operation to the status report. Thus, if there is any error or loss of the status report, it can be detected and recovered by the HARQ process.
3. The HARQ process on the transmitting side may reach its maximum retransmission limit in certain cases, for example, if the link adaptation techniques used are not enough strong compared to the actual radio link quality. In this case, the HARQ process on the transmitting side can notify the transmitting side of the AM RLC entity and the AM RLC entity can retransmit the AMD PDU, as shown in Section 7.3.1.4. Since both HARQ and ARQ entities are located in the eNodeB or in the UE, such interaction between them can be carried out quickly. It may be noted that such interaction between HARQ and ARQ functions was not possible in the HSPA because of implementing the RLC operation at the radio network controller (RNC) instead of NodeB.

4. The same sequence number and reception buffer are used for both the ARQ operation in the RLC layer and the HARQ operation in the MAC layer. Thus, unlike the HSPA, the HARQ reordering does not need an additional sequence number and reception buffer. Moreover, reordering is performed only at the RLC, whereas in the HSPA, reordering is performed at both the RLC and MAC layers. In addition, the two reordering operations increase delay in the HSPA. This benefit has again been possible in LTE because of locating both the HARQ and ARQ entities in the eNodeB.
5. The HARQ process treats the retransmission by the AM RLC entity as new data and thus does not attempt to combine this retransmission with any earlier transmission.
6. If an ACK in the HARQ process is misinterpreted as a NACK, then the correctly received transport block can be retransmitted creating a duplicate RLC PDU. The receiving RLC entity detects duplicate RLC PDUs, if there are any, and discards them.
7. Some HARQ processes may require repeated retransmissions while other HARQ processes may not. Thus, some transport blocks can be more delayed than the others. This may result in the reception of out-of-sequence RLC PDUs. The receiving RLC entity reorders out-of-sequence RLC PDUs.

### 20.3.3 HARQ Information on the PDCCH

The PDCCH carries the following HARQ-related information while allocating resources for downlink and uplink transmissions.

1. Each HARQ process is identified with a unique 3-bit-long HARQ process identifier (HARQ ID). The HARQ ID is explicitly signaled on the PDCCH only in the case of downlink transmission as shown in Section 5.1.3.2.
2. The incremental redundancy HARQ process has transport block to be used for transmission or retransmission, indexed with a redundancy version. The redundancy version specifies a starting point in the buffer to start reading out bits as shown in Section 3.3. The redundancy version is signaled on the PDCCH for both downlink and uplink transmissions.
3. The PDCCH includes a 1-bit New Data Indicator (NDI) field for both downlink and uplink transmissions. For downlink transmission, the NDI toggles for every new transmission and then the UE stores it as new data. The NDI does not toggle in case of retransmissions and the UE combines the received data with the data currently stored in the soft buffer for this transport block. For uplink transmission, the NDI on DCI format 0 does not toggle instructing retransmission and toggles instructing new transmission.

### 20.3.4 Types of HARQ Retransmissions

The retransmissions in HARQ process can be adaptive or nonadaptive:

- *Adaptive HARQ:* The transmitter may change some or all of the transmission attributes in each retransmission as compared to the initial transmission because of variations in the radio channel conditions. The retransmissions better fit in the radio channel conditions, but signaling the transmission attributes for each retransmission is required. The transmission attributes that are changed may include the MCS and resource block allocation in frequency. However, the HARQ retransmissions must use the same transport block size as the initial transmission. The transmission attributes for the redundancy versions are explicitly signaled in each retransmission. For the downlink data transfer, the HARQ retransmissions are always adaptive. For the uplink data transfer, the HARQ retransmissions can be adaptive only in certain cases.

- *Nonadaptive HARQ:* The retransmissions either use the same transmission attributes or change the attributes according to a predefined rule. Thus, signaling the transmission attributes in each retransmission is avoided. For the uplink data transfer, the HARQ retransmissions are usually nonadaptive but can be adaptive sometimes.

The retransmissions in HARQ process can be synchronous or asynchronous:

- *Synchronous:* The retransmissions of a packet occur at predefined times relative to the initial transmission. Thus, the receiver can realize to which HARQ process the retransmission belongs and this obviates the need to signal the HARQ ID. For uplink data transfer, the HARQ retransmissions are always synchronous. These HARQ retransmissions can be either adaptive or nonadaptive.
- *Asynchronous:* The retransmissions of a packet occur at any time relative to the initial transmission. Thus, to let the receiver realize to which HARQ process the retransmission belongs, signaling the HARQ ID is required. However, asynchronous HARQ allows more flexibility in scheduling. For the downlink data transfer, the HARQ retransmissions are always asynchronous. These HARQ retransmissions are adaptive in LTE although, in general, the asynchronous transmissions can be either adaptive or nonadaptive.

### 20.3.5 HARQ for the Downlink Transmission

The retransmissions in the HARQ process are always asynchronous and adaptive. Since the retransmission is asynchronous, it can occur at any time. The eNodeB retransmits when it wants after the reception of the HARQ ACK/NACK. The interval between the retransmission and the initial transmission or between retransmissions does not have to be 8 ms and can be more than that. Therefore, there is full scheduling freedom but with the overhead of signaling on the PDCCH. The PDCCH must include the HARQ ID to indicate the particular HARQ process. Since the retransmission is adaptive, the PDCCH includes new resource block allocation, MCS, and redundancy versions. The transport block size cannot be changed.

The UE sends the HARQ ACK/NACK on the PUCCH or the PUSCH on subframe $n + 4$ for either dynamic or semipersistent downlink transmission on the PDSCH on subframe $n$. If the feedback is a NACK, the retransmission is scheduled dynamically for either dynamic or semipersistent transmission (i.e., the HARQ retransmissions cannot exercise semipersistent scheduling).

### 20.3.6 HARQ for the Uplink Transmission

The uplink retransmissions in the HARQ process are always synchronous. Thus, the HARQ retransmission has a fixed schedule in time. The retransmission always occurs 8 ms after the initial transmission or after the last retransmission (i.e., if the initial transmission occurs on subframe $n$, its retransmission occurs on subframe $n + 8$). The uplink HARQ has a fixed gap of 4 TTI between the uplink transmission and the ACK/NACK on the PHICH (i.e., if the initial transmission occurs on subframe $n$, its ACK/NACK is sent on subframe $n + 4$). Since the HARQ retransmission is always synchronous, signaling the HARQ ID is not required. After the UE retransmits its data, the eNodeB will send an ACK/NACK for the retransmitted data.

The uplink retransmissions are basically nonadaptive, but there may be adaptive retransmissions in certain cases. For the usual nonadaptive case, the retransmission occurs at the same set of resource blocks in frequency as the initial transmission or the last retransmission (i.e., it has a fixed schedule in frequency). Thus, signaling the resource block allocation is not needed. Also, the Redundancy Versions (RVs) use a fixed pattern; RV0, RV1, RV2 and RV3 are used for initial transmission, 1st retransmission, 2nd retransmission and 3rd retransmission, respectively. Thus, signaling the redundancy versions is also not needed. As a result, no signaling except the NACK on the PHICH may be used for the retransmission in uplink. The DCI format 0 on PDCCH may or may not be used to schedule a retransmission. However, the DCI format 0 can be used, for example, if the eNodeB does not receive retransmission after sending the NACK. Then the eNodeB can assume that the UE is missing the NACK on the PHICH, so it may schedule the retransmission using DCI format 0. If DCI format 0 is used, then it can indicate either nonadaptive or adaptive retransmission. In both nonadaptive and adaptive cases, the HARQ retransmission remains synchronous (i.e., it obeys the fixed schedule in time). So, signaling the HARQ ID is not required on DCI format 0.

In both nonadaptive and adaptive cases, the DCI format 0 overrides the ACK/NACK on the PHICH. The New Data Indicator (NDI) field on DCI format 0 now indicates, instead of ACK/NACK, whether a retransmission or a new transmission is to be performed as explained in Section 20.3.3. Thus, when an ACK is received on the PHICH, the UE does not clear the transmission buffer right away. The UE rather waits and checks if a subsequent DCI format 0 schedules retransmission or not.

The adaptive retransmission has to use DCI format 0 on PDCCH for the indication of retransmission. The adaptive retransmission can change the scheduling of resources in frequency. So, it may be used, for example, to change the scheduling in frequency domain for the sake of advantage in uplink frequency resources allocation for other purposes. The adaptive retransmission can also use any redundancy version and there is no predefined value for it. The redundancy version is specified on DCI format 0 as explained in Section 5.1.3.2.

The reaction of the UE to various indications from the eNodeB is summarized in Table 20.10.

The eNodeB configures the maximum limit in total number of HARQ transmissions for all the HARQ processes in uplink using the MaxHARQ-Tx field. Appendix A.2 shows how the eNodeB sends the MaxHARQ-Tx field to the UE while

**Table 20.10** Reaction of the UE to Indications from the eNodeB

| HARQ Feedback on the PHICH | Resource Grant on the PDCCH | Reaction of the UE | Type of Retransmission |
|---|---|---|---|
| ACK or NACK | New transmission | New transmission according to the resource grant on the PDCCH | N/A |
| ACK or NACK | Retransmission | Retransmission according to the resource grant on the PDCCH | Synchronous and nonadaptive or adaptive |
| ACK | None | No retransmission but the transmission buffer is not immediately cleared | N/A |
| NACK | None | Retransmission | Synchronous and nonadaptive |

setting up or modifying radio bearers. The value of MaxHARQ-Tx can be 1, 2, 3, 4, 5, 6, 7, 8, 10, 12, 16, 20, 24, or 28.

The TTI bundling can be optionally used in the uplink as shown in Section 19.2.2.4. If the retransmission of a TTI bundle is needed, the whole TTI bundle is retransmitted. Then the maximum number of parallel HARQ processes is 4 instead of 8. The four transport blocks of the retransmitted TTI bundle are sent in consecutive subframes like the original TTI bundle. The same HARQ ID is used for each of these transport blocks. The HARQ retransmission of the TTI bundle is nonadaptive and the redundancy version changes in a predetermined manner. The eNodeB waits for the reception and decoding of the whole TTI bundle before sending the HARQ ACK/NACK. If the last TTI in a TTI bundle is subframe $n$, then the HARQ ACK/NACK is transmitted in subframe $n + 4$. However, since a measurement gap may stop transmission in certain TTIs, it may happen that the last TTI of the TTI bundle has no transmission. In this case, the HARQ ACK/NACK for the TTI bundle is still sent four subframes after the last TTI despite the fact that it has no transmission.

# CHAPTER 21

# Data Transfer Session Setup

A data transfer session enables an application (e.g., browsing, e-mail, chat, file transfer, file sharing, voice, video, online gaming, and so forth). The data transfer session requires the availability of the EPS bearers that fulfill the QoS requirements for the particular application. When the UE is in the ECM-IDLE state, radio bearers, the UE-associated logical S1-connection, and S1-U bearers do not exist. Thus, all existing EPS bearers remain in a dormant state. In order to set up a data transfer session in the ECM-IDLE state, a service request procedure is invoked. The service request procedure establishes radio and S1 bearers for dormant bearers (i.e., it reactivates the default EPS bearer and the existing dedicated EPS bearers). The service request procedure can be classified as follows:

1. *UE-triggered service request procedure:* This procedure is invoked when the UE is in the ECM-IDLE state and needs to initiate the transmission of the uplink user data or signaling information.
2. *Network-triggered service request procedure:* This procedure is invoked when the MME needs to communicate with the UE in the ECM-IDLE state. In the ECM-IDLE state, the serving GW finds no downlink user plane TEID in its context. At this moment, if the serving GW receives downlink data for the UE, it needs a service request to reestablish the dormant bearers. Another typical example of network-triggered service request procedure is the case when the MME or the HSS initiates the detach procedure with the UE in the ECM-IDLE state as shown in Section 15.3.2.

If the application level requirements cannot be satisfied by the default EPS bearer and existing dedicated EPS bearers, then an additional dedicated EPS bearer needs to be established. A dedicated EPS bearer carries traffic flows that use a specific QoS treatment between the UE and the PDN GW. Therefore, a dedicated EPS bearer can be established with a specific QoS and TFT to support the application level requirements, provided that the user's subscription allows it. The dedicated EPS bearer establishment is triggered by the network, but it can be triggered based on the request from the UE. The dedicated EPS bearer setup can be classified as follows:

1. UE-initiated dedicated EPS bearer setup;
2. Network-initiated dedicated EPS bearer setup.

The default EPS bearers support only best effort data transfer, and for certain applications (e.g., VoIP) the default EPS bearer does not fulfill the QoS requirements. If, for example, there is only a default EPS bearer available and a VoIP call needs to be set up, then a dedicated EPS bearer needs to be established. In this case, the default EPS bearer can be used for the negotiation required to set up the VoIP call. At the end of the negotiation, the network can initiate a dedicated EPS bearer setup to support the VoIP call. Alternatively, the UE can initiate a dedicated EPS bearer setup, but this requires that the UE has received proper indications from the network of the support for the existing VoIP call procedure beforehand.

## 21.1 Service Request Procedure

The typical service request procedure is described next assuming that the data transfer can be initiated using the existing EPS bearer and that an additional EPS bearer establishment is not required.

### 21.1.1 UE-Triggered Service Request

The typical UE-triggered service request procedure, shown in Figure 21.1, is described in the following sections.

#### 21.1.1.1 NAS Signaling Connection Establishment and Service Request

The UE is in the ECM-IDLE state when the service request procedure is invoked, so it first needs to set up a way to communicate with the network in order to send a request for the tracking area update. The UE initiates the establishment of NAS signaling connection between the UE and the MME. The establishment of the NAS signaling connection is comprised of RRC connection establishment, MME selection and UE-associated logical S1-connection establishment which were explained in Sections 14.1.1, 14.2 and 14.3.2, respectively.

During this procedure, when the UE sends an RRCConnectionRequest message to the eNodeB, it uses its S-TMSI in the UE-Identity field and it sets the Establishment-Cause field to mobile originated data transfer.

When the UE sends an RRCConnectionSetupComplete message to the eNodeB, the DedicatedInfoNAS IE in this message contains SERVICE REQUEST as the initial NAS message. The RegisteredMME IE in the RRCConnectionSetupComplete message indicates the MME with which the UE is registered. The eNodeB forwards the SERVICE REQUEST message to the MME and thus establishes a UE-associated logical S1-connection. The eNodeB sends the INITIAL UE MESSAGE to the MME that contains the SERVICE REQUEST message. The SERVICE REQUEST message is integrity protected. The SERVICE REQUEST message includes the key set identifier (KSI) value of the current EPS security context and the five least significant bits of the NAS COUNT value. It also includes a short message authentication code (MAC) for its integrity protection.

## 21.1 Service Request Procedure

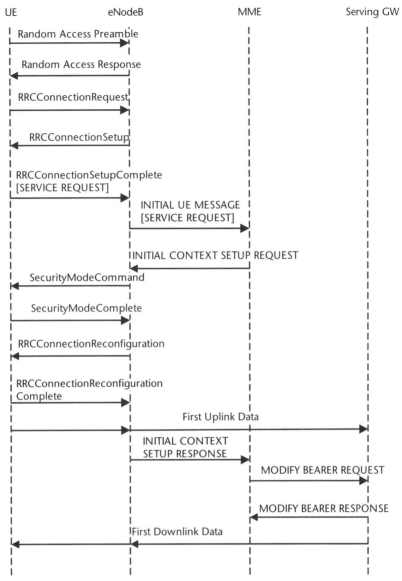

**Figure 21.1** UE-triggered service request when downlink data arrives.

### 21.1.1.2 Authentication and Security Activation

The authentication that executes the EPS AKA procedure and the establishment of a new NAS security context may be performed. If the integrity check of SERVICE REQUEST message has failed, then authentication step is mandatory; otherwise, it is optional. These procedures were explained in Sections 12.2 and 12.3.

### 21.1.1.3 Initial Context Setup

The MME derives the UE context based on the S-TMSI. The MME determines the existing EPS bearers from the UE context. The MME establishes E-RAB for these dormant EPS bearers as follows.

*Initial Context Setup Request*
The MME sends an INITIAL CONTEXT SETUP REQUEST message to the eNodeB. This message was explained in Section 15.1.7.4. The E-RAB ID IE of the INITIAL CONTEXT SETUP REQUEST message contains the E-RAB ID for the E-RABs to be established corresponding to the dormant EPS bearers. The E-RAB Level QoS Parameters IE gives the QoS parameters for the dormant EPS bearers. The INITIAL CONTEXT SETUP REQUEST message also includes the serving GW address, uplink S1-TEID, EPS bearer QoS, security context, the MME signaling connection ID, and the handover restriction list. Upon receipt of the INITIAL CONTEXT SETUP REQUEST message, the eNodeB establishes the UE context and stores the security context, the MME signaling connection ID, the EPS bearer QoS, and the S1-TEID in the UE context.

*Radio Bearer Establishment*
The eNodeB establishes SRB2 and DRBs corresponding to the active EPS bearers. The radio bearer establishment is described here:

1. The security function is activated in the AS layer as explained in Section 12.4.1.
2. The eNodeB determines the QoS of the radio bearers to be established based on the EPS bearer QoS. The eNodeB sends an RRCConnectionReconfiguration message to the UE. This message contains the RadioResourceConfigDedicated IE, which sets up SRB2 and DRBs. The RadioResourceConfigDedicated IE includes detailed configurations of layer 1 and layer 2 for the radio bearers. The RRCConnectionReconfiguration message also includes the MeasConfig IE for the configuration of measurements and reporting. The eNodeB is allowed to send the RRCConnectionReconfiguration message before it receives the SecurityModeComplete message (i.e., before the AS security activation is completed).
3. If the UE had information stored for any EPS bearers but the network does not set up a radio bearer for them, then the UE removes these EPS bearers. Thus, the EPS bearers are synchronized between the UE and the network.
4. The UE sends RRCConnectionReconfigurationComplete message to the eNodeB to confirm the successful completion of the establishment of the radio bearer corresponding to the dedicated EPS bearer.
5. Now the UE can start the uplink data transfer. The eNodeB uses the serving GW address and the uplink S1-TEID provided in INITIAL CONTEXT SETUP REQUEST message for uplink traffic.

*Initial Context Setup Response*
The eNodeB sends the INITIAL CONTEXT SETUP RESPONSE message to the MME indicating the successful establishment of the UE context and the radio bearers. The INITIAL CONTEXT SETUP RESPONSE message contains the list of accepted EPS bearers. It also contains the address of the eNodeB and the TEID used to be used by the serving GW for downlink traffic at the S1-U interface.

The eNodeB sends the INITIAL CONTEXT SETUP FAILURE message to the MME if the UE context setup is not successful.

### 21.1.1.4 Bearer Modify

The MME sends a MODIFY BEARER REQUEST message to the serving GW and thus indicates the establishment of the radio bearer along with the eNodeB user plane information. The MODIFY BEARER REQUEST message contains the eNodeB address, the downlink S1 TEID, and the delay downlink packet notification request. The serving GW sends a MODIFY BEARER RESPONSE message to the MME as acknowledgment. The MODIFY BEARER RESPONSE message contains the EPS bearer identity.

At this stage, the serving GW can start the transmission of downlink data. The serving GW may first transmit its buffered downlink data.

## 21.1.2 Network-Triggered Service Request

The network-triggered service request procedure is described here for the case when the serving GW receives downlink data for the UE and it needs to reestablish the dormant bearers as shown in Figure 21.2:

1. The serving GW buffers the downlink data and identifies the MME that is serving the UE.
2. The serving GW sends a DOWNLINK DATA NOTIFICATION message to the MME.

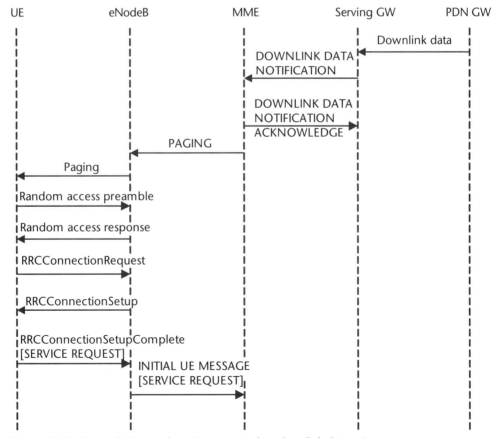

**Figure 21.2** Network-triggered service request when downlink data arrives.

3. The MME sends a DOWNLINK DATA NOTIFICATION ACK message to the serving GW. If the serving GW receives more downlink data packets for the UE, the serving GW keeps buffering them, but it does not send additional DOWNLINK DATA NOTIFICATION messages to the MME.
4. In the ECM-IDLE state, the MME is aware of the UE location with an accuracy of the TAI list of the UE. The MME sends a PAGING message to all eNodeBs that belong to the area covered by the current TAI list of the UE. The MME includes the NAS ID for paging, TAIs in the current TAI list of the UE, both IMSI and S-TMSI of the UE, the paging DRX length, and the paging priority indication.
5. The eNodeBs calculate the paging frame and paging occasion based on the IMSI and send paging messages to the particular UE as explained in Chapter 13. If the MME has sent a paging priority indication and the eNodeB finds a congestion situation, it may send paging messages according to the indicated priority.

Upon reception of the paging message, the UE performs an RRC connection establishment using contention-based random access and sends a SERVICE REQUEST message. Thus, the next steps in the network-triggered service request procedure are identical to all the steps in the UE-triggered service request procedure, explained in Section 21.1.1. The sixth step here will be the first step of the UE-triggered service request procedure. During this procedure, when the UE sends an RRCConnectionSetupComplete message, it may not include the RegisteredMME IE in this message, as the eNodeB is aware of the MME.

## 21.2 Dedicated EPS Bearer Setup

A dedicated EPS bearer can only be established with a PDN if there is already a default EPS bearer established with that PDN. If the UE is in the ECM-IDLE state when a dedicated EPS bearer needs to be established, then a service request procedure, explained in Section 21.1, is performed first. The dedicated EPS bearer context activation procedure may also take place as a part of the attach procedure. Then the dedicated EPS bearer context activation takes place together with the default EPS bearer context activation.

### 21.2.1 Network-Initiated Dedicated EPS Bearer Setup

The network initiated dedicated EPS bearer setup procedure, shown in Figure 21.3, is comprised of the steps shown in the following sections.

#### 21.2.1.1 Configuring a Dedicated EPS Bearer

The dedicated EPS bearer is configured as follows:

1. The PDN GW uses the QoS policy to assign the EPS bearer QoS. The PDN GW sends a CREATE BEARER REQUEST message to the serving GW. This message contains the IMSI, PTI, EPS bearer QoS, TFT, S5/S8

## 21.2 Dedicated EPS Bearer Setup

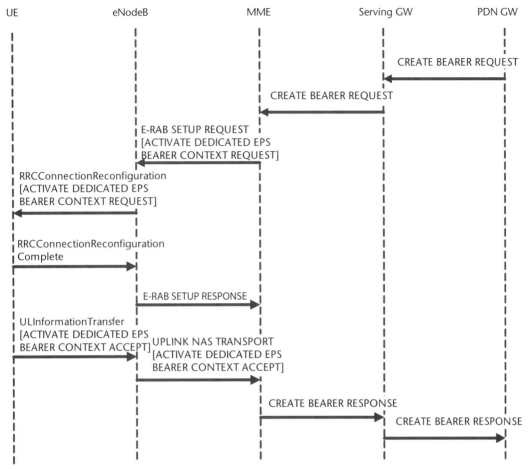

**Figure 21.3** Network-initiated dedicated EPS bearer setup.

TEID, charging ID, and linked bearer identity (LBI). The LBI is the EPS bearer identity (EBI) of the default EPS bearer for the PDN connection and it is included to indicate the default EPS bearer for the particular PDN connection.

2. The serving GW sends the CREATE BEARER REQUEST message to the MME. This message contains the IMSI, PTI, EPS bearer QoS, TFT, S1-TEID, and LBI.
3. The MME assigns an EBI for the dedicated EPS bearer. The MME sends an E-RAB SETUP REQUEST message to the eNodeB and thus instructs allocate resources on the LTE-Uu and S1 interfaces for the EPS bearer. The E-RAB SETUP REQUEST message contains the ACTIVATE DEDICATED EPS BEARER CONTEXT REQUEST message. It also contains S1-TEID.

#### 21.2.1.2 Radio Bearer Establishment

The eNodeB establishes the DRB corresponding to the active EPS bearers:

1. The eNodeB determines the QoS of the radio bearers to be established based on the EPS bearer QoS. The eNodeB sends an RRCConnectionReconfiguration message to the UE. This message contains a RadioResourceConfigDedicated IE that includes detailed configurations of layer 1 and layer 2 for the DRB. The RRCConnectionReconfiguration message also includes the MeasConfig IE for the configuration of measurements and reporting. In addition, the RRCConnectionReconfiguration message includes a DedicatedInfoNAS IE, which contains the ACTIVATE DEDICATED EPS BEARER CONTEXT REQUEST message. This message contains the following major IEs:
   - *EBI:* This IE contains the EBI that the MME allocated to the EPS bearer. The UE stores the EBI of the dedicated EPS bearer.
   - *Traffic flow template (TFT):* This IE specifies the TFT parameters and operations for the dedicated EPS bearer. It may contain packet filters. If it contains packet filters for the uplink direction, then the UE uses the uplink packet filter to determine the mapping of traffic flows to the radio bearer.
   - *EPS quality of service:* This IE specifies the EPS quality of service assigned for the EPS bearer.
   - *Linked EBI:* This IE indicates the default EPS bearer for the particular PDN connection. The UE links the dedicated bearer to the default bearer indicated by the LBI.
   - *Procedure transaction identity (PTI):* The PTI identifies a procedure and allows the distinction of different bidirectional messages flows.
2. The UE sends an RRCConnectionReconfigurationComplete message to the eNodeB to confirm the successful completion of the establishment of the radio bearer corresponding to the dedicated EPS bearer.

### 21.2.1.3 Dedicated EPS Bearer Setup Accept

The acceptance of the dedicated EPS bearer setup is notified as follows:

1. The eNodeB sends an E-RAB SETUP RESPONSE message to the MME and acknowledges the bearer activation. This message contains the EBI and the S1-TEID. The UE sends an ACTIVATE DEDICATED EPS BEARER CONTEXT ACCEPT message using the ULInformationTransfer message. The eNodeB forwards this message to the MME using the UPLINK NAS TRANSPORT message. The ACTIVATE DEDICATED EPS BEARER CONTEXT ACCEPT message contains the EBI. Thus, the UE acknowledges the activation of the dedicated EPS bearer context.
2. The MME sends a CREATE BEARER RESPONSE message to the serving GW and acknowledges the bearer activation. This message contains the EBI and the S1-TEID.
3. The serving GW sends CREATE BEARER RESPONSE message to the PDN GW and acknowledges the bearer activation. This message contains the EBI and the S5/S8-TEID.

Data transfer can now take place using the dedicated EPS bearer.

### 21.2.2 UE-Requested Dedicated EPS Bearer Setup

The dedicated EPS bearer setup may take place at the request of the UE. In this case, either the UE-requested bearer resource allocation procedure or the UE-requested bearer resource modification procedure is invoked. In the case of the UE-requested bearer resource allocation procedure, the UE first seeks a dedicated EPS bearer setup using the following steps as shown in Figure 21.4:

1. The UE sends BEARER RESOURCE ALLOCATION REQUEST message to the MME seeking the allocation of bearer resources for the new dedicated bearer. The BEARER RESOURCE ALLOCATION REQUEST message contains the following IEs:
   - *TFT:* This IE specifies the TFT parameters and operations for the dedicated EPS bearer.
   - *EPS QoS:* This IE specifies the requested EPS QoS for the EPS bearer.
   - *LBI:* The LBI is the EBI of the default EPS bearer for a PDN connection. This IE includes the LBI to indicate the default EPS bearer for the particular PDN connection.
   - *PTI:* The PTI identifies a procedure and allows the distinction between different bidirectional messages flows. The PTI is released when the procedure is completed.
2. The MME checks if the LBI matches with the EBI of any of the active default EPS bearers. The MME selects the serving GW that is used for the particular default EPS bearer indicated by the LBI. The MME sends the BEARER RESOURCE COMMAND message to the selected serving GW. This message contains the IMSI, LBI, PTI, and EPS QoS.
3. The serving GW selects the PDN GW that is used for the particular default EPS bearer indicated by the LBI. The serving GW sends the BEARER RESOURCE COMMAND message to the PDN GW. This message contains the IMSI, LBI, PTI, and EPS QoS.

After the UE seeks a dedicated EPS bearer setup, the network establishes a dedicated EPS bearer as shown in Section 21.2.1 (i.e., the steps shown in Section 21.2.1 are performed).

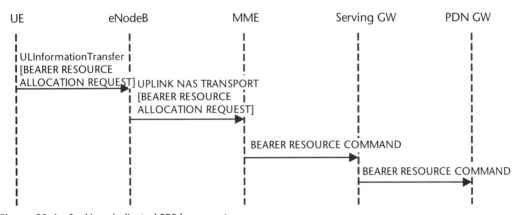

**Figure 21.4** Seeking dedicated EPS bearer setup.

## 21.3 Voice Services

The major source of revenue for most cellular operators is still voice services. However, LTE does not support the circuit-switched (CS) service and it employs end-to-end IP connection, so the voice transfer does not find any superb solution, especially in the initial days of deployment. Some operators even plan to deliver LTE for only data with no voice in the first 1 or 2 years. The key problems of voice implementation are as follows:

- The voice transfer has to use VoIP since LTE uses all-IP technology. The VoIP typically uses RTP/UDP/IP and requires large overhead. The header compression explained in Section 8.1.3 helps mitigate this problem.
- Since no dedicated CS channel is available, frequent resource allocation is required, causing high control overhead. The semipersistent scheduling explained in Section 19.2 helps mitigate this problem.
- The high data rate support in LTE makes the Maximum Allowable Path Loss (MAPL) somewhat small. The TTI bundling explained in Section 19.2.2.4 helps mitigate this problem in the uplink.

The voice calls can be classified as Mobile Originated (MO) calls and Mobile Terminated (MT) calls. If the UE under consideration initiates the call (i.e., if it is the caller party), then the call is regarded as an MO call. On the other hand, if the UE under consideration, receives the call (i.e., if it is the called party, then the call is regarded as an MT call).

### 21.3.1 Voice Service Solutions

A number of solutions are defined for voice services in LTE and each of them has its own advantages and disadvantages:

1. *IP Multimedia System (IMS):* The IMS uses an architecture framework to deliver fully IP-based multimedia services. The IMS is viewed as the ultimate voice solution for LTE. Many operators will use other options after the initial LTE rollout, but they envision the implementation of IMS in time. The recent additions in the specifications allow ongoing voice calls to be handed over between the IMS and a CS network (e.g., GSM and UMTS). Single Radio Voice Call Continuity (SRVCC) is a feature to support for this transfer.
2. *Voice over LTE via Generic Access (VoLGA):* The VoLGA packetizes voice information and transfers using the LTE network (i.e., CS data and signaling are tunneled over IP). VoLGA has been accepted by many operators as an interim solution until IMS is implemented. VoLGA is basically supported by the industry consortium known as the VoLGA Forum. VoLGA uses a new entity, the VoLGA access network controller (VANC). The VANC is a 3GPP Generic Access Network Controller (GANC), which has been modified to support CS services over LTE. The VANC introduces an overlay access between the UE and the MSC. A security tunnel is established

between the UE and the VANC. The UE registers with the MSC using the security tunnel. Thereafter, voice calls can be set up using the security tunnel. VoLGA allows the handover of ongoing voice calls to the CS network [e.g., GSM, UMTS-based on a single radio voice call continuity (SRVCC)].

3. *CS Fallback (CSFB):* The CSFB establishes CS voice calls by allowing the UE to fall back to 2G/3G. CSFB has also been accepted by many operators as an interim solution until the IMS is implemented. The CSFB procedure is explained in Section 21.3.2.

4. *Over-the-top voice:* The over-the-top voice facilitates voice services over LTE with the help of external voice service providers.

A comparison among different solutions for voice services is shown in Table 21.1.

**Table 21.1** Comparison Among Voice Service Solutions

| Solutions for Voice Services | Advantages | Disadvantages |
|---|---|---|
| IMS | Full IP-based multimedia services. | Complex solution |
| | Operational costs are minimized | Large investment is required. |
| | | Major changes are required in network |
| VoLGA | Simpler solution | The implementation for VoLGA is more complicated than CSFB. Addition of VANC is required. |
| | Small investment is required. | |
| | Small call setup delay. | |
| | MSC upgrade is not required. | The investment for VoLGA is larger than CSFB. |
| | Good migration path to IMS | |
| | Femtocells are supported | The VoLGA standards are not accepted by 3GPP. |
| | Supports IMS RCS and combinational voice/data services | New client software may have difficulties with compatibility. |
| | All CS services including SMS can be used over LTE | There can be problem in roaming with a network that does not use VoLGA. |
| CSFB | It utilizes the existing 2G/3G network including using the existing operations support systems (OSSs) and billing systems. | Longer call setup delay. |
| | | Increased signaling because of switching between RATs. |
| | No new network elements need to be added. | Some modifications are required in the core network including an upgraded MSC and SGSN interface. Extensive testing is required to ensure the proper operation after modification. However, the modifications are small compared to the other voice service options. |
| | The implementation is relatively cheaper. | |
| | | LTE functionalities are not available during voice calls. |
| | | The UE has to support 2G/3G. |
| | | Significant likelihood of call drop. |
| | | Femtocells are not supported. |
| Over-the-top voice | Simple solution. | There is no direct control over the quality. |
| | Special voice applications may be offered. | Handover of ongoing voice calls to GSM or UMTS is not feasible. |

### 21.3.2 CSFB Procedure

If there is registration with 2G/3G, the UE can fallback to UTRAN/GERAN for CSFB operation. As shown in Section 15.1.1.1, during attach procedure, the UE sends ATTACH REQUEST message to the MME. The UE can select Combined EPS/IMSI Attach for the EPS Attach Type IE in the ATTACH REQUEST message and thus, the UE requests for registration with 2G/3G in addition to registration with EPS. As shown in Section 15.1.7.4, the EPS Attach Result IE in the ATTACH ACCEPT message indicates the type of accepted registration, whether it is registration with 2G/3G in addition to EPS or registration with EPS only.

The S3 interface, shown in Figure 1.1, can enable idle mode signaling reduction (ISR). ISR is activated while inter-RAT cell reselection takes place and does not occur during the initial attach procedure. ISR allows the UE to remain simultaneously registered in both the MME and the SGSN. Thus, if ISR is activated, the UE is assigned both the GUTI and the P-TMSI. The UE is paged on both E-UTRAN and UTRAN/GERAN for MT CSFB calls. Thus, when the serving GW receives downlink data for the UE as shown in Section 21.1.2, the serving GW sends a DOWNLINK DATA NOTIFICATION message to the MME as well as the SGSN. The tracking area update or routing area update is not required following the inter-RAT cell reselection or handover unless the tracking area or the routing area has been changed. Thus, in practice, coordination with the geographical boundaries between tracking areas and location/routing areas of UTRAN/GERAN is expected in order to obviate the necessity to execute the update procedure following inter-RAT cell reselection or handover. However, in specification, there is no restriction in the relationship between tracking areas and location areas and location/routing areas of UTRAN/GERAN.

On the other hand, if ISR is not activated, the tracking area update is invoked following the inter-RAT cell reselection or handover from UTRAN/GERAN to E-UTRAN. Similarly, the location area update and/or routing area update are invoked following the inter-RAT cell reselection or handover from E-UTRAN to UTRAN/GERAN.

The CSFB voice call is established using one of the following methods:

1. *PS handover:* A PS handover to UTRAN/GERAN is performed. Thereafter, a CS voice call is set up at UTRAN/GERAN.
2. *Redirection:* The redirection is typically used when the PS handover is not supported. In this case, the RRC connection is released with redirection to UTRAN/GERAN. Then the UE reselects to the UTRAN/GERAN cell and establishes the RRC connection. Thereafter, a CS voice call is set up at the UTRAN/GERAN. When the RRC connection is released, the RRCConnectionRelease message includes the RedirectedCarrierInfo IE to inform the UE of the UTRAN/GERAN carrier frequencies for redirection. The RedirectedCarrierInfo IE includes the ARFCN-ValueUTRA IE, which contains the UTRAN carrier frequencies, and the CarrierFreqsGERAN IE, which contains groups of GERAN carrier frequencies.

The redirection procedure requires a much higher call setup delay compared to the PS handover method. The PS handover requires a delay only for the measurements and handover procedure, whereas the redirection requires a delay for the RRC connection release, tuning to the UTRAN/GERAN cell, measurements of UTRAN/GERAN cell, reading system information, and the RRC connection setup at the UTRAN/GERAN cell. Hence, in Release 9, the RRCConnectionRelease message includes additional information about the redirected UTRAN/GERAN cells, including system information. This addition allows the call setup delay in the redirection procedure to be almost equal to the delay in PS handover. Release 9 introduces the following additional IEs in the RRCConnectionRelease message:

1. *CellInfoListUTRA-FDD:* This IE includes the CellInfoUTRA-FDD IE for each of the redirected UTRAN cells. The CellInfoUTRA-FDD IE includes the following IEs:
   - *PhysCellId:* It contains the primary scrambling code (PSC) of the cell.
   - *UTRA-BCCH-Container:* It contains the system information blocks (SIBs) broadcast in the cell.
2. *CellInfoListGERAN:* This IE includes CellInfoGERAN IE for each of the redirected GERAN cells. The CellInfoGERAN IE includes the following IEs:
   - *PhysCellIdGERAN:* It contains the base station identify code (BSIC) of the cell.
   - *CarrierFreqGERAN:* It contains the carrier frequency for the cell.
   - *SystemInfoListGERAN:* It contains the system information (SI) messages broadcast in the cell.

At the end of the voice call, the UE returns to E-UTRAN, and there is no specification for this return procedure. The inter-RAT handover or cell reselection procedure can be used for the return. If the EPS service was suspended during the CS service, it is resumed after the return.

### 21.3.2.1 MO Call Setup

The CSFB MO call setup, shown in Figure 21.5, includes the following steps:

1. If the UE is in the ECM-IDLE state when the MO call is triggered, then it first needs to set up a way to communicate with the network. Therefore, the UE initiates the establishment of NAS signaling connection between the UE and the MME. The establishment of the NAS signaling connection is comprised of RRC connection establishment, MME selection and UE-associated logical S1-connection establishment which were explained in Sections 14.1.1, 14.2 and 14.3.2, respectively. During this procedure, when the UE sends the RRCConnectionSetupComplete message to the eNodeB, the DedicatedInfoNAS IE in this message contains an EXTENDED SERVICE REQUEST as the initial NAS message. The eNodeB forwards the EXTENDED SERVICE REQUEST message using the INITIAL UE MESSAGE.
   On the other hand, if the UE is in the ECM-CONNECTED state, then there is already a NAS signaling connection existing between the UE and the MME. Thus, the UE directly sends an EXTENDED SERVICE REQUEST message to the eNodeB using a ULInformationTransfer

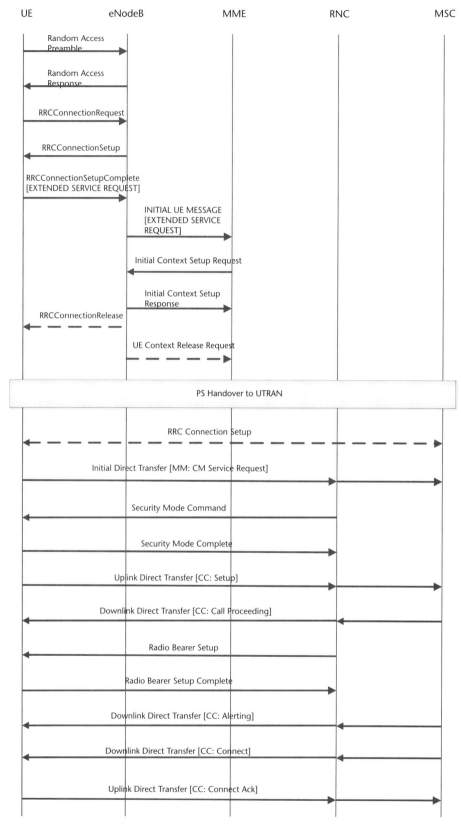

**Figure 21.5** CSFB MO call setup procedure.

message. The eNodeB forwards the EXTENDED SERVICE REQUEST message to the MME using an UPLINK NAS TRANSPORT message.

The EXTENDED SERVICE REQUEST message includes the Service Type IE, which indicates the MO CSFB call. This message can also optionally include the EPS Bearer Context Status IE, which indicates the active EPS bearer contexts.

2. If the UE was in the ECM-IDLE state, the MME sends an Initial Context Setup Request message to the eNodeB and the eNodeB establishes a UE context. On the other hand, if the UE was in the ECM-CONNECTED state, the MME sends an UE Context Modification Request message to the eNodeB. Both the Initial Context Setup Request message and the UE Context Modification Request message include the CS Fallback Indicator IE and this IE is set to "CS Fallback required," instructing that the UE should fallback to 2G/3G. Both these messages also include location area identification (LAI) of the UE. The eNodeB acknowledges by sending the Initial Context Setup Response message or the UE Context Modification Response message to the MME depending on if the UE was in the ECM-IDLE state or the ECM-CONNECTED state, respectively.

3. The eNodeB optionally solicits a measurement report from the UE in order to determine the target UTRAN for the PS handover or redirection. If measurement reports from the UE are used, then it is UE-assisted handover. Conversely, if measurement reports from the UE are not used, then it is blind handover. The handover types are explained in Section 23.3.1.2.

4. If the PS handover method is used, then the eNodeB initiates the PS handover to UTRAN. The handover procedure is explained in Section 23.3.4. On the other hand, if redirection is used, then the eNodeB releases the RRC connection by sending the RRCConnectionRelease message to the UE. The eNodeB sends the UE Context Release Request message to the MME. The MME releases the UE Context in the eNodeB as well as all eNodeB-related information in the S-GW. Then the UE reselects to UTRAN and establishes the RRC connection at UTRAN.

5. The MO CS voice call is set up at UTRAN with the necessary steps as shown in Figure 21.5.

### 21.3.2.2 MT Call Setup

The CSFB MT call setup, shown in Figure 21.6, includes the following steps:

1. When an MT call arrives at the MSC, the MSC sends a Paging Request message to the MME over the SGs interface indicating that there is an incoming CS call. This message includes the IMSI, TMSI, location information, and priority indication.

2. If the UE is in the ECM-IDLE state, then the MME sends the Paging message to the UE via eNodeBs. If the Paging Request message includes the TMSI, the MME uses the TMSI to derive the S-TMSI and uses the S-TMSI in a Paging message as the UE identity. If this message includes the IMSI, the MME uses the IMSI in the Paging message as the UE identity. If the MME has the TAI list of the UE stored, the MME sends the Paging message to all

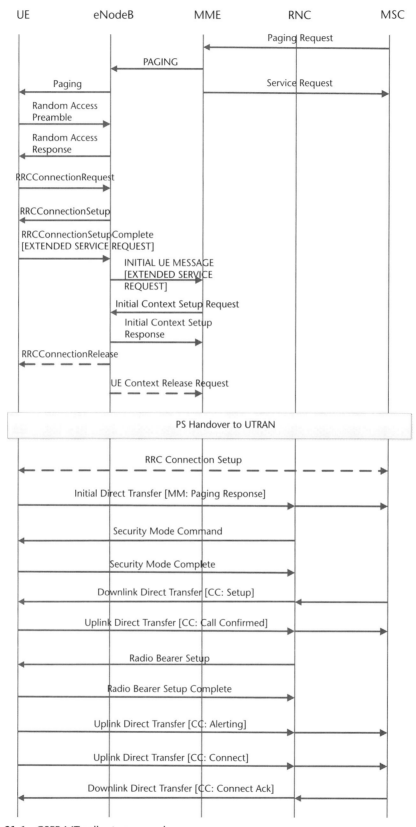

**Figure 21.6** CSFB MT call setup procedure.

eNodeBs that belong to the area covered by the TAI list. If the MME does not have a stored TAI list of the UE, the MME may use the location information received in the Paging Request message from the MSC. The eNodeBs send the paging message to the UE. Upon reception of the Paging message, the UE initiates the establishment of the NAS signaling connection between the UE and the MME. The establishment of the NAS signaling connection is comprised of RRC connection establishment, MME selection and UE-associated logical S1-connection establishment which were explained in Sections 14.1.1, 14.2 and 14.3.2, respectively. During this procedure, when the UE sends the RRCConnectionSetupComplete message to the eNodeB, the DedicatedInfoNAS IE in this message contains an EXTENDED SERVICE REQUEST as the initial NAS message. The eNodeB forwards the EXTENDED SERVICE REQUEST message using the INITIAL UE MESSAGE.

On the other hand, if the UE is in the ECM-CONNECTED state, then there is already a NAS signaling connection existing between the UE and the MME. Then, in order to indicate the requirement of the CS service, the MME sends a CS SERVICE NOTIFICATION message to the eNodeB using the DLInformationTransfer message. The eNodeB forwards the CS SERVICE NOTIFICATION message to the UE using the DOWNLINK NAS TRANSPORT message. In response, the UE sends the EXTENDED SERVICE REQUEST message to the eNodeB using the ULInformationTransfer message. The eNodeB forwards the EXTENDED SERVICE REQUEST message to the MME using UPLINK NAS TRANSPORT message.

The Service Type IE in the EXTENDED SERVICE REQUEST message indicates the MT CSFB call.

3. The MME sends the Service Request message to the MSC over the SGs interface indicating whether the UE was in the ECM-IDLE state or the ECM-CONNECTED state. The Service Request message precludes the MSC retransmitting the Paging Request message.

4. If the UE was in the ECM-IDLE state, the MME sends an Initial Context Setup Request message to the eNodeB and the eNodeB establishes a UE context. On the other hand, if the UE was in the ECM-CONNECTED state, the MME sends an UE Context Modification Request message to the eNodeB. Both the Initial Context Setup Request message and the UE Context Modification Request message include the CS Fallback Indicator IE, which is set to "CS Fallback required," instructing that the UE should fall back to 2G/3G. Both these messages also include the LAI of the UE.

The eNodeB acknowledges by sending the Initial Context Setup Response message or the UE Context Modification Response message to the MME depending on if the UE was in the ECM-IDLE state or the ECM-CONNECTED state, respectively.

5. The eNodeB optionally solicits a measurement report from the UE in order to determine the target UTRAN for the PS handover or redirection. If measurement reports from the UE are used, then it is a UE-assisted handover. Conversely, if measurement reports from the UE are not used, then it is a blind handover. The handover types are explained in Section 23.3.1.2.

6. If the PS handover method is used, then the eNodeB initiates the PS handover to UTRAN. The handover procedure is explained in Section 23.3.4. On the other hand, if redirection is used, then the eNodeB releases the RRC connection by sending an RRCConnectionRelease message to the UE. The eNodeB sends UE Context Release Request message to the MME. The MME releases the UE context in the eNodeB as well as all eNodeB-related information in the S-GW. Then the UE reselects to UTRAN and establishes the RRC connection at UTRAN.
7. The MT CS voice call is set up at UTRAN with the necessary steps as shown in Figure 21.6.

## 21.4 Short Message Service (SMS)

The SMS is also a good source of revenue for the operators and it has been supported using the CS infrastructure in different 2G/3G cellular technologies. Although LTE uses all-IP, it supports SMS.

### 21.4.1 SMS Solutions

The following solutions are available to support SMS.

1. *IMS:* Like voice, the IMS is viewed as the ultimate solution for SMS. A gateway called an IP short message gateway (IP-SM-GW) is defined for interoperability between SMS and IMS messaging service. The IMS supports SMS in one of the following ways:
    - *Page-mode messaging:* The message is sent like a notification without requiring any answer. The SIP protocol can be used if the size of the message is not larger than 1,500 bytes. If the size is larger than 1,500 bytes, then the Message Session Relay Protocol (MSRP) is used.
    - *Session-mode messaging:* A session is established for conversation and the message is sent using the session. The MSRP is used in this case.
    - OMA SIMPLE IM: This is an instant messaging (IM) service released by the Open Mobile Alliance (OMA) for the IMS.
2. *Interim voice solutions:* The interim voice solutions (e.g., VoLGA and CSFB) support SMS in addition to voice, so when VoLGA or CSFB is used for voice services, SMS can also be supported.
3. *SMS over SGs:* The SMS over the SGs interface can be used if the IMS is not yet supported and thus, it is treated as an interim solution. The SMS over SGs uses the CS domain infrastructure, but the UE does not require a fallback to 2G/3G. The SMS over the SGs interface procedure is explained in Section 21.4.2.

### 21.4.2 SMS over the SGs Procedure

The UE is required to have registration in the CN domain for SMS over the SGs interface. As shown in Section 15.1.1.1, during the attach procedure, the UE sends an ATTACH REQUEST message to the MME. The UE can indicate its intention for

## 21.4 Short Message Service (SMS)

SMS over the SGs interface but not CSFB operation in the Additional Update Type IE of the ATTACH REQUEST message. As shown in Section 15.1.7.4, if registration has been accepted for only SMS over the SGs interface, it is indicated in the Additional Update Result IE of the ATTACH ACCEPT message.

The registration for SMS over the SGs interface can also be performed during the tracking area update procedure. As shown in Section 22.1, the UE can indicate its intention for SMS over the SGs interface but not CSFB operation in the Additional Update Type IE of the TRACKING AREA UPDATE REQUEST message. As shown in Section 22.5, if registration has been accepted for only SMS over the SGs interface, it is indicated in the Additional Update Result IE of the TRACKING AREA UPDATE ACCEPT message.

### 21.4.2.1 Mobile Originated (MO) SMS

The MO SMS procedure, shown in Figure 21.7, includes the following steps:

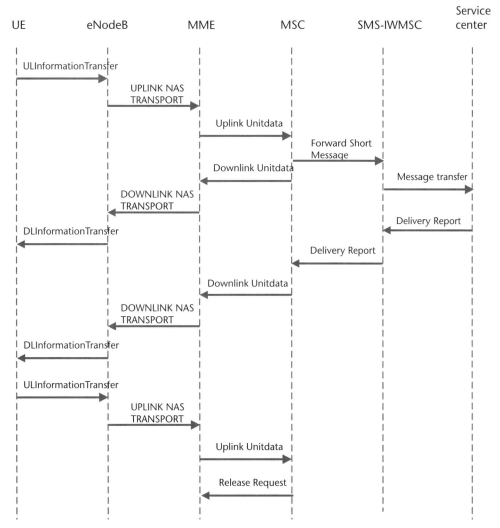

**Figure 21.7** MO SMS procedure.

1. If the UE is in the ECM-CONNECTED state, the UE sends the message for SMS to the eNodeB encapsulating the message in the DedicatedInfoNAS IE of the ULInformationTransfer message. The eNodeB forwards the message for SMS to the MME using the UPLINK NAS TRANSPORT message. If the UE is in the ECM-IDLE state, the UE needs to invoke the UE-triggered service request procedure first and send a SERVICE REQUEST message to the MME as shown in Section 21.1.1. Thereafter, the UE sends the message for SMS in the same way as it does for the ECM-CONNECTED state.
2. The MME sends an Uplink Unitdata message to the MSC over the SGs interface. This message encapsulates the message for SMS.
3. The MSC forwards the message to the SMS interworking MSC (SMS-IWMSC), which is capable of receiving a short message and submitting it to the appropriate service center.
4. The MSC acknowledges receipt of the message for SMS to the UE. For this purpose, the MSC sends a Downlink Unitdata message to the MME. The MME sends a DOWNLINK NAS TRANSPORT message to the eNodeB and the eNodeB sends a DLInformationTransfer message to the UE.
5. The SMS-IWMSC forwards the message for SMS to the service center.
6. The service center sends delivery report to the SMS-IWMSC.
7. The SMS-IWMSC forwards the delivery report to the MSC.
8. The MSC sends the Downlink Unitdata message to the MME. This message encapsulates the delivery report.
9. The MME sends DOWNLINK NAS TRANSPORT message to the eNodeB and the eNodeB sends the DLInformationTransfer message to the UE. These messages contain the delivery report.
10. The UE acknowledges the receipt of the delivery report by sending an ULInformationTransfer message. The eNodeB sends an UPLINK NAS TRANSPORT message the MME.
11. The MME sends an Uplink Unitdata message to the MSC, which receives acknowledgment of the receipt of the delivery report.
12. The MSC sends a Release Request message to the MME. This message indicates to the MME that no more NAS messages need to be transferred to the UE.

### 21.4.2.2 Mobile Terminated (MT) SMS

The MT SMS procedure, shown in Figure 21.8, includes the following steps:

1. The service center sends the message for SMS to the SMS-IWMSC.
2. The SMS-IWMSC forwards the message for SMS to the MSC.
3. The MSC sends a Paging Request message to the MME over the SGs interface that there is an incoming SMS. This message includes the IMSI, TMSI, location information, and SMS indicator.
4. If the UE is in the ECM-IDLE state, then the MME sends a Paging message to the UE via eNodeBs. The UE invokes the UE-triggered service request procedure and sends a SERVICE REQUEST message to the MME

## 21.4 Short Message Service (SMS)

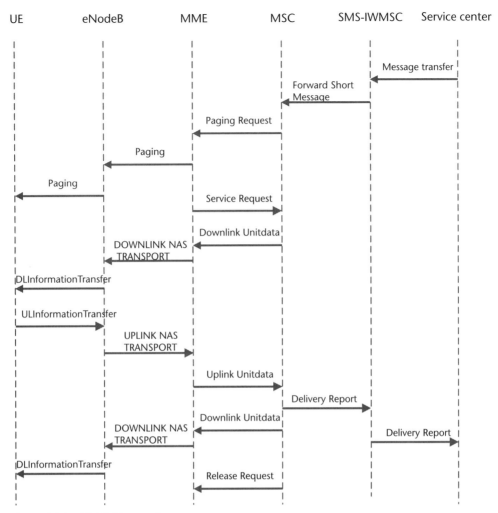

**Figure 21.8** MT SMS procedure.

as shown in Section 21.1.1. If the UE is in the ECM-CONNECTED state, then this step is not required and is skipped.

5. The MME sends a Service Request message to the MSC over the SGs interface indicating whether the UE was in the ECM-IDLE state or the ECM-CONNECTED state.
6. The MSC sends a Downlink Unitdata message to the MME. This message encapsulates the message for SMS.
7. The MME sends a DOWNLINK NAS TRANSPORT message to the eNodeB and the eNodeB sends a DLInformationTransfer message to the UE. These messages encapsulate the message for SMS.
8. The UE acknowledges the receipt of the message for SMS by sending a delivery report using the ULInformationTransfer message. The eNodeB forwards the delivery report to the MME using the UPLINK NAS TRANSPORT message.
9. The MME sends an Uplink Unitdata message to the MSC over the SGs interface. This message encapsulates the delivery report.

10. The MSC forwards the delivery report to the SMS-IWMSC.
11. The SMS-IWMSC forwards the delivery report to the service center.
12. The MSC sends a Downlink Unitdata message to the MME. This message acknowledges the receipt of the delivery report.
13. The MME sends DOWNLINK NAS TRANSPORT message to the eNodeB and the eNodeB sends a DLInformationTransfer message to the UE, which receives acknowledgment of the receipt of the delivery report.
14. The MSC sends a Release Request message to the MME. This message indicates to the MME that no more NAS messages need to be transferred to the UE.

# CHAPTER 22

# Tracking Area Update

The tracking area update (TAU) registers the UE in the tracking area of the current cell. It is similar with routing area update used in UMTS. The typical cases when the TAU procedure is invoked are:

1. *UE mobility:* The TAU procedure takes place due to the mobility of the UE in both the ECM-IDLE state and the ECM-CONNECTED state. In the ECM-IDLE state, there is no NAS signaling connection between the UE and the MME, but the MME is aware of the UE location with an accuracy of the TAI list of the UE. This awareness allows the MME to page the UE only within the area covered by the current TAI list. In order to keep up this awareness, after the completion of cell reselection, if the UE finds that it has moved to an area covered by a different TAI list, the UE updates the MME with its current location using the TAU procedure.

    In the ECM-CONNECTED state, after the completion of handover action, the UE performs the TAU procedure provided that the UE finds that the tracking area has been changed and the new TAI is not one of the TAIs in the TAI list currently possessed by the UE. In this case, the TAU procedure involves a change of the serving MME. However, the serving gateway (GW) may or may not be changed.

2. *Periodic update:* In the ECM-IDLE state, the UE performs the TAU procedure periodically in order to notify the network about its availability. For this purpose, the UE maintains a periodic tracking area update timer known as T3412. Every time T3412 expires, the UE performs the TAU procedure and T3412 is restarted. However, if the UE is attached for emergency bearer services when T3412 expires, the UE performs the detach procedure with the network instead of the TAU procedure.

    The network can assign and update the value of T3412 during the attach procedure and the TAU procedure using the T3412 Value IE in the ATTACH ACCEPT message and the TRACKING AREA UPDATE ACCEPT message, respectively. The T3412 Value IE was explained in Section 15.1.7.4. Its default duration is 54 minutes. The network may set the value of T3412 to zero or indicate a deactivation of T3412. In these cases, the UE does not perform the periodic TAU procedure.

    If the UE moves out of coverage, it can no longer perform the periodic TAU procedure. Therefore, if the UE does not perform the periodic TAU procedure at its due time, the MME starts a timer called the mobile reachable timer. When the mobile reachable timer expires without any communication with the UE, the MME infers that the UE has moved out of coverage. The default value of the mobile reachable timer is 4 minutes greater than T3412, but if the UE is attached for emergency bearer services, the value of this timer is set equal to T3412. When the mobile reachable timer expires without any periodic TAU, the MME starts another timer called the implicit detach timer. This is because the MME wants to ensure that the UE has been out of coverage for a long enough duration before it detaches the UE. At this moment, the MME does not page the UE if downlink data arrives for the UE. The MME rather sends a Downlink Data Notification Reject message to the serving GW if it receives a Downlink Data Notification message from the serving GW. If the implicit detach timer expires and the UE does communicate with the network, then the MME infers that the UE has been out of coverage for a long period, so it performs an implicit detach with the UE.

The signaling in the TAU procedure consumes resources. In order to reduce this consumption, Release 10 introduces the assignment of a longer value of T3412 by including the T3412 Extended Value IE in both the ATTACH ACCEPT message and the TRACKING AREA UPDATE ACCEPT message. It may be noted that this increases the duration of both the periodic TAU timer and the mobile reachable timer. If a network failure occurs or the UE moves out of coverage, the longer value of T3412 may increase the delay to detect them. The use of the T3412 Extended Value IE in the ATTACH ACCEPT message and TRACKING AREA UPDATE ACCEPT message was explained in Section 15.1.7.4 and will be explained in Section 22.5, respectively.

3. *Change in UE properties:* The UE performs the TAU procedure in the case of certain changes in the properties of the UE, for example, when the UE changes its network capability information or it changes the UE-specific DRX parameter.
4. *MME load rebalancing:* The UE performs the TAU procedure during the MME load rebalancing procedure. The MME load rebalancing will be explained in Section 25.3.
5. *NAS signaling connection recovery:* The UE performs the TAU procedure when the NAS signaling connection is lost and the UE needs to reestablish the NAS signaling connection.

The TAU procedure may or may not involve a change of MME depending on the current location of the UE. Also, if the MME is changed, then the TAU procedure may or may not involve a change of the serving GW. The new MME determines if the serving GW needs to be changed. The MME changes the serving GW if the same serving GW cannot continue serving or if another serving GW can serve more effectively. The TAU procedure includes the steps one after another explained in the following sections and shown in Figure 22.1. If the TAU procedure is invoked in the ECM-CONNECTED state, then it does not require the steps shown in Sections 22.2, 22.3, and 22.4 and they are skipped.

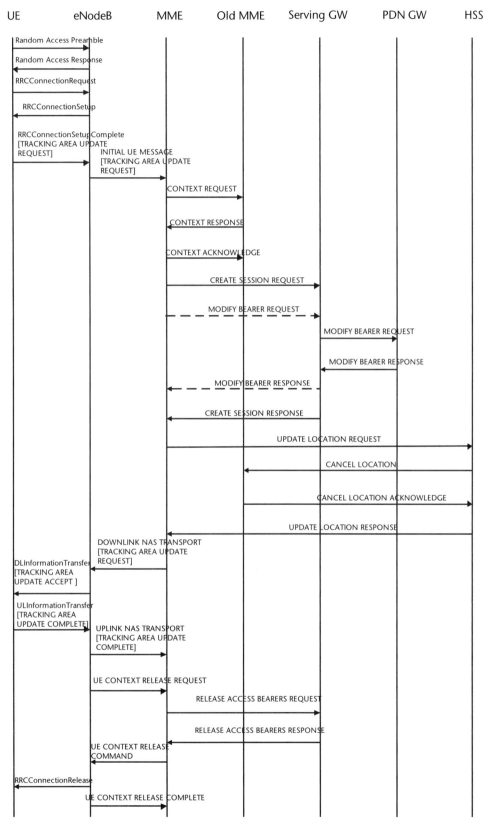

**Figure 22.1** TAU procedure.

## 22.1 NAS Signaling Connection Establishment and TAU Request

If the UE is in the ECM-IDLE state when the TAU procedure is invoked, then it first needs to set up a way to communicate with the network in order to send a request for the TAU. Thus, the UE initiates the establishment of the NAS signaling connection between the UE and the MME. The establishment of the NAS signaling connection is comprised of RRC connection establishment, MME selection and UE-associated logical S1-connection establishment which were explained in Sections 14.1.1, 14.2 and 14.3.2, respectively. During this procedure, when the UE sends an RRCConnectionRequest message to the eNodeB, it uses its S-TMSI in the UE-Identity field and it sets the EstablishmentCause field to the mobile originated signaling.

When the UE sends an RRCConnectionSetupComplete message to the eNodeB, the DedicatedInfoNAS IE in this message contains the TRACKING AREA UPDATE REQUEST as the initial NAS message. The eNodeB forwards the TRACKING AREA UPDATE REQUEST message to the MME and thus establishes the UE-associated logical S1 connection. The eNodeB sends an INITIAL UE MESSAGE message to the MME that contains the TRACKING AREA UPDATE REQUEST message.

On the other hand, if the UE is in the ECM-CONNECTED state, then there is already a NAS signaling connection existing between the UE and the MME, so the UE directly sends the TRACKING AREA UPDATE REQUEST message to the eNodeB using the ULInformationTransfer message. The eNodeB forwards the TRACKING AREA UPDATE REQUEST message to the MME using an UPLINK NAS TRANSPORT message.

The TRACKING AREA UPDATE REQUEST message includes the following major IEs:

1. *Last visited registered TAI:* This IE contains the TAI with which the UE registered most recently. The UE includes this information in order to help the MME produce a good list of TAIs for the TAI list.
2. *EPS bearer context status:* If the UE wants to indicate which EPS bearer contexts are active and which are not, then this IE is included. This IE indicates the state of each EPS bearer context identifying them by their EPS bearer identities.
3. *EPS update type:* This IE specifies the scenario for the requested TAU procedure. It can be any of the following types:
    - *TA updating:* It indicates that the TAU procedure is invoked for any reason other than the expiry of T3412. The most common reason is the change of the tracking area of the UE and the fact that the new TAI is not one of the TAIs in the TAI list currently possessed by the UE.
    - *Periodic updating:* It indicates that the TAU procedure is invoked because the periodic tracking area updating timer (i.e., T3412) has expired.
    - *Combined tracking area/location area updating:* It indicates that Location Area Update (LAU) procedure is invoked in addition to the TAU, since the UE attaches to both EPS and 2G/3G.
    - *Combined tracking area/location area updating with IMSI attach:* It indicates that the UE attaches to EPS and it wants to attach to 2G/3G.

The EPS Update Type IE also includes an active flag. The active flag indicates if the UE requests for the activation of the radio bearer and the S1 bearer for all the active EPS bearers along with the current TAU procedure. The active EPS bearers are already indicated by the EPS Bearer Context Status IE. Such activation typically occurs if the UE has some uplink user or signaling data available when the TAU procedure is invoked.

4. *Additional update type:* This IE is used to indicate if the UE requests for SMS service over the SGs interface but not the circuit-switched fallback (CSFB) operation. If this request is accepted, then the UE does not fall back to UTRAN/GERAN but the UE registers for exchange of the SMS over the SGs interface as explained in Section 21.4.2.
5. *Voice domain preference and UE's usage setting:* This IE is used to indicate if the UE supports CSFB and SMS over the SGs interface or if the UE supports IMS for voice.
6. *UE radio capability information update needed:* This IE indicates if the MME needs to be updated with new UE radio capability information.
7. *DRX parameter:* The UE may change the UE-specific DRX parameters, which may even be the reason for performing the TAU. This IE indicates the change of the UE-specific DRX parameters.
8. *UE network capability:* This IE includes the ciphering and integrity protection algorithms supported by the UE unless the UE is performing a periodic TAU.
9. *Old GUTI:* This IE contains the most recent GUTI possessed by the UE at the time of request for a TAU.
10. *NAS key set identifier, $KSI_{ASME}$:* If the UE has valid NAS key set identifier, $KSI_{ASME}$, it is included here. If a valid security context does not exist, then it is indicated that no key is available.

## 22.2 UE Context Update

If the UE is in the ECM-IDLE state when the TAU procedure is invoked and also if the MME has been changed, then the new MME is updated with the UE context from the old MME as shown next. This updating is not required if the MME has not been changed.

1. The new MME derives the old MME from the old GUTI IE in the TRACKING AREA UPDATE REQUEST message. The new MME sends a CONTEXT REQUEST message to the old MME in order to retrieve the user information. This message includes old GUTI, F-TEID of the new MME at the S10 interface, the UE Validated flag, and the TRACKING AREA UPDATE REQUEST message. The UE Validated field flag indicates that the new MME has validated the integrity protection of the TRACKING AREA UPDATE REQUEST message based on the EPS-cached security context. The old MME validates the CONTEXT REQUEST message using the TRACKING AREA UPDATE REQUEST message and it sends an error message to the new MME if the integrity check fails. The old MME uses

the F-TEID of the new MME at the S10 interface to send signaling to the new MME.

2. The old MME sends a CONTEXT RESPONSE message to the new MME. This message includes the IMSI, IMEI, MSISDN, unused EPS authentication vectors, $KSI_{ASME}$, $K_{ASME}$, EPS bearer contexts, subscribed UE AMBR in uplink and downlink, serving GW signaling address and TEIDs, UE core network capability, UE-specific DRX parameters, and F-TEID of the old MME at the S10 interface. The new MME uses the F-TEID of the old MME at the S10 interface to send signaling to the old MME. The EPS bearer contexts include the EPS bearer identity (EBI), uplink TFT, downlink TFT, and QoS parameters for each EPS bearer.

3. The new MME performs authentication that executes the EPS AKA procedure and establishes a new NAS security context as explained in Sections 12.2 and 12.3. If the integrity check of the TRACKING AREA UPDATE REQUEST message has failed in a previous step, then this authentication step is mandatory; otherwise, it is optional.

4. The new MME constructs an MM context for the UE. The MM context contains mobility management and UE security parameters. The new MME sends a CONTEXT ACKNOWLEDGE message to the old MME. The old MME considers that the information in the gateways and HSS is no longer valid.

## 22.3 Bearer Update

If the UE is in the ECM-IDLE state when the TAU procedure is invoked, then the bearer context is updated, which takes place as follows:

1. The new MME considers the EPS bearers indicated in the aforementioned CONTEXT RESPONSE message for establishment. The new MME checks on which bearers are intended to be kept by the UE based on the EPS Bearer Context Status IE of the TRACKING AREA UPDATE REQUEST message. Then the new MME initiates the establishment of the appropriate EPS bearers according to the indicated order. The MME deactivates the EPS bearers that cannot be established.

2. The new MME determines if the serving GW needs to be changed. If the serving GW remains unchanged, then the new MME sends a MODIFY BEARER REQUEST message to the serving GW for each PDN connection. This message contains the new MME address, TEID, and serving network identity. The PDN GW address is indicated in the bearer contexts.

    If the serving GW is changed, the new MME sends a CREATE SESSION REQUEST message to the serving GW for each PDN connection. This message contains the IMSI, bearer contexts, MME address and TEID, type, protocol type over S5/S8, RAT type, and serving network. The bearer contexts are included for each EPS bearer and the bearer context includes the PDN GW address. The Type field indicates to the serving GW to send the Create Session Request to the PDN GW. The Protocol Type over S5/S8 field indicates to the serving GW if GTP or PMIP should be used over the S5/S8 interface. The RAT type indicates if there is any change in the RAT.

4. The serving GW sends a MODIFY BEARER REQUEST message to each PDN GW for each PDN connection and informs the PDN GW if there are any changes (e.g., if the RAT type has been changed). This message contains the serving GW address and the F-TEID of the serving GW for both the control plane and the user plane.
5. The PDN GW updates its bearer contexts. The PDN GW sends a MODIFY BEARER RESPONSE message to the serving GW. This message contains the TEID for uplink traffic.
6. The serving GW updates its bearer contexts. If the serving GW is not changed, then the serving GW sends a MODIFY BEARER RESPONSE message to the MME. This message contains the serving GW address and the TEID for uplink traffic. On the other hand, if the serving GW is changed, the serving GW sends a CREATE SESSION RESPONSE message to the MME. This message contains the serving GW address and the TEIDs for the control plane and the user plane and the PDN GW TEIDs for uplink traffic and the control plane.

## 22.4 Updating HSS

If the UE is in the ECM-IDLE state when the TAU procedure is invoked and if the MME has been changed, then the HSS is updated as follows:

1. The new MME checks if there is already subscription data of the UE available based on its GUTI. If there are no subscription data available, then the new MME sends an UPDATE LOCATION REQUEST message to the HSS and thus informs the HSS of the change of the MME. This message contains the new MME identity, MME capabilities, IMSI of the UE, ULR-flags, and IMS voice over PS sessions. The ULR-flags indicate that the update location is sent from an MME and the MME registration has to be updated in the HSS. The MME Capabilities field indicates regional access restriction functionality supported by the MME. The Homogenous Support of IMS over PS Sessions field indicates if the IMS is supported for voice homogeneously in all tracking areas under the serving MME.
2. The HSS updates its MME registration. It also sends a CANCEL LOCATION message to the old MME. This message includes the IMSI of the UE and cancellation type. The Cancellation Type field is set to the update procedure. However, if the HSS rejects the location update request of the MME, the MME rejects the TAU procedure. This may occur, for example, if the subscription of the UE does not allow access to this tracking area.
3. The old MME deletes MM and bearer contexts. It also sends a CANCEL LOCATION ACK message to the HSS as an acknowledgment of its release. This message includes the IMSI of the UE.

4. After cancellation of the old MME context, the HSS sends an UPDATE LOCATION ACK message to the new MME. This message includes the IMSI and the subscription data of the UE.

## 22.5 Tracking Area Update Accept

If the MME duly accepts the TAU request, it notifies the UE by sending a TRACKING AREA UPDATE ACCEPT message. This is performed whether the UE is in the ECM-IDLE state or the ECM-CONNECTED state when the TAU procedure is invoked and also whether or not the MME has been changed. The MME sends a TRACKING AREA UPDATE ACCEPT message to the eNodeB using a DOWNLINK NAS TRANSPORT message. Thus, the MME responds to the TRACKING AREA UPDATE REQUEST message. The eNodeB forwards the TRACKING AREA UPDATE ACCEPT message to the UE by including it in the NAS-DedicatedInformation IE of the DLInformationTransfer message. The TRACKING AREA UPDATE ACCEPT message includes the following major IEs:

- *EPS update result:* This IE contains the type of the TAU that has been accepted. This IE can represent the TA updated, the combined TA/LA updated, the TA updated and ISR activated, the combined TA/LA updated, or the ISR activated.
- *GUTI:* This IE is included if the MME assigns a new GUTI to the UE. Then this IE contains the new GUTI.
- *TAI list:* If the MME assigns a new TAI list to the UE, this IE is included and contains the new TAI list. Thus, the allocation of GUTI and/or TAI list (i.e., GUTI reallocation) may be performed as a part of the TAU procedure. The GUTI reallocation can also occur independently as explained in Section 15.2.
- *Additional update result:* This IE is included in response to the Additional Update Type IE in the TRACKING AREA UPDATE REQUEST message. It is used to indicate if the registration in the CN domain has been accepted for only SMS service over the SGs interface or the UE can fallback to UTRAN/GERAN for circuit-switched fallback (CSFB) operation as explained in Section 21.4.2.
- *T3412 value:* This IE assigns the value for T3412. This IE was explained in Section 15.1.7.4.
- *T3412 extended value:* This IE is introduced in Release 10. This IE assigns a longer value for T3412. This IE was explained in Section 15.1.7.4. When this IE is included in the TRACKING AREA UPDATE ACCEPT message, the UE applies its value instead of the value of the T3412 Value IE for T3412. If this IE is not included, the UE applies the value of the T3412 Value IE for T3412. If this IE is included, the T3412 Value IE is invariably included, but it may happen that neither this IE nor T3412 Value IE is included. Then the UE applies the stored value from the most recent ATTACH ACCEPT message or TRACKING AREA UPDATE ACCEPT message.

- *EPS bearer context status:* This IE indicates which EPS bearer contexts have been activated. This IE indicates the state of each EPS bearer identifying them by their EPS bearer identities.
- *Equivalent PLMNs:* If the MME assigns a new list of equivalent PLMNs to the UE, this IE is included and contains the new list.
- *EPS network feature support:* This IE is included to inform the UE of the support of certain features, if there are any.

If a new GUTI has been assigned in the TRACKING AREA UPDATE ACCEPT message, then the UE sends a TRACKING AREA UPDATE COMPLETE message to the MME to acknowledge the received GUTI.

If the network cannot accept the TAU request, it sends the TRACKING AREA UPDATE REJECT message to the UE instead of the TRACKING AREA UPDATE ACCEPT message. The TRACKING AREA UPDATE REJECT message includes the EMM Cause IE, which depicts the cause of rejection.

## 22.6 S1 Release

The S1 release procedure is performed at this stage, provided that the active flag in the EPS Update Type IE of the TRACKING AREA UPDATE REQUEST message was not set. The S1 release procedure was explained in Section 14.4. If the active flag was set, the NAS signaling connection is kept. Moreover, the radio bearer and the S1 bearer are established for the additional active EPS bearers, if needed, in conjunction with the TAU procedure.

# CHAPTER 23

# Change of Cell

LTE supports mobility (i.e., the service is continued as the user moves out of the coverage of one cell and enters the coverage of another cell). The serving cell is updated automatically with the movement of the UE. The decision for the change of cell depends on the relative radio link quality between the current cell and the neighbor cells at the position of the UE. In the RRC_IDLE state, if the UE changes the cell on which it is camped, the cell reselection procedure is performed. On the other hand, in the RRC_CONNECTED state, if the serving cell changes, then the handover procedure is performed. The handover can also be performed to move a user from a heavily loaded cell to a lightly loaded cell for the purpose of load balancing.

## 23.1 Neighbor Cells

The neighbor cells around the UE can be classified as follows:

1. *Intrafrequency cells:* If the current cell and the neighbor cell operate on the same carrier frequency, then the neighbor cell is called an intrafrequency cell. Intrafrequency measurements are predominant in a system with frequency reuse = 1. There are three different scenarios for intrafrequency cells as shown in Table 23.1.
2. *Interfrequency cells:* If the current cell and the neighbor cell operate on a different carrier frequency, then the neighbor cell is called an interfrequency cell. There are three different scenarios for interfrequency cells as shown in Table 23.2.
3. *Inter-RAT cells:* The neighbor cell operates on a radio access technology (RAT) other than LTE (e.g., GERAN, UTRAN, and CDMA2000). LTE service is not likely to be available everywhere before long, so switching to a cell on another RAT will be necessary where the coverage of LTE finishes. In addition, switching to another RAT can be performed in order to use the circuit-switched (CS) service, particularly for voice.

**Table 23.1** Scenarios for Intrafrequency Cells

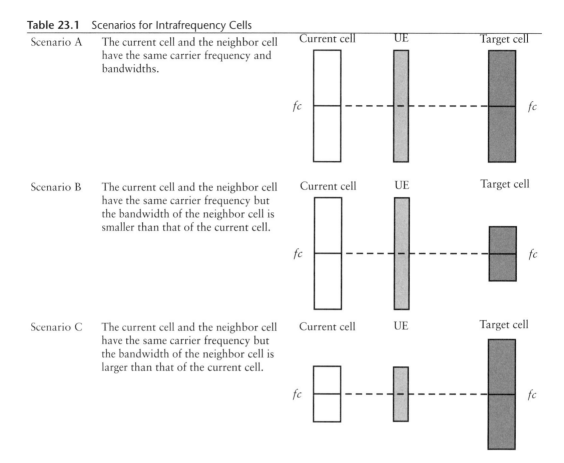

| | |
|---|---|
| Scenario A | The current cell and the neighbor cell have the same carrier frequency and bandwidths. |
| Scenario B | The current cell and the neighbor cell have the same carrier frequency but the bandwidth of the neighbor cell is smaller than that of the current cell. |
| Scenario C | The current cell and the neighbor cell have the same carrier frequency but the bandwidth of the neighbor cell is larger than that of the current cell. |

## 23.2 Cell Reselection

While a UE is camped on a cell in the RRC_IDLE state, it attempts to operate on the best-quality RF carrier available at its present location. Therefore, it keeps on checking on the neighbor cells around itself. If it finds a neighbor cell better than the serving cell such that the cell reselection conditions are fulfilled, the UE itself selects the new cell without notifying the network; this is known as cell reselection. Since there are no resources allocated to the UE in the RRC_IDLE state, notifying the network is of no use. Such notification procedure would unnecessarily waste wireless resources and power. The cell reselection can involve a change of RAT.

The eNodeB does not provide the UE with any list of neighbor cells for the UEs to consider for cell reselection. The UE rather searches for candidate neighbor cells, performs measurement on the detected cells, and performs cell reselection, provided that the necessary conditions are satisfied. It may be noted that certain SIB types specify neighbor cells, but it is merely to assign different parameter values to the specific neighbor cells and not to provide any list of neighbor cells for measurement.

The frequencies for interfrequency and inter-RAT cell reselection are specified on SIB types as shown next, and the UE uses only these frequencies. In addition, the SIB types assign relative priorities to these frequencies using the

## 23.2 Cell Reselection

**Table 23.2** Scenarios for Interfrequency Cells

| Scenario D | The current cell and the neighbor cell have different carrier frequencies and the bandwidth of the neighbor cell is smaller than that of the current cell. The bandwidth of the neighbor cell lies within the bandwidth of the current cell. | 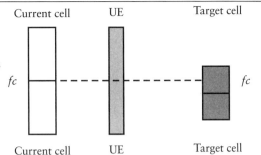 |
|---|---|---|
| Scenario E | The current cell and the neighbor cell have different carrier frequencies and the bandwidth of the neighbor cell is larger than that of the current cell. The bandwidth of the current cell lies within the bandwidth of the neighbor cell. | 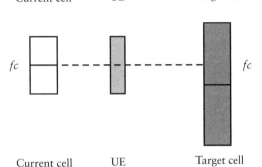 |
| Scenario F | The current cell and the neighbor cell have different carrier frequencies and their bandwidths are not overlapping. | 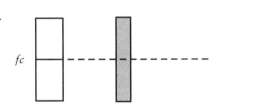<br /> |

CellReselectionPriority IE. Such priorities help in spreading users among different colocated frequencies in the same cell. The CellReselectionPriority IE can have any integer value from 0 through 7 and its higher value is interpreted as a higher priority. The E-UTRAN and any other RAT cannot be assigned the same priority. The priority setting can be skipped by omitting the CellReselectionPriority IE for any of these frequencies and, if skipped, the UE does not evaluate that particular frequency. The eNodeB provides the priority of the current E-UTRAN frequency using the CellReselectionPriority IE on SIB type 3.

- *E-UTRAN frequencies:* SIB type 5 contains the list of E-UTRAN frequencies and their CellReselectionPriority IE. It can accommodate 8 E-UTRAN frequencies at the maximum.
- *UTRAN frequencies:* SIB type 6 contains the list of UTRAN frequencies and their CellReselectionPriority IE. It can accommodate 16 UTRAN frequencies at the maximum.

- *GERAN frequencies:* SIB type 7 contains the list of GERAN frequencies and their CellReselectionPriority IE. The GERAN frequencies are grouped for this purpose. SIB type 7 can include 16 such groups at the maximum.

Specific dedicated priorities can be optionally assigned to a UE when the UE enters the RRC_IDLE state from the RRC_CONNECTED state using the RRCConnectionRelease message. These dedicated priorities are deleted when the UE enters the RRC_CONNECTED state again or when the PLMN selection is performed. The RRCConnectionRelease message includes the IdleModeMobilityControlInfo IE. The IdleModeMobilityControlInfo IE optionally includes the list of E-UTRAN frequencies, UTRAN frequencies, and GERAN frequencies along with the CellReselectionPriority IE to set their priorities. The CellReselectionPriority IE can have any integer value from 0 through 7, and its higher value is interpreted as higher priority. The priorities of the RRCConnectionRelease message override the priorities of SIB type 5, SIB type 6, and SIB type 7. The priorities can also be assigned when the UE enters the RRC_IDLE state from another RAT. The UE applies the priority information on the RRCConnectionRelease message for the T320 timer period after the RRC connection release. The RRCConnectionRelease message can include the T320 timer value and the value can be 5, 10, 20, 30, 60, 120, or 180 minutes. When the UE receives the RRCConnectionRelease message, it starts the T320 timer. When the T320 timer expires, the UE starts applying cell reselection priority information provided in the SIB types.

SIBs are also used to broadcast many other cell reselection-related parameters, and SIB type 3 is the most important among these SIBs. If SIB type 3 is not broadcast, the UE does not perform cell reselection.

### 23.2.1 Measurement of Neighbor Cells

The UE performs periodic measurements of the neighbor cells along with the periodic measurement of the serving cell so that it can compare cells existing around itself and operate on the cell of the best quality. The UE first identifies a neighbor cell using the cell search procedure and then initiates the periodic measurements of that neighbor cell.

The UE uses the parameter Srxlev of the serving cell and the neighbor cells for comparison purposes. Srxlev of the serving cell is denoted as $S_{ServingCell}$ and Srxlev of a neighbor cell is denoted as $S_{nonServingCell,x}$. Srxlev of the serving cell is determined as described in Section 9.1.1. Srxlev of the neighbor cells are determined as follows:

1. *Intrafrequency cells and interfrequency cells:* Srxlev of intrafrequency cells and interfrequency cells are calculated as follows:

   $$Srxlev = Qrxlevmeas - Qrxlevmin - Pcompensation$$

   where Pcompensation = MAX $[P_{EMAX} - P_{PowerClass}, 0]$
   The parameters in these equations are explained here:
   - *Qrxlevmeas:* It is the reference signal received power (RSRP) in dBm measured by the UE in the cell.

- *Qrxlevmin:* Qrxlevmin is the minimum required RSRP for reselection to the cell. The Q-Rxlevmin field on SIB type 3 specifies the value of Qrxlevmin to be used for intrafrequency cells. SIB type 5 contains the Q-Rxlevmin field individually for each of the interfrequencies to specify the value of Qrxlevmin. The actual value of Qrxlevmin is the Q-Rxlevmin field value × 2 [dBm]. The value of Q-Rxlevmin can be any integer between –70 and –22.
- $PE_{MAX}$: $PE_{MAX}$ is the maximum allowed uplink transmission power broadcast by the eNodeB. The P-Max field on SIB type 3 specifies the value of PEMAX to be used for intrafrequency cells. SIB type 5 contains the P-Max field individually for each of the interfrequencies to specify the value of $PE_{MAX}$. The value of P-Max can range between – 30 dBm and 33 dBm.
- $P_{PowerClass}$: $P_{PowerClass}$ is the maximum RF output power of the UE according to the particular power class of the UE.

  SIB type 3 and SIB type 5 contain the following additional parameters for measurement of intrafrequency cells and interfrequency cells, respectively.
- *AllowedMeasBandwidth:* It gives the maximum allowed measurement bandwidth on the E-UTRA carrier frequency in terms of the number of resource blocks, which can be 6, 15, 25, 50, 75, or 100.
- *PresenceAntennaPort1:* This is a 1-bit field. If it is set to true, then it indicates that the UE should assume the presence of at least two antenna ports in all neighboring cells of the E-UTRA frequency. Thus, in this case, both antenna port 0 and antenna port 1 are available. For RSRP determination, the UE may use cell-specific reference signals R1 (on antenna port 1) in addition to R0 (on antenna port 0) if the UE can reliably detect the reference signals from the second antenna port.

  SIB type 4 assigns different parameter values to the intrafrequency neighbor cells. It can specify 16 intrafrequency cells as such at the maximum. SIB type 5 assigns different parameter values to interfrequencies and to cells on these intrafrequencies. It can specify 8 E-UTRAN frequencies at the maximum and 16 interfrequency cells at the maximum on each of the E-UTRAN frequencies.

2. *UTRAN cells:* Srxlev of UTRAN cells are calculated as follows:

$$Srxlev = RSCP - Qrxlevmin - MAX\begin{pmatrix} \text{P-MaxUTRA} - \\ \text{Maximum Output} \\ \text{Power of the UE, 0} \end{pmatrix}$$

Received Signal Code Power (RSCP) is measured by the UE over Common Pilot Channel (CPICH) of the UTRAN cell in dBm. SIB type 6 contains the Q-RxLevMin IE and the P-MaxUTRA IE to provide other parameter values in the equation as follows:
- Qrxlevmin specifies the minimum required received signal level at the UE in the UTRAN cell in dBm. The value of Qrxlevmin is calculated as:

(Q-RxLevMin IE value $\times$ 2 + 1) dBm. The IE value can be between $-60$ and $-13$.
- P-MaxUTRA specifies the maximum allowed uplink transmission power in the UTRAN cell in dBm. SIB type 6 contains the P-MaxUTRA IE, which gives the value of P-MaxUTRA. The value can be between $-50$ dBm and 33 dBm.

3. *GERAN cells*: Srxlev of GERAN cells are calculated as follows:

$$\text{Srxlev} = \text{RxLeV} - \text{Q-RxLevMin} - \text{MAX}\begin{pmatrix} \text{P-MaxGERAN} - \\ \text{Maximum Output} \\ \text{Power of the UE, 0} \end{pmatrix}$$

RxLev is the RSSI measured by the UE over the control channel of the GERAN cell in dBm. SIB type 7 contains the Q-RxLevMin IE and the P-MaxGERAN IE to provide other parameter values in the equation as follows:
- Q-RxLevMin specifies the minimum required received signal level at the UE in the GERAN cell in dBm. The value of Q-RxLevMin is calculated as: (the Q-RxLevMin IE value $\times$ 2 $-$ 115) dBm. The IE value can be any integer between 0 and 45.
- P-MaxGERAN specifies the maximum allowed uplink transmission power in the GERAN cell in dBm. SIB type 7 contains the P-MaxGERAN IE, which gives the value of P-MaxGERAN. The value can be between 0 dBm and 39 dBm.

### 23.2.1.1 Intrafrequency Measurement Conditions

The UE performs intrafrequency measurements in the following cases:

1. SIB type 3 contains $S_{intrasearch}$, and $S_{ServingCell}$ is less than or equal to $S_{intrasearch}$ (i.e., the UE does not perform intrafrequency measurements if $S_{ServingCell}$ is greater than $S_{intrasearch}$).
2. $S_{intrasearch}$ is not sent on SIB type 3.

SIB type 4 contains the list of blacklisted intrafrequency neighboring cells. The UE does not consider any blacklisted cells as candidates for cell reselection even if detected.

### 23.2.1.2 Interfrequencies and Inter-RAT Measurement Conditions

1. If an E-UTRAN interfrequency or inter-RAT frequency has a priority higher than the priority of the current E-UTRA frequency, then the UE performs measurements of the E-UTRAN interfrequency or inter-RAT frequency.
2. If an E-UTRAN interfrequency or an inter-RAT frequency has equal or lower priority than the reselection priority of the current E-UTRA frequen-

cy, then the UE performs measurements of the E-UTRAN interfrequency or the inter-RAT frequency in the following cases:

- SIB type 3 contains $S_{nonintrasearch}$, and $S_{ServingCell}$ is less than or equal to $S_{nonintrasearch}$. The value of $S_{nonintrasearch}$ is given as the ReselectionThreshold IE, which can be any integer up to 31. The value of $S_{nonintrasearch}$ is calculated as: (ReselectionThreshold IE value × 2) dB.
- SIB type 3 does not contain $S_{nonintrasearch}$.

#### 23.2.1.3 Aggressive Measurements

If the UE finds that the serving cell does not fulfill the cell selection criterion for a particular period, then the UE initiates the measurements of all neighbor cells indicated by the serving cell regardless of any measurement rules currently restricting the measurement activities. This particular period depends on the DRX cycle length as shown in Table 23.3.

### 23.2.2 Cell Reselection Criteria

The UE can reselect a new cell if more than 1 second has elapsed since the UE camped on the current serving cell. The cell reselection takes place when both these conditions are met:

1. The target neighbor cell is suitable and fulfills the following S criteria:
   - Intrafrequency cells and interfrequency cells: Srxlev > 0.
   - UTRAN cells: Srxlev > 0 as well as Squal > 0 where Squal = Ec/Io – Qqualmin.
   
   Ec/Io is measured by the UE over CPICH of the UTRAN cell in decibels. Qqualmin specifies the minimum required quality level in the UTRAN cell in decibels. SIB type 6 contains the value of Qqualmin. Its value can be between – 24 dB and 0 dB.
   - GERAN cells: Srxlev > 0.
2. Certain cell reselection conditions are found to be fulfilled for a neighbor cell for a particular duration, $Treselection_{RAT}$. $Treselection_{RAT}$ is broadcast on different SIB types for various RATs as shown next. $Treselection_{RAT}$ can be any integer number of seconds up to 7 seconds.
   - Intrafrequency cells: $Treselection_{RAT}$ is expressed as $Treselection_{EUTRAN}$. SIB type 3 contains the T-ReselectionEUTRA field, which gives the value of $Treselection_{EUTRAN}$.

Table 23.3 Period to Trigger Aggressive Measurement

| DRX Cycle Length (sec) | Period to Trigger Measurement (sec) |
|---|---|
| 0.32 | 1.28 (4 DRX cycles) |
| 0.64 | 2.56 (4 DRX cycles) |
| 1.28 | 2.56 (2 DRX cycles) |
| 2.56 | 5.12 (2 DRX cycles) |

- Interfrequency cells: TreselectionRAT is expressed as Treselection$_{EUTRAN}$. SIB type 5 contains the T-ReselectionEUTRA field for each of the E-UTRA frequencies, which gives the value of TreselectionEUTRAN for the E-UTRA frequencies.
- UTRAN cells: TreselectionRAT is expressed as Treselection$_{UTRAN}$. SIB type 6 contains the T-ReselectionUTRA field for each of the UTRAN frequencies, which gives the value of TreselectionUTRAN for the UTRAN frequencies.
- GERAN cells: TreselectionRAT is expressed as Treselection$_{GERAN}$. SIB type 7 contains the T-ReselectionGERAN field for each of the GERAN frequencies, which gives the value of Treselection$_{GERAN}$ for the GERAN frequencies.

The cell reselection conditions are specified in three different ways depending on the various relative priorities among different types of cells. These three ways are applied in the order of priority as shown here:

- *Priority 1:* Interfrequency frequency or inter-RAT frequency of higher priority than the serving frequency;
- *Priority 2:* Intrafrequency or equal priority interfrequency;
- *Priority 3:* Interfrequency frequency or inter-RAT frequency of lower priority than the serving frequency.

### 23.2.2.1 Priority 1: Interfrequency Frequency or Inter-RAT Frequency of Higher Priority Than the Serving Frequency

The UE reselects to a cell on a E-UTRAN frequency or inter-RAT frequency of a higher priority than the serving frequency if it finds $S_{nonServingCell,x}$ (i.e., Srxlev of the neighbor cell of a higher-priority E-UTRAN frequency or inter-RAT frequency) to stay greater than the threshold value, $Thresh_{x, high}$, for the duration Treselection$_{RAT}$. Since the cell reselection condition does not incorporate the Srxlev of the serving cell, it is possible that the Srxlev of the neighbor cell is lower than the Srxlev of the serving cell when the cell reselection occurs. The ReselectionThreshold IE gives the value of $Thresh_{x, high}$. The ReselectionThreshold IE is broadcast on SIB type 5 for interfrequencies, on SIB type 6 for UTRAN frequencies, and on SIB type 7 for GERAN frequencies. The value of $Thresh_{x, high}$ is calculated as: (ReselectionThreshold IE value × 2) dB. The ReselectionThreshold IE can be any integer up to 31.

### 23.2.2.2 Priority 2: Intrafrequency or Equal Priority Interfrequency

The UE considers the reselection to neighbor cells of the current E-UTRAN frequency and equal priority interfrequency neighbor cells evaluating the rank of the serving cell and the neighbor cells. In order to avoid too many cell reselections, a hysteresis is applied to the rank of the serving cell. For reselecting a new cell, the cell has to be better than the serving cell by the hysteresis that is configurable by the network. In addition, an amount of offset is applied to the rank of the neighbor cells, which allows biasing the reselection toward particular cells. The ranks of the serving cell and neighbor cells are calculated frequently as follows:

The rank of the serving cell, $R_s = Q_{meas,s} + Q_{Hyst}$

## 23.2 Cell Reselection

The rank of a neighbor cell, $R_n = Q_{meas,n} - Qoffset$

The UE reselects to a new intrafrequency and equal priority interfrequency cell if the new cell stays higher ranked than the serving cell for duration Treselection-$_{RAT}$. If more than one neighbor cell is higher ranked, then the UE reselects to the highest ranked cell. A separate Treselection$_{RAT}$ timer is started for each cell that becomes better ranked than the serving cell. If the Treselection$_{RAT}$ timer expires for any cell when it is the highest ranked, the UE triggers cell reselection to that cell. The different terms in the rank calculation are:

1. $Q_{meas,s}$ *and* $Q_{meas,n}$: $Q_{meas,s}$ and $Q_{meas,n}$ are the RSRP measured by the UE on the serving cell and on the neighbor cell, respectively.
2. $Q_{Hyst}$: $Q_{Hyst}$ is the hysteresis to be applied to the rank of the serving cell. SIB type 3 contains the parameter $Q_{Hyst}$. The value of $Q_{Hyst}$ can be 0, 1, 2, 3, 4, 5, 6, 8, 10, 12, 14, 16, 18, 20, 22, or 24 dB.
3. *Qoffset*: The value of Qoffset is determined as follows:
   - Intrafrequency: SIB type 4 contains the Q-OffsetCell field for each of the intrafrequency cells listed. This field gives the Qoffset for the particular intrafrequency cell. The value of the Q-OffsetCell can be −24, −22, −20, −18, −16, −14, −12, −10, −8, −6, −5, −4, −3, −2, −1, 0, 1, 2, 3, 4, 5, 6, 8, 10, 12, 14, 16, 18, 20, 22, or 24 dB.
   - Interfrequency: SIB type 5 contains the Q-OffsetFreq field for each of the E-UTRAN frequencies. The value of Q-OffsetFreq is given by the Q-OffsetRange IE. SIB type 5 also contains the Q-OffsetCell field for each of the interfrequency cells listed and the value of the Q-OffsetCell is also given by the Q-OffsetRange IE. The value of the Q-OffsetRange IE can be −24, −22, −20, −18, −16, −14, −12, −10, −8, −6, −5, −4, −3, −2, −1, 0, 1, 2, 3, 4, 5, 6, 8, 10, 12, 14, 16, 18, 20, 22, or 24 dB. The Qoffset for a particular interfrequency cell is calculated as the value of the Q-OffsetFreq for the E-UTRAN frequency plus the value of the Q-OffsetCell for the cell.

### 23.2.2.3 Priority 3: Interfrequency Frequency or Inter-RAT Frequency of Lower Priority Than the Serving Frequency

When the conditions for reselecting to the aforementioned first two types of cells are not found to be satisfied, the UE can reselect to a cell on an E-UTRAN frequency or inter-RAT frequency of a lower priority than the serving frequency, provided the following condition is met.

$S_{ServingCell}$ (i.e., Srxlev of the serving cell) stays lower than the threshold, Thresh$_{serving, low}$, as well as $S_{nonServingCell,x}$ (i.e., Srxlev of the neighbor cell of a lower-priority E-UTRAN frequency or inter-RAT frequency) stays greater than the threshold, Thresh$_{x, low}$ for the duration Treselection$_{RAT}$. The different terms in this condition are described here:

- Thresh$_{serving, low}$: SIB type 3 contains the ThreshServingLow field and its value is given by the ReselectionThreshold IE. The value of Thresh$_{serving, low}$ is calculated as: (ReselectionThreshold IE value × 2) dB. The ReselectionThreshold IE can be any integer up to 31.

- $Thresh_{x,\ low}$: SIB type 5 contains the $Thresh_x$, low field for each of the E-UTRA frequencies. SIB type 6 contains the $Thresh_x$, low field for each of the UTRAN frequencies. SIB type 7 contains the $Thresh_x$, low field for each of the GERAN frequencies in terms of the ReselectionThreshold IE. The value of $Thresh_x$, low is calculated as: (ReselectionThreshold IE value × 2) dB.

### 23.2.2.4 Reduction of $Q_{hyst}$ and $Treselection_{RAT}$ for Speedy UEs

LTE attempts to ensure good performance even at a high speed. When the UE moves at a very high speed, it passes through many cells in short period of time. Thus, it requires changes of serving cells too frequently. In this case, the good quality of radio link can be maintained if the UE can switch the serving cell as soon as the signal from the new cell starts getting better than the signal from the old cell. This is achieved by allowing less stringent cell reselection conditions through the reduction of the values of $Q_{hyst}$ and $Treselection_{RAT}$. Also, the higher the speed of the UE is, the less stringent the cell reselection conditions should be. Therefore, apart from the normal-mobility state supporting UEs at a low speed, the high-mobility state and medium-mobility state are defined in order to improve performance for UEs at a very high speed and a medium-high speed, respectively. Thus, the following three different states of mobility are defined for the UE:

1. High-mobility state;
2. Medium-mobility state;
3. Normal-mobility state.

The UE determines its appropriate mobility state based on the number of cell reselections during a certain time as shown next. However, consecutive reselections between the same two cells do not reflect the speed of the UE in a particular direction. Thus, the UE does not take consecutive reselections between the same two cells into account while detecting the mobility state:

- *High-mobility state:* The UE enters a high-mobility state if there are more than $N_{CR\_H}$ number of cell reselections in the $T_{CRmax}$ duration.
- *Medium-mobility state:* The UE enters a medium-mobility state if there are more than a $N_{CR\_M}$ number of cell reselections but less than a $N_{CR\_H}$ number of cell reselections in the $T_{CRmax}$ duration.
- *Normal-mobility state:* The UE stays in a normal-mobility state if the parameters $T_{CRmax}$, $N_{CR\_H}$, $N_{CR\_M}$, and $T_{CRmaxHyst}$ are not sent in the system information. When they are sent, the UE enters the normal-mobility state from the medium-mobility state or the high-mobility state if neither the medium-mobility state nor the high-mobility state is detected during the time period $T_{CRmaxHyst}$.

SIB type 3 contains the SpeedStateReselectionPars IE, which includes the MobilityStateParameters IE. The MobilityStateParameters IE includes the following fields for the detection of the mobility state:

- T-Evaluation: This field gives the value of $T_{CRmax}$.
- T-HystNormal: This field gives the value of $T_{CRmaxHyst}$.
- N-CellChangeMedium: This field gives the value of $N_{CR\_M}$.
- N-CellChangeHigh: This field gives the value of $N_{CR\_H}$.

The values of $T_{CRmax}$ and $T_{CRmaxHyst}$ can be 30, 60, 120, 180, or 240 seconds. The values of $N_{CR\_M}$ and $N_{CR\_H}$ can be any integer up to 16.

When the normal-mobility state is detected, the UE does not modify any parameters. However, if the high-mobility state or the medium-mobility state is detected, the UE modifies the values of $Q_{hyst}$ and Treselection as shown here:

- *$Q_{hyst}$*: The hysteresis, $Q_{hyst}$, applied to the rank of the serving cell, is reduced for speedy UEs. Thus, a lesser difference between the signal levels of the serving cell and the new cell can satisfy the cell reselection conditions leading to a quicker cell reselection. The SpeedStateReselectionPars IE includes Q-HystSF IE (speed-dependent ScalingFactor for $Q_{hyst}$). Q-HystSF includes the SF-High and SF-Medium fields. They both take on only negative values. Their values can be –6, –4, –2, or 0 dB. When the high-mobility state or the medium-mobility state is detected, $Q_{hyst}$ is decreased by adding SF-High or SF-Medium, respectively, to $Q_{hyst}$.
- *Treselection$_{RAT}$*: The cell reselection takes place when certain conditions are found to be fulfilled for the duration, Treselection$_{RAT}$. Therefore, Treselection$_{RAT}$ is reduced by a scaling factor for speedy UEs in order to allow a quicker cell reselection. Different SIB types provide the values of TreselectionRAT for various RATs. The same different SIB types provide the scaling factors for various RATs. Thus, the scaling factor is given by SIB type 3, SIB type 5, SIB type 6, and SIB type 7 for intrafrequency, interfrequency, UTRAN, and GERAN cells, respectively. The value of each of these scaling factors is given by the SpeedStateScaleFactors IE. The SpeedStateScaleFactors IE includes the SF-High and SF-Medium fields. SF-High or SF-Medium can take on the values 0.25, 0.5, 0.75, or 1. When the high-mobility state or the medium-mobility state is detected, TreselectionRAT for the particular RAT is decreased by multiplying by the corresponding SF-High or SF-Medium, respectively.

## 23.3 HANDOVER

In the RRC_CONNECTED state, the mobility of the UE is supported through the only procedure, handover between eNodeBs. The handover is always hard handover (i.e., it uses the break-before-make approach). Since soft handover is not supported, only a single radio link can exist between the UE and the network.

### 23.3.1 Types of Handover

The handover can be classified based on the involvement of a change of RAT as follows:

1. *Intra-E-UTRAN handover:* The handover takes place between two E-UTRAN cells.
2. *Inter-RAT handover:* The handover takes place between an E-UTRAN cell and a cell of a RAT other than LTE (e.g., GERAN, UTRAN, and CDMA2000).

Each of these handover types can be further classified as shown in the following sections.

#### 23.3.1.1 Intra-E-UTRAN Handover

The intra-E-UTRAN handover can be classified based on the interface used for handover initiation as follows. However, from the perspective of the UE, these two types of handover appear to be the same.

1. *X2-based handover:* If there is an X2 interface interconnecting the source eNodeB and the target eNodeB, then the handover procedure typically uses the X2 interface. In this case, the source eNodeB and the target eNodeB belong to the same MME. The X2-based handover procedure differs slightly in the two following cases:
   - The handover does not change the serving GW.
   - The handover changes the serving GW.
2. *S1-based handover:* The handover between E-UTRAN cells takes place using the S1 interfaces that interconnect the source eNodeB and the target eNodeB with their MMEs, typically when the X2 interface cannot be used. This may happen in the following cases:
   - The X2 interface does not exist between the source eNodeB and the target eNodeB
   - A change of the MME and/or a change of the serving GW is required and the source eNodeB is configured to use the S1 interface because of this change. In this case, the source eNodeB and the target eNodeB may belong to different tracking areas.
   - The target eNodeB rejects the attempt of the source eNodeB to initiate handover using the X2 interface.
   
   The S1-based handover may involve a change of the MME or not, depending on the change of the location of the UE. Also, if the MME changes, then the S1-based handover may or may not involve a change of the serving GW.

The intra-E-UTRAN handover can also be classified based on the scenario for handover as follows:

1. *UE-assisted handover:* This is the typical type of handover. It provides for network-controlled and UE-assisted mobility. The UE performs

measurements of the radio link quality of serving and neighbor cells and sends measurement reports to the network. The network considers the radio link quality reported and makes the decision for handover. This is often referred to as *backward handover* because it uses exchange of information over the radio link with the old eNodeB.

2. *Blind handover:* The network makes a decision of handover without depending on any measurement reports from the UE. The blind handover can be quickly performed due to the avoidance of any measurements, so it can be used when the delay in handover is undesired. However, the radio link quality at the new cell is uncertain. The blind handover typically occurs to move a user from a heavily loaded cell to a lightly loaded cell for the purpose of load balancing.

3. *Radio link failure (RLF) handover:* The RLF handover was defined in Section 14.1.3.2. This handover procedure is UE-based.

    In order to improve the success of the RLF handover, multiple target eNodeBs can be prepared simultaneously for handover. Then source eNodeB performs handover preparation phase, explained in Section 23.3.3.2, with multiple potential target eNodeBs instead of only one. The source eNodeB performs the handover execution phase with only one target eNodeB. If there is failure in the signaling related to handover because of a poor radio link, the UE selects a suitable E-UTRA cell from the prepared ones. Then the UE can establish an RRC connection with this cell promptly. Since the RRCConnectionReestablishmentRequest message includes the C-RNTI of the UE and PCI of the source cell as shown in Section 14.1.4.1, the new target eNodeB can detect the source eNodeB. Then this new target eNodeB can obtain the UE context from the source eNodeB. The ReestablishmentCause field in the RRCConnectionReestablishmentRequest message indicates that handover failure has triggered the RRC connection reestablishment procedure. After successful handover with the new target eNodeB, it sends the UE CONTEXT RELEASE message to the source eNodeB. Then the source eNodeB may indicate cancellation to the other prepared eNodeBs.

    A variant of the RLF handover is the forward handover. The forward handover procedure is also UE-based. It is performed when the serving eNodeB has a radio link too poor to allow it to duly receive measurement reports from the UE. Then, unlike the typical RLF handover, the serving eNodeB fails to prepare any target eNodeB for handover. Despite no target eNodeB prepared for handover, the UE can choose and access an eNodeB on its own and thus conduct forward handover.

### 23.3.1.2 Inter-RAT Handover

The inter-RAT handover takes place between an E-UTRAN cell and a cell of GERAN, UTRAN, or CDMA2000. The source eNodeB uses the S1 interface to initiate the inter-RAT handover. The inter-RAT handover can be classified as follows:

1. *UE-assisted handover:* It is network-controlled and UE-assisted handover. The UE performs measurements of the radio link quality of the serving cell and inter-RAT neighbor cells and sends measurement reports to the

network. The network considers the radio link quality reported and makes the decision for handover.
2. *Blind handover:* The network makes a decision of handover without depending on any measurement reports from the UE. The blind handover typically occurs for the purpose of circuit-switched fallback (CSFB) because the reduction of call setup delay is crucial in CSFB.
3. *RLF handover:* The RLF handover can also be inter-RAT. In this case, the UE selects a cell using another RAT (e.g. GERAN or UTRAN).

The inter-RAT handover procedure differs slightly in the following two cases:
- The handover does not change the serving GW.
- The handover changes the serving GW.

### 23.3.2 Measurement and Reporting

The network needs to be aware of the radio link quality of the serving cell as well as the neighbor cells at the position of the UE in order to make the decision for handover. This requires that the UE measures the radio link quality of serving and neighbor cells and sends the measurement results to the network.

#### 23.3.2.1 Configuration Procedure

The eNodeB configures how the UE would perform the measurement and send measurement reports, and, for this purpose, the eNodeB sends the MeasConfig IE on the RRCConnectionReconfiguration message. This configuration involves the following basic items:

1. *Measurement object:* The eNodeB specifies and configures what the UE would measure; this is known as measurement object. Each measurement object configured is provided an ID. The eNodeB can specify 32 measurement objects at the maximum. A particular measurement object specifies the measurement configuration for any one and only one of the following items:
   - The serving E-UTRA frequency for intrafrequency measurements.
   - A nonserving E-UTRA frequency for interfrequency measurements. The measurement object also specifies measurement bandwidth and other parameters because they are not known to the UE for the nonserving E-UTRA carrier.
   - A particular UTRAN frequency and a set of cells on the UTRAN frequency for UTRAN measurements.
   - A set of GERAN frequencies for GERAN measurements.
   
   For intrafrequency and interfrequency measurements, the measurement object specifies neighbor cells but merely to assign cell-specific offset values to the neighbor cells and not to provide any list of neighbor cells for measurement. The UE rather searches for candidate neighbor cells and performs measurement on the detected cells. However, for inter-RAT measurements, the measurement object provides a neighbor cell list (NCL) for measurement.

2. *Measurement reporting:* After the UE performs the measurement of the radio link quality of neighbor cells, it sends the measurement results to the network. This is known as measurement reporting. The eNodeB configures the measurement reporting procedure for the measurement objects. The eNodeB defines a number of different measurement reporting configurations for this purpose. Each of these measurement reporting configurations is provided an ID. The eNodeB can specify 32 measurement reporting configurations at the maximum.
3. *Measurement identity:* The eNodeB can specify certain measurement configurations to apply. A particular measurement configuration is identified by its ID. The eNodeB can specify multiple measurement identities for the measurement of multiple configurations. A measurement configuration links one measurement object with one reporting configuration. Multiple measurement identities can use the same measurement reporting ID and thus link more than one measurement object with the same reporting configuration. Also, multiple measurement identities can use the same measurement object ID and thus link more than one reporting configuration with the same measurement object. The eNodeB can specify 32 measurement configurations at the maximum.

The UE performs measurements of neighbor cells only when the RSRP of the serving cell goes below the threshold (i.e., the UE does not measure neighbor cells if the RSRP of the serving cell is higher than the threshold regardless of any other measurement configurations). The MeasConfig IE includes the S-Measure field, which gives the threshold RSRP value. The value of the S-Measure field is given by the RSRP-Range IE according to Table 23.4. The eNodeB can avoid setting any threshold RSRP value for a serving cell and allow the UE to perform measurements of the neighbor cells at all times. This is done by setting value 0 for S-Measure.

### 23.3.2.2 Measurement Object

When the eNodeB adds or modifies a measurement object, it includes the MeasObjectToAddModList IE in the MeasConfig IE. The MeasObjectToAddModList

**Table 23.4** RSRP Values

| RSRP-Range IE Value | RSRP Value |
|---|---|
| 0 | RSRP < −140 dBm |
| 1 | −140 dBm ≤ RSRP < −139 dBm |
| 2 | −139 dBm ≤ RSRP < −138 dBm |
| 3 | −138 dBm ≤ RSRP < −137 dBm |
| ... | ... |
| ... | ... |
| ... | ... |
| 95 | −46 dBm ≤ RSRP < −45 dBm |
| 96 | −45 dBm ≤ RSRP < −44 dBm |
| 97 | −44 dBm ≤ RSRP |

IE includes the MeasObjectToAddMod IE for each measurement object. The MeasObjectToAddMod IE includes the MeasObjectId IE, which contains the ID for the measurement object. A new ID is assigned for new measurement objects. The ID can be any integer number up to 32. The MeasObjectToAddMod IE also includes MeasObjectEUTRA or MeasObjectUTRA or MeasObjectGERAN IE; these are explained here:

1. *MeasObjectEUTRA:* This IE is used to configure an intrafrequency or interfrequency measurement object. This IE includes the following IEs:
   - CarrierFreq: This field gives the EARFCN of the E-UTRA carrier frequency.
   - AllowedMeasBandwidth: The function of this field is explained in Section 23.2.1..
   - PresenceAntennaPort1: The function of this field is explained in Section 23.2.1.
   - NeighCellConfig: This field is explained in Section 24.3.
   - OffsetFreq: This field gives the value of OffsetFreq, which has to be applied in case of Events A3, A4, and A5 as shown in Section 23.3.2.3. The value of OffsetFreq is given by the Q-OffsetRange IE. The Q-OffsetRange IE can be −24, −22, −20, −18, −16, −14, −12, −10, −8, −6, −5, −4, −3, −2, −1, 0, 1, 2, 3, 4, 5, 6, 8, 10, 12, 14, 16, 18, 20, 22, or 24 dB.
   - CellForWhichToReportCGI: This IE gives the PCI of the cells for which the UE is required to try to acquire the ECGI for the purpose of ANRF. ANRF is explained in Section 25.4.
   - CellsToAddModList: This IE is used to assign cell-specific offset values to the neighbor cells. When it adds or modifies the offset assignment, it includes CellsToAddMod IE for the particular cell. The CellsToAddMod includes the following fields:
     - *CellIndex:* Each neighbor cell is provided an index. An index may also indicate a range of neighbor cells. The index can be an integer up to 32.
     - *PhysCellId:* It gives the physical cell identifier (PCI) of the cell.
     - *CellIndividualOffset:* This field gives the value of CellIndividualOffset, which is to be applied to the cell in case of Events A3, A4, and A5 as shown in Section 23.3.2.3.
   
   If any neighbor cell is removed from the offset assignment, then the MeasObjectEUTRA IE includes the CellsToRemoveList IE, which contains the CellIndex of the cells to be removed.

2. *MeasObjectUTRA:* This IE is used to configure a UTRAN frequency as the measurement object. It includes the following IEs with regard to the UTRAN frequency:
   - CarrierFreq: This field gives the UARFCN of the UTRAN carrier frequency.
   - OffsetFreq: This field gives the value of OffsetFreq, which is to be applied to the UTRAN carrier frequency in case of Events B1 and B2 as shown in Section 23.3.2.3. The offset is given by Q-OffsetRangeInterRAT IE whose value can be between −15 dB and 15 dB.

- CellForWhichToReportCGI: This field gives the Primary Scrambling Code (PSC) of the UTRAN cells for which the UE is required to try to acquire the global cell identity for the purpose of ANRF.
- CellsToAddModList: This IE is used to specify the cells to be measured on the UTRAN frequency. It includes the CellsToAddModListUTRA-FDD IE for FDD cells and the cellsToAddModListUTRA-TDD IE for the TDD cells. The CellsToAddModListUTRA-FDD IE includes the following fields:
  - *CellIndex*: Each neighbor cell is provided an index. The index can be an integer up to 32.
  - *PhysCellIdUTRA-FDD*: It gives the PSC of the UTRAN cell, which can be between 0 and 511.

  When the MeasObjectEUTRA IE removes any neighbor UTRAN cells, it includes the CellsToRemoveList IE, which contains the CellIndex of the cells to be removed.

3. *MeasObjectGERAN*: This IE is used to configure a group of GERAN frequencies as the measurement object. It includes the following IEs:
   - CarrierFreqsGERAN: This IE gives the ARFCNs of the group of GERAN carrier frequencies.
   - OffsetFreq: This field gives the value of OffsetFreq, which is to be applied to the GERAN carrier frequencies in case of Events B1 and B2 as shown in Section 23.3.2.3. The offset is given as the Q-OffsetRangeInterRAT IE whose value can be between −15 dB and 15 dB.
   - CellForWhichToReportCGI: This field gives the base station identity code (BSIC) of the GERAN cells on this group of GERAN carrier frequencies for which the UE is required to try to acquire the global cell identity for the purpose of ANRF.

The MeasConfig IE includes the MeasObjectToRemoveList IE in order to leave a measurement object out. The MeasObjectToRemoveList IE contains the MeasObjectId IEs for the measurement objects to be removed.

#### 23.3.2.3 Measurement Reporting

A measurement reporting configuration is identified by its ID. The UE triggers measurement reporting in either of two ways:

1. *Event-triggered:* The UE begins sending measurement reports when any of the particular events takes place. The UE is considered to enter a particular event when certain conditions are met. Similarly, the UE is considered to leave a particular event when certain conditions are met. As long as the event remains activated, the UE keeps on sending measurement reports one after another at a certain interval until it reaches a maximum number of measurement reports. The network specifies entering and leaving conditions for the various events, the interval between measurement reports, and the maximum number of measurement reports to be sent.

   There are five types of events for measurement reporting of E-UTRA cells as shown in Table 23.6:

- *Event A1:* The serving cell becomes better than a threshold.
- *Event A2:* The serving cell becomes worse than a threshold.
- *Event A3:* The neighbor cell becomes better than the serving cell by an offset.
- *Event A4:* The neighbor cell becomes better than a threshold.
- *Event A5:* The serving cell becomes worse than a threshold, threshold 1, and the neighbor cell becomes better than another threshold, threshold 2.

There are two types of events for the measurement reporting of inter-RAT cells as shown in Table 23.7:

- *Event B1:* The neighbor cell becomes better than a threshold.
- *Event B2:* The serving cell becomes worse than a threshold and the neighbor cell becomes better than another threshold.

2. *Periodical:* The UE sends measurement reports regardless of the satisfaction of any conditions. The UE keeps on sending the measurement reports one after another at a certain interval until it reaches a maximum number of measurement reports. The network specifies the interval between measurement reports and the maximum number of measurement reports to be sent.

When the eNodeB adds or modifies a measurement reporting configuration, it includes the ReportConfigToAddModList IE in the MeasConfig IE. The ReportConfigToAddModList IE includes the ReportConfigToAddMod IE for each reporting configuration. The ReportConfigToAddMod IE includes the ReportConfigId IE, which specifies the ID of the measurement reporting configuration. A new ID is assigned for new measurement reporting. The ID can be any integer number up to 32.

*E-UTRA Measurement Reporting*
If the measurement object is an E-UTRA frequency, then the ReportConfigToAddMod IE includes the ReportConfigEUTRA IE in order to configure measurement reporting.

*Parameters for Both Event-Triggered and Periodical Measurement Reporting*
The ReportConfigEUTRA IE includes the following fields that are applied to both event-triggered and periodical measurement reporting cases:

- *TriggerType:* This field specifies if the measurement reporting is event-triggered or periodical.
- *TriggerQuantity:* This field specifies if the UE would send measurement reports in terms of RSRP or RSRQ.
- *ReportQuantity:* This field can have the value "SameAsTriggerQuantity" or "Both." If its value is set to "SameAsTriggerQuantity," then the UE sends measurement reports in terms of RSRP or RSRQ as specified by TriggerQuantity. If the value is set to "Both," then the UE sends measurement reports in terms of RSRP and RSRQ.
- *ReportAmount:* This field specifies the maximum number of measurement reports to be sent by the UE. Its value can be 1, 2, 4, 8, 16, 32, 64, or infinity.

When the value of ReportAmount is infinity, the UE keeps on sending measurement reports until any reporting condition changes (e.g., the leaving condition for an event is met).

- *ReportInterval:* This field gives the length of the interval between measurement reports. It can be 120, 240, 480, 640, 1,024, 2,048, 5,120, or 10,240 ms or 1, 6, 12, 30, or 60 minutes.
- *MaxReportCells:* This field specifies the maximum number of neighbor cells to be included in the measurement report. It can be any integer up to 8.

*Event-Triggered Measurement Reporting*
The ReportConfigEUTRA IE includes the following fields in the case of event-triggered measurement reporting to specify the criteria for sending measurement reports in addition to the aforementioned common fields:

- *TimeToTrigger:* This field specifies the period during which the specific criteria for a particular event need to be met in order to trigger the measurement reporting. Its value can be 0, 40, 64, 80, 100, 128, 160, 256, 320, 480, 512, 640, 1,024, 1,280, 2,560, or 5,120 ms.
- *Hysteresis:* This IE specifies the value of hysteresis, which is used for the determination of the entering condition and leaving condition for a particular event-triggered measurement reporting. The value of hysteresis is calculated as: (the IE value × 0.5) dB, where the IE value can be any integer up to 30.
- *Threshold:* This IE gives the threshold value for a particular event-triggered measurement reporting as follows.
  - *A1-Threshold:* The value of threshold for Event A1.
  - *A2-Threshold:* The value of threshold for Event A2.
  - *A4-Threshold:* The value of threshold for Event A4.
  - *A5-Threshold:* The value of threshold 1 for Event A5.
  - *A5-Threshold:* The value of threshold 2 for Event A5.
  
  All the threshold values are given using the following IEs:
  - *RSRP-Range:* This IE is used for the value of threshold if the measurement reports are made in terms of RSRP. Its value can be any integer up to 97 and the value corresponds to the RSRP value as shown in Table 23.4.
    - *RSRQ-Range:* This IE is used for the value of threshold if the measurement reports are made in terms of RSRQ. Its value can be any integer up to 34 and the value corresponds to the RSRQ value as shown in Table 23.5.
- *A3-Offset:* This IE gives the offset value to be used in the case of Event A3. The value of offset is calculated as: (the IE value × 0.5) dB.
- *ReportOnLeave:* This 1-bit field indicates if the UE would send a measurement report when the leaving condition for an event is met.

The conditions to be fulfilled for entering and leaving event-triggered measurement reporting are shown in Table 23.6.

**Table 23.5** RSRQ Values

| RSRQ-Range IE Value | RSRQ Value |
|---|---|
| 0 | RSRQ < –19.5 dB |
| 1 | –19.5 dB ≤ RSRQ < –19 dB |
| 2 | –19 dB  RSRQ < –18.5 dB |
| 3 | –18.5 dB ≤ RSRQ < –18 dB |
| ... | ... |
| ... | ... |
| ... | ... |
| 32 | –4 dB ≤ RSRQ < –3.5 dB |
| 33 | –3.5 dB ≤ RSRQ < –3 dB |
| 34 | –3 dB ≤ RSRQ |

*Periodical Measurement Reporting*
In order to configure periodical measurement reporting, the ReportConfigEUTRA IE includes the Purpose IE in addition to the aforementioned common fields. The Purpose IE indicates if this measurement reporting is for the purpose of handover decision or ANRF explained in Section 25.4. The Purpose IE can take on any of the following values:

- *ReportStrongestCells:* The measurement reporting is for the purpose of handover decision. The UE keeps on sending the measurement reports one after another periodically until it reaches the specified number of measurement reports.
- *ReportCGI:* The measurement reporting is for the purpose of ANRF to discover the identity of certain cells. The MeasObjectEUTRA IE in the MeasConfig IE includes the CellForWhichToReportCGI IE, which contains the PCI of the cells for which the UE is required to attempt for the acquisition of the ECGI.
  Since, in this case, the purpose is reporting the identity of certain cells and not the radio link quality, the UE sends measurement report only once. Thus, the value of ReportAmount is considered 1.

*Inter-RAT Measurement Reporting*
If the measurement object is an inter-RAT frequency, then the ReportConfigToAddMod IE includes the ReportConfigInterRAT IE in order to configure measurement reporting.

*Parameters for Both Event-Triggered and Periodical Measurement Reporting*
The ReportConfigInterRAT IE includes the following fields that are applied to both event-triggered and periodical measurement reporting cases:

- *TriggerType:* This field specifies if the measurement reporting is event-triggered or periodical.

## 23.3 HANDOVER

**Table 23.6** The Conditions for Event-Triggered EUTRA Measurement Reporting

| Event Type | Basis for the Event | Entering Condition | Leaving Condition | Parameter Specification for the Condition |
|---|---|---|---|---|
| Event A1 | The UE enters the event when the serving cell becomes better than a threshold. | Serving cell measurement − Hysteresis > Threshold | Serving cell measurement + Hysteresis < Threshold | ReportConfigEUTRA IE specifies hysteresis and threshold. |
| Event A2 | The UE enters the event when the serving cell becomes worse than a threshold. | Serving cell measurement + Hysteresis < Threshold | Serving cell measurement − Hysteresis > Threshold | ReportConfigEUTRA IE specifies hysteresis and threshold. |
| Event A3 | The UE enters the event when the neighbor cell becomes better than the serving cell by an offset. | Neighbor cell measurement + OffsetFreq for neighbor cell + CellIndividualOffset for neighbor cell − Hysteresis > Serving cell measurement + OffsetFreq for serving cell + CellIndividualOffset for serving cell + A3-Offset | Neighbor cell measurement + OffsetFreq for neighbor cell + CellIndividualOffset for neighbor cell + Hysteresis < Serving cell measurement + OffsetFreq for serving cell + CellIndividualOffset for serving cell + A3-Offset | MeasObjectEUTRA IE specifies OffsetFreq and CellIndividualOffset. ReportConfigEUTRA IE specifies hysteresis and A3-Offset. |
| Event A4 | The UE enters the event when the neighbor cell becomes better than a threshold. | Neighbor cell measurement + OffsetFreq for neighbor cell + CellIndividualOffset for neighbor cell − Hysteresis > Threshold | Neighbor cell measurement + OffsetFreq for neighbor cell + CellIndividualOffset for neighbor cell + Hysteresis < Threshold | MeasObjectEUTRA IE specifies OffsetFreq and CellIndividualOffset. ReportConfigEUTRA IE specifies hysteresis and threshold. |
| Event A5 | The UE enters the event when the serving cell becomes worse than a threshold, Threshold 1, and the neighbor cell becomes better than another threshold, Threshold 2. | Serving cell measurement + Hysteresis < Threshold 1 and Neighbor cell measurement + OffsetFreq for neighbor cell + CellIndividualOffset for neighbor cell − Hysteresis > Threshold 2 | Serving cell measurement − Hysteresis > Threshold 1 or Neighbor cell measurement + OffsetFreq for neighbor cell + CellIndividualOffset for neighbor cell + Hysteresis < Threshold 2 | MeasObjectEUTRA IE specifies OffsetFreq and CellIndividualOffset. ReportConfigEUTRA IE specifies hysteresis, Threshold 1, and Threshold 2. |

- *ReportAmount:* This field specifies the maximum number of measurement reports to be sent by the UE. Its value can be 1, 2, 4, 8, 16, 32, 64, or infinity. When the value of ReportAmount is infinity, the UE keeps on sending measurement reports until any reporting condition changes (e.g., the leaving condition for an event is met).
- *ReportInterval:* This field gives the length of the interval between measurement reports. It can be 120, 240, 480, 640, 1,024, 2,048, 5,120, or 10,240 ms or 1, 6, 12, 30, or 60 minutes.
- MaxReportCells: This field specifies the maximum number of neighbor cells to be included in the measurement report. It can be any integer up to 8.

*Event-Triggered Measurement Reporting*
The ReportConfigInterRAT IE includes the following fields in the case of event-triggered measurement reporting to specify the criteria for sending measurement reports in addition to the aforementioned common fields:

- *TimeToTrigger:* This field specifies the period during which the specific criteria for a particular event needs to be met in order to trigger the measurement reporting. Its value can be 0, 40, 64, 80, 100, 128, 160, 256, 320, 480, 512, 640, 1,024, 1,280, 2,560, or 5,120 ms.
- *Hysteresis:* This IE specifies the value of hysteresis that is used for the determination of the entering condition and leaving condition for the particular event-triggered measurement reporting that is currently defined. The value of hysteresis is calculated as: (IE value × 0.5) dB, where the IE value can be any integer up to 30.
- *Threshold:* This IE gives the threshold value for the particular event-triggered measurement reporting currently defined as follows:
    - *B1-ThresholdUTRA:* The value of threshold for UTRAN cells for Event B1.
    - *B1-Threshold GERAN:* The value of threshold for GERAN cells for Event B1.
    - *B2-Threshold1:* The value of threshold 1 for Event B2.
    - *B2-Threshold2UTRA:* The value of threshold 2 for UTRAN cells for Event B2.
    - *B2-Threshold2GERAN:* The value of threshold 2 for GERAN cells for Event B2.
  
  The value of B2-Threshold1 is given using the following IEs:
    - *RSRP-Range:* This IE is used for the value of threshold if the measurement reports are made in terms of RSRP as shown in Table 23.4.
    - *RSRQ-Range:* This IE is used for the value of threshold if the measurement reports are made in terms of RSRQ as shown in Table 23.5.
  
  The value of B1-ThresholdUTRA and B2-Threshold2UTRA are given using the following IEs:
    - *UTRA-RSCP:* This IE is used for the value of threshold if the measurement reports of UTRAN cells are made in terms of RSCP over CPICH. The actual value of the threshold is: (IE value − 115) dBm. The IE value can be any integer between − 5 and 91.
    - *UTRA-EcN0:* This IE is used for the value of threshold if the measurement reports of UTRAN cells are made in terms of Ec/No over CPICH. The actual value of the threshold is (IE value − 49)/2 dB. The IE value can be any integer between 0 and 49.
  
  The value of B1-Threshold GERAN and B2-Threshold2GERAN are given using the ThresholdGERAN IE, which gives the value of threshold in terms of RxLev, which is the RSSI over the control channel of the GERAN cell. The actual value of the threshold is (IE value − 110) dBm. The IE value can be any integer between 0 and 63.

The conditions to be fulfilled for entering and leaving event-triggered measurement reporting are shown in Table 23.7.

*Periodical Measurement Reporting*
In order to configure periodical measurement reporting, the ReportConfigInterRAT IE includes the Purpose IE in addition to the aforementioned common fields. The Purpose IE indicates if this measurement reporting is for the purpose of handover decision or to discover neighbor cells as a part of the ANRF explained in Section 25.4. The Purpose IE can take on any of the following values:

1. *ReportStrongestCells:* The measurement reporting is for the purpose of handover decision. The UE keeps on sending the measurement reports one after another periodically until it reaches the specified number of measurement reports.
2. *ReportCGI:* The measurement reporting is for the purpose of ANRF to discover the identity of certain inter-RAT cells as follows:
    - *UTRAN cells:* The MeasObjectUTRA IE of the MeasConfig IE includes the CellForWhichToReportCGI IE, which contains the PSC of the UTRAN cells for which the UE is required to attempt for the acquisition of the global cell identity. The UE also attempts to acquire the location area code (LAC), the routing area code (RAC), and the list of additional PLMN identities of the UTRAN cell, if multiple PLMN identities are broadcast.
    - *GERAN cells:* The MeasObjectGERAN IE of MeasConfig IE includes the CellForWhichToReportCGI IE, which contains the BSIC of the GERAN cells for which the UE is required to attempt for acquisition of the cell global identification (CGI) of the GERAN cell. The UE also attempts

Table 23.7 The Conditions for Event-Triggered Inter-RAT Measurement Reporting

| Event Type | Basis for the Event | Entering Condition | Leaving Condition | Parameter Specification for the Condition |
|---|---|---|---|---|
| Event B1 | The UE enters the event when the inter-RAT neighbor cell becomes better than a threshold. | Neighbor cell measurement + OffsetFreq for Neighbor cell – Hysteresis > Threshold | Neighbor cell measurement + OffsetFreq for neighbor cell + Hysteresis < Threshold | MeasObjectUTRA IE specifies OffsetFreq for UTRAN cells. MeasObjectGERAN IE specifies OffsetFreq for GERAN cells. ReportConfigInterRAT IE specifies hysteresis and threshold. |
| Event B2 | The UE enters the event when the serving cell becomes worse than a threshold and the inter-RAT neighbor cell becomes better than another threshold. | Serving cell measurement + Hysteresis < Threshold 1 and Neighbor cell measurement + OffsetFreq for neighbor cell – Hysteresis > Threshold 2 | Serving cell measurement – Hysteresis > Threshold 1 or Neighbor cell measurement + OffsetFreq for neighbor cell + Hysteresis < Threshold 2 | MeasObjectUTRA IE specifies OffsetFreq for UTRAN cells. MeasObjectGERAN IE specifies OffsetFreq for GERAN cells. ReportConfigInterRAT IE specifies hysteresis, Threshold 1, and also Threshold 2 for UTRAN and GERAN cells. |

to acquire the RAC of the GERAN cell. System Information type 3 contains the MCC, MNC, LAC, and RAC.

Since, in this case, the purpose is reporting the identity of certain cells and not the radio link quality, the UE sends measurement report only once. Thus, the value of ReportAmount is considered to be 1.

3. *ReportStrongestCellsForSON:* This is used for the network to detect and report the strongest UTRAN cell for the purpose of self-optimizing networks (SON). It is currently not used for the GERAN network.

Since, in this case, the purpose is to report a detected UTRAN cell, the UE sends the measurement report only once and reports only the cell detected. Thus, both the values of MaxReportCells and ReportAmount are considered to be 1.

*Reduction of TimeToTrigger for Speedy UEs*

When the UE moves at a very high speed in the RRC_CONNECTED state, it passes through many cells in a short period of time. Thus, it requires handovers too frequently, so, similar with the case explained in Section 23.2.2.4, the good quality of radio link can be maintained if the handover takes place as soon as the signal from the new cell starts becoming better than the signal from the old cell. Therefore, the UE should trigger measurement reporting quicker, and this is achieved by reducing the value of TimeToTrigger for event-triggered measurement reporting. The values of TimeToTrigger in both the ReportConfigInterRAT IE and the ReportConfigEUTRA IE are reduced and the reduction is made the same way as was shown in Section 23.2.2.4, using the same IEs of MobilityStateParameters and SpeedStateScaleFactors.

The UE determines its appropriate mobility state based on the number of cell reselections during certain time as shown next. However, consecutive handovers between the same two cells do not reflect the speed of the UE in a particular direction, so the UE does not take consecutive reselections between the same two cells into account while detecting the mobility state.

- *High-mobility state:* The UE enters a high-mobility state if there are more than $N_{CR\_H}$ numbers of handovers in the $T_{CR_{max}}$ duration.
- *Medium-mobility state:* The UE enters the medium-mobility state if there are more than $N_{CR\_M}$ numbers of handovers but less than $N_{CR\_H}$ numbers of handovers in the $T_{CRmax}$ duration.
- *Normal-mobility state:* The UE stays in the normal-mobility state if the parameters $T_{CRmax}$, $N_{CR\_H}$, $N_{CR\_M}$, and $T_{CRmaxHyst}$ are not sent. When they are sent, the UE enters the normal-mobility state from the medium-mobility or high-mobility state if neither the medium-mobility state nor the high-mobility state is detected during the time period $T_{CRmaxHyst}$.

The MeasConfig IE contains the SpeedStatePars IE, which includes the MobilityStateParameters IE.

The MobilityStateParameters IE includes the T-Evaluation, T-HystNormal, N-CellChangeMedium, and N-CellChangeHigh fields for the detection of the mobility state as explained in Section 23.2.2.4.

The SpeedStatePars IE includes the TimeToTrigger-SF IE whose value is given by SpeedStateScaleFactors IE. The SpeedStateScaleFactors IE is explained in Section 23.2.2.4. When the normal-mobility state is detected, the UE does not modify TimeToTrigger, but when the high-mobility state or medium-mobility state is detected, TimeToTrigger is decreased by multiplying by SF-High or SF-Medium, respectively, to allow the intended quicker triggering of measurement reporting.

### 23.3.2.4 Measurement Gap

In the case of measurements of interfrequency and inter-RAT cells, the UE needs to tune its receiver to a different carrier frequency. This is facilitated by configuring a certain pause in uplink and downlink data transmission on all subcarriers and allowing the UE to perform interfrequency or inter-RAT measurements in the pause period. Such pause periods or gaps in transmission are called measurement gaps. The UE does not need any measurement gaps for the measurements of intrafrequency cells. Typically, the measurement gaps may be configured only when the radio link quality of the serving cell goes below certain threshold level. Unlike UMTS, only one measurement gap configuration is used by the UE at a time.

The UE indicates if measurement gaps are required for interfrequency and inter-RAT bands during UE capability information transfer as shown in Section 15.1.7.4. During the measurement gaps, the UE does not transmit any data. If there is uplink grant allocating resources in a measurement gap, then the UE processes the grant but it does not transmit in the allocated uplink resources. Also, during the measurement gaps, the UE does not transmit the Sounding Reference Signal (SRS), CQI/PMI/RI, and HARQ feedback. If the schedule of HARQ retransmission falls within a measurement gap, the HARQ retransmission does not take place. The UE also does not transmit in the subframe immediately after the measurement gap.

The eNodeB configures a single measurement gap pattern that has a periodical repetition of constant gap durations. In other words, a fixed length of measurement gap is repeated at fixed time intervals. The length of the measurement gap is called the Measurement Gap Length (MGL). The time interval at which the measurement gaps are repeated is called the Measurement Gap Repetition Period (MGRP).

Since the data transmission is paused during measurement gaps, a frequent scheduling of measurement gaps can reduce throughput more, but this may be justified depending on the number of interfrequency or inter-RAT measurements required. Therefore, two types of measurement gap patterns with different periodicity are defined allowing a trade-off. The two measurement gap patterns are differentiated with their IDs as shown in Table 23.8.

The MeasConfig IE includes the MeasGapConfig IE, which is used to set up, release, and configure measurement gaps. When it sets up measurement gap, it

**Table 23.8** Measurement Gap Pattern

| Gap Pattern ID | Measurement Gap Length (MGL) | Measurement Gap Repetition Period (MGRP) |
|---|---|---|
| 0 | 6 ms (6 subframes) | 40 ms (4 radio frames) |
| 1 | 6 ms (6 subframes) | 80 ms (8 radio frames) |

configures the gap by indicating the gap pattern ID and the value of GapOffset for the particular measurement gap pattern as explained here:

1. *Measurement Gap Pattern ID 0:* The measurement gaps are 6 ms or 6 subframes long and they are repeated in every 4 radio frames. The value of GapOffset can be any integer between 00 and 39. The last decimal digit of GapOffset is a number between 0 and 9 and this gives the subframe number in a radio frame where each gap starts. The first decimal digit of GapOffset is a number between 0 and 3 and this gives the radio frame number in each chunk of 4 radio frames where each gap starts. The chunks of 4 radio frames start with a System Frame Number (SFN) divisible by 4 [i.e., (SFN of the first radio frame in the chunks of 4 radio frames) mod 4 = 0].

2. *Measurement Gap Pattern ID 1:* The measurement gaps are 6 ms or 6 subframes long and they are repeated in every 8 radio frames. The value of GapOffset can be any integer between 00 and 79. The last decimal digit of GapOffset is a number between 0 and 9 and this gives the subframe number in a radio frame where each gap starts. The first decimal digit of GapOffset is a number between 0 and 7 and this gives the radio frame number in each chunk of 8 radio frames where each gap starts. The chunks of 8 radio frames start with an SFN divisible by 8 [i.e., (SFN of the first radio frame in the chunks of 4 radio frames) mod 8 = 0].

When the UE measures any LTE carrier, it can surely capture the Primary Synchronization Signal (PSS) and the Secondary Synchronization Signal (SSS) in the 6-ms-long gap. This is because eNodeB transmits PSS and SSS on the first 0.5-ms slot of first and sixth subframes of each radio frame, and thus, PSS and SSS repeat every 5 ms. The 6-ms-long measurement gap also provides a margin for the receiver to tune to the different LTE carrier and return to the serving LTE carrier. The UE also performs channel estimation based on the cell-specific reference signals and determines RSRP and RSRQ. The 6-ms-long measurement gap includes enough reference symbols for channel estimation.

### 23.3.2.5 Configuring Measurement Triggering

The eNodeB specifies a number of measurement configurations for the UE to apply and each measurement configuration is identified by its ID. A measurement configuration links one measurement object with one reporting configuration. When the eNodeB adds or modifies a measurement configuration, the MeasConfig IE includes the MeasIdToAddModList IE. The MeasIdToAddModList IE includes the MeasIdToAddMod IE for each measurement configuration.

The MeasIdToAddMod IE includes the following IEs to link a particular measurement object with a particular measurement reporting configuration:

- *MeasId:* This field contains the ID for the measurement configuration. The ID can be any integer number up to 32.
- *MeasObjectId:* This field gives the ID of the measurement object to be linked with a particular measurement reporting configuration. The ID of the measurement object is given by MeasObjectId.

- *ReportConfigId:* This field gives the ID of the measurement reporting configuration to be linked with a particular measurement object. The ID of the measurement reporting is given by ReportConfigId.

The MeasConfig IE includes the MeasIdToRemoveList IE in order to leave a measurement configuration out. The MeasIdToRemoveList IE contains the MeasId IE of the measurement configuration to be removed.

### 23.3.2.6 Updating UE with Measurement Configurations

The UE uses the VarMeasConfig IE to specify the currently required measurements and reporting for the UE itself based on instructions from the eNodeB. The UE stores and maintains the VarMeasConfig IE, which includes the key description of the measurements and reporting to be performed. The VarMeasConfig IE is not transmitted over the air. The UE determines the exact measurements to be performed based on the MeasConfig IE and then it updates the VarMeasConfig IE accordingly. The VarMeasConfig IE includes the following IEs:

- *MeasIdList:* This IE includes the MeasId of all measurement configurations for which the UE is required to perform measurements. Thus, when the MeasIdToAddModList IE adds a new measurement configuration, the UE adds the MeasId of the MeasIdToAddModList IE to this MeasIdList IE. Similarly, when the MeasIdToRemoveList IE removes a measurement configuration, the UE removes the MeasId of the MeasIdToRemoveList IE from this MeasIdList IE.
- *MeasObjectList:* This IE consists of the MeasObjectId of all measurement objects for which the UE is required to perform measurements. Thus, when the MeasObjectToAddModList IE adds a measurement object, the UE adds the MeasObjectId of MeasObjectToAddModList IE to this MeasObjectList IE. Similarly, when MeasObjectToRemoveList IE removes a measurement object, the UE removes the MeasObjectId of MeasObjectToRemoveList IE from this MeasObjectList IE.
- *ReportConfigList:* This IE consists of the ReportConfigId of all measurement reporting which the UE is required to use to send measurement reports. Thus, when ReportConfigToAddModList IE adds a measurement reporting configuration, the UE adds the ReportConfigId of ReportConfigToAddModList IE to this ReportConfigList IE. Similarly, when the ReportConfigToRemoveList IE removes a measurement reporting configuration, the UE removes the ReportConfigId of ReportConfigToRemoveList IE from this ReportConfigList IE.
- *S-Measure:* This IE is updated with the information of the same IE in the MeasConfig IE.
- *QuantityConfig:* This IE is updated with the information of the same IE in the MeasConfig IE.
- *SpeedStatePars:* This IE is updated with the information of the same IE in the MeasConfig IE.

### 23.3.2.7 Performing Measurements

Layer 1 of the UE performs measurements on the E-UTRA downlink cell-specific signal, determines RSRP and RSRQ, and reports the results to Layer 3. In the RRC_CONNECTED state, a measurement period of 200 ms is defined for intrafrequency measurements. In the 200 ms, the physical layer of the UE can perform RSRP and RSRQ measurements of at least 8 identified intrafrequency cells and report the results to Layer 3 when no measurement gaps are activated.

Layer 3 of the UE performs Layer 3 filtering of the measured results received from the physical layer before using the results for evaluation or reporting. Layer 3 filtering removes the effect of fast fading and Layer 1 measurement errors. Layer 3 filtering is performed for the measurements of the serving cell, intrafrequency, interfrequency, and inter-RAT neighbor cells.

The newly filtered measured result, $F_n$, is generated as follows:

$$F_n = (1-a) \times F_{n-1} + a \times M_n$$

where

$M_n$ = The newly received measurement result from the physical layer;

$F_{n-1}$ = The last filtered measurement result;

$a = 1/2^{(k/4)}$, where $k$ is called the filter coefficient. The relative influence of the recent measurement and older measurements is controlled by the filter coefficient, $k$.

The MeasConfig IE contains the QuantityConfig IE, which gives the value of the filter coefficient, $k$. The value of $k$ can be 0, 1, 2, 3, 4, 5, 6, 7, 8, 9, 11, 13, 15, 17, or 19. If $k$ is set to 0, then Layer 3 filtering is not performed.

The filter input rate is implementation dependent. The Layer 3 filtering period should be chosen to remove the effect of fast fading but not shadowing. It should be adapted depending on the degree of correlation present in samples. For example, at the high speed of the UE, the samples are not highly correlated. Therefore, it would be more accurate to have a shorter filtering period for slow-speed users in order to follow the shadowing. Also, the filter is adapted to preserve its time characteristics at different input rates.

### 23.3.2.8 Sending Measurement Reports

The UE sends measurement results using a MeasurementReport message to the eNodeB according to the measurement reporting configuration. The MeasurementReport message includes the MeasResults IE to carry the measurement results. The MeasResults IE includes the MeasId field indicating the ID of the measurement configuration for which the reporting is made. The MeasResults IE conveys the measurement results for different types of cells as follows:

1. *Serving cell*: The MeasResults IE includes the MeasResultPCell IE in order to report the measurement results for the serving cell in terms of either

RSRP or RSRQ using the following IEs. The MeasResultPCell IE includes following fields:
- *RSRPResult:* This IE gives the value of the RSRP of serving cell using the RSRP-Range IE as shown in Table 23.4.
- *RSRQResult:* This IE gives the value of the RSRQ of serving cell using the RSRQ-Range IE as shown in Table 23.5.

2. *Intrafrequency and interfrequency neighbor cells:* The MeasResults IE includes MeasResultListEUTRA IE in order to report the measurement results for intrafrequency and interfrequency neighbor cells. The MeasResultListEUTRA IE includes following fields:
   - *PhysCellId:* This contains the PCI of the neighbor cell.
   - *RSRPResult:* This IE gives the value of the RSRP of the neighbor cells using the RSRP-Range IE as shown in Table 23.4.
   - *RSRQResult:* This IE gives the value of the RSRQ of the neighbor cells using the RSRP-Range IE as shown in Table 23.5.

   When the Purpose IE in the ReportConfigEUTRA IE is ReportCGI, the UE sends the identities of the neighbor cells for the purpose of the ANRF. This is performed using the following IEs:
   - *CellGlobalId:* It contains the ECGI of the cells which is a globally unique identifier of the cell.
   - *TrackingAreaCode:* It contains the tracking area code (TAC) to which the cell belongs.
   - *PLMN-IdentityList:* It contains PLMNs from the PLMN-IdentityList of the neighbor cell.

3. *UTRAN neighbor cells:* The MeasResults IE includes the MeasResultListUTRA IE in order to report the measurement results for the UTRAN cells. The MeasResultListUTRA IE includes the following fields:
   - *PhysCellId:* This contains the PSC of the UTRAN neighbor cell.
   - *UTRA-RSCP:* This gives the value of the RSCP over the CPICH of the UTRAN cells. The actual value of the RSCP is (IE value – 115) dBm. The IE value can be any integer between –5 and 91.
   - *UTRA-EcN0:* This gives the value of Ec/No over the CPICH of the UTRAN cells. The actual value of Ec/No is (IE value – 49)/2 dB. The IE value can be any integer between 0 and 49.

   When the Purpose IE in the ReportConfigInterRAT IE is ReportCGI, the UE sends the identities of the UTRAN neighbor cells for the purpose of the ANRF. This is performed using the following IEs:
   - *CellGlobalId:* It contains the globally unique identification of the UTRAN cell. It is constructed from the MCC, MNC, and a 28-bit-long cell identity that is unique in the PLMN.
   - *LocationAreaCode:* It contains the LAC to which the cell belongs.
   - *RoutingAreaCode:* It contains the RAC to which the cell belongs.
   - *PLMN-IdentityList:* It contains PLMNs from the PLMN-IdentityList of the neighbor cell.

4. *GERAN neighbor cells:* The MeasResults IE includes the MeasResultGERAN IE in order to report the measurement results for the GERAN cells. The MeasResultGERAN IE includes the following fields:

- *CarrierFreq:* This contains the ARFCN of the GERAN carrier frequency.
- *PhysCellId:* This contains the BSIC of the GERAN neighbor cell.
- RSSI: This gives the value of the RSSI over the control channel of the GERAN cell.

When the Purpose IE in the ReportConfigInterRAT IE is ReportCGI, the UE sends the identities of the GERAN neighbor cells for the purpose of the ANRF. This is performed using the following IEs:

- *CellGlobalId:* It contains the CGI of the GERAN cell. It is constructed from the MCC, MNC, LAC, and a 16-bit-long cell identity that is unique in the LAC.
- *RoutingAreaCode:* It contains the RAC to which the cell belongs.

### 23.3.3 Intra-E-UTRAN Handover Procedure

This section explains the general procedure for the X2-based handover between E-UTRAN cells without a change of the serving GW. This type of handover is expected to take place commonly. The whole handover procedure can be considered as a composition of the following phases:

1. Measurements and reporting;
2. Handover decision;
3. Handover preparation;
4. Handover execution;
5. Handover completion.

The measurements and reporting phase was explained in Section 23.3.2. The other four phases are explained in the following sections. The exchange of messages is shown in Figure 23.1.

#### 23.3.3.1 Handover Decision

The network makes the decision for handover. The actual RRM algorithm for making this decision depends on the implementation at eNodeB, but it involves the following points:

- The source eNodeB determines the radio link quality of the serving cell and neighbor cells at the position of the UE based on the measurement reports from the UE. The source eNodeB checks if the signal from the target cell is sufficiently better than the signal from the serving cell continuously for a certain period before a decision for handover is made. In order to check on the continuity of the signal condition, the source eNodeB relies on consecutive multiple measurement reports.
- Appropriate hysteresis and offset values need to be determined.
- The criteria for the decision of handover must be consistent with the criteria for cell reselection. Otherwise, switching between the RRC_IDLE and RRC_CONNECTED states might cause switching back and forth between cells.

## 23.3 HANDOVER

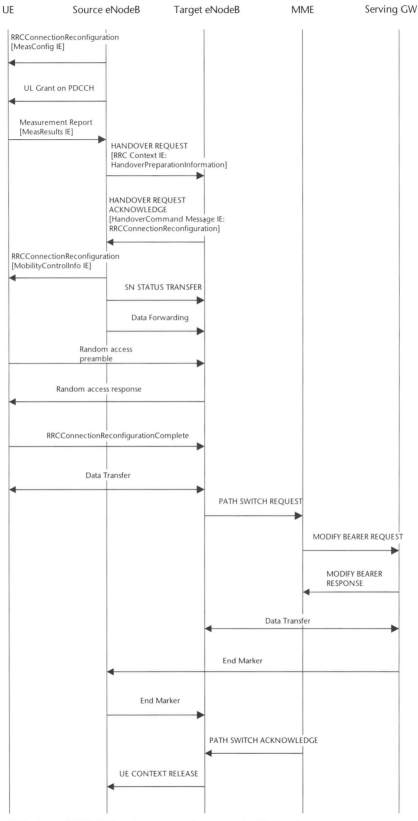

**Figure 23.1** Intra-E-UTRAN handover procedure over the X2 interface.

- The source eNodeB determines the traffic load situation at eNodeBs. The source eNodeB may consider load balancing among cells in making the handover decision. The inhomogeneous load distributions among cells may keep some cells overloaded and fail to satisfy the users, whereas other cells may have a few users with free resources. An appropriate offset used in the decision of handover can prioritize the lightly loaded cells and thus improve homogeneity in load distribution, better fulfilling the user requirement.

#### 23.3.3.2 Handover Preparation

The source eNodeB determines a target eNodeB for handover and thereafter, it conducts a negotiation with the target eNodeB as shown next.

*Handover Request*

The source eNodeB sends a HANDOVER REQUEST message to the target eNodeB. This message also includes the following information:

- ECGI of the target cell.
- GUMMEI of the serving MME.
- UE X2 signaling context reference at source eNodeB.
- UE Context Information: This includes the UE-AMBR, UE security capabilities, AS security information, and UE S1 signaling context reference.
- E-RABs To Be Setup List: This includes the E-RAB contexts of the E-RABs that will need to be set up at the target eNodeB if downlink data forwarding takes place. This includes the QoS profiles of the bearers uplink GTP tunnel endpoint at the serving GW.
- RRC Context IE: This IE contains an RRC message that is the HandoverPreparationInformation message. The HandoverPreparationInformation message includes the following major IEs:
  - UE-CapabilityRAT-ContainerList: This IE contains the radio access capabilities of the UE as explained in Section 15.1.7.4.
  - AS-Config: The AS-Config IE contains information about RRC configuration information in the source cell.
  - RRM-Config: This IE includes the UE-InactiveTime IE, which indicates how long the UE has not received or transmitted any user data.
  - AS-Context: The IE includes the ReestablishmentInfo IE, which contains information needed for the RRC connection reestablishment.

*Handover Request Acknowledge*

The target eNodeB determines if the resources can be granted depending on the E-RAB QoS information received. The admission control function at the target eNodeB can be used in this purpose. If the target eNodeB is ready for allocation of the resources, then it sends the HANDOVER REQUEST ACKNOWLEDGE to the source eNodeB.

The HANDOVER REQUEST ACKNOWLEDGE message contains the E-RAB context of the E-RABs that are accepted for setting up at the target eNodeB. It also includes uplink and downlink GTP tunnel endpoints to be used for forwarding of uplink and downlink data. It may also optionally include the E-RABs that have not been admitted. Moreover, the HANDOVER REQUEST ACKNOWLEDGE message contains the Target eNodeB To Source eNodeB Transparent Container IE. This IE transparently contains the HandoverCommand message, which is an RRC message. The HandoverCommand message includes the HandoverCommandMessage IE. The HandoverCommandMessage IE contains the entire RRCConnectionReconfiguration message, which the source eNodeB will directly send to the UE.

### 23.3.3.3 Handover Execution

The handover execution phase includes the following steps.

*Handover Instruction*

The source eNodeB sends the RRCConnectionReconfiguration message to the UE as an instruction for handover action. Earlier, the target eNodeB fully created this RRCConnectionReconfiguration message and forwarded to the source eNodeB using the Target eNodeB To Source eNodeB Transparent Container IE of the HANDOVER REQUEST ACKNOWLEDGE message.

The RRCConnectionReconfiguration message includes the MobilityControlInfo IE, which specifies how the UE would act at the target eNodeB. The MobilityControlInfo IE includes the following parameters:

- *TargetPhysCellId:* This IE gives the PCI of the target cell.
- *CarrierFreqEUTRA:* This IE gives the downlink EARFCN and uplink EARFCN used at the target cell.
- *CarrierBandwidthEUTRA:* This IE gives both the downlink and uplink bandwidth of the target cell in terms of the number of resource blocks, which can be 6, 15, 25, 50, 75, or 100.
- *NewUE-Identity:* This IE assigns a new C-RNTI to the UE for use at the target cell.
- *RACH-ConfigDedicated:* This IE is typically included. In this case, the UE performs the noncontention-based random access procedure using the parameters specified in this IE as explained in Chapter 11. On the other hand, if the target eNodeB does not assign a dedicated random access preamble, this RACH-ConfigDedicated IE is not included in the MobilityControlInfo IE and the UE performs the contention-based random access procedure at the target eNodeB.
- *RadioResourceConfigCommon:* As shown in Appendix A.3, this IE includes detailed physical layer parameters and random access parameters to be used at the target cell.

When the UE receives the RRCConnectionReconfiguration message it starts timer T304.

*Proper Delivery of Packets*
The RLC and MAC layers are reset after handover. Since the RLC layer is reset, it can no longer deliver the PDCP SDUs in order. Also, there can be data missing due to the interruption of service. The RLC context is not transferred during handover, so if there are any remaining RLC PDUs, they are flushed. The PDCP layer is reestablished after handover and it offers in-order delivery of packets, duplicate detection, and necessary retransmission.

In order to avoid the loss of data, the source eNodeB forwards data to the target eNodeB over the X2 interface temporarily during the handover action until the data transfer begins between the UE and the target eNodeB. Such data forwarding is not required in the case of the HSPA because the RLC layer is implemented in a central node RNC that can keep data for retransmission after handover. The actions for proper delivery of packets are explained here for different bearers:

1. *Data radio bearer (DRB) of the RLC acknowledged mode:* The RLC acknowledged mode is used for error-sensitive applications, as shown in Section 7.3. In this case, the handover is required to be lossless and, for this purpose, the source eNodeB performs two steps during handover:
   - *Notification of the PDCP SN status:* The source eNodeB notifies the target eNodeB about the uplink and downlink PDCP SN status by sending an SN STATUS TRANSFER message. This message contains the following information:
     - *Downlink PDCP SN transmitter status:* This includes the next PDCP SN that the target eNodeB needs to assign to a new SDU.
     - *Uplink PDCP SN receiver status:* This includes the PDCP SN of the first missing UL PDCP SDU, which is called the First Missing PDCP SN (FMS). This may also include a bit map of the receiver status for the UL PDCP SDUs that are out of sequence and need to be retransmitted by the UE at the target cell.
   - *Data transfer:* After sending the SN STATUS TRANSFER message, the source eNodeB forwards the data that are not yet transferred successfully to the target eNodeB over the X2 interface. This takes place as follows for downlink and uplink data:
     - *Downlink data:* The source eNodeB does not abort ongoing RLC transmissions to the UE as it starts data forwarding to the target eNodeB. The source eNodeB forwards the following downlink data to the target eNodeB: PDCP PDUs that are not yet acknowledged and their PDCP SNs. Secondly, IP packets that have been received over the S1 interface but not yet transmitted to the UE; these data do not have any PDCP SN.
     - *Uplink data:* The source eNodeB forwards uplink data to the target eNodeB only when the target eNodeB has requested it to do so. The source eNodeB forwards the uplink PDCP SDUs received in-sequence to the serving GW until it sends the SN STATUS TRANSFER message to the target eNodeB. When the source eNodeB sends the SN STATUS TRANSFER message, it stops forwarding the uplink PDCP SDUs to the serving GW and initiates forwarding them to the target eNodeB. The source eNodeB forwards the uplink PDCP SDUs received

out of sequence to the target eNodeB if the target eNodeB has requested it to do so. The source eNodeB discards these PDCP SDUs if it has not received such a request from the target eNodeB.

After handover, the UE and target eNodeB perform actions explained in Section 8.1.5.

2. *DRB of the RLC unacknowledged mode:* The source eNodeB forwards only IP packets that have been received over the S1 interface but not yet transmitted to the UE to the target eNodeB. These data do not have any PDCP SNs. Thus, the source eNodeB does not send an SN STATUS TRANSFER message to the target eNodeB for the RLC unacknowledged mode. The PDCP layer resets the SN and HFN values after handover. The target eNodeB and the UE transmit all PDCP SDUs that are not yet transmitted. No PDCP status report is sent.

3. *Signaling radio bearer (SRB):* Data forwarding is not applied. The PDCP layer resets the SN and HFN values after handover. If there is any PDCP SDU, it is discarded.

*Downlink Synchronization with Target eNodeB*
After the reception of the RRCConnectionReconfiguration message, the UE performs downlink synchronization with the target eNodeB as explained in Section 10.1. The UE does not delay the handover to the target eNodeB for delivering the HARQ/ARQ responses to the source eNodeB.

*Transmission of Random Access Preamble*
The UE sends the random access preamble on the PRACH to perform either non-contention-based random access or contention-based random access at the target eNodeB depending on the allocation of dedicated random access preamble using the RACH-ConfigDedicated IE in the MobilityControlInfo IE.

*Random Access Response*
The target eNodeB sends a random access response (RAR) message on the PDSCH and associated indication on the PDCCH as a part of random access procedure explained in Chapter 11. The RAR message includes uplink grant-allocating resources on the PUSCH for the UE. The Temporary C-RNTI field in the RAR message contains the C-RNTI of the UE, which the source eNodeB earlier sent to the UE using the NewUE-Identity IE on RRCConnectionReconfiguration message in the handover instruction step. If noncontention-based random access occurs, the UE considers it as the successful completion of the random access procedure as the RAR message is received.

*Handover Complete Indication*
The UE sends an RRCConnectionReconfigurationComplete message to the target eNodeB notifying that it has successfully accessed the target cell. The UE uses the uplink resources allocated by the RAR message for this purpose. The RRCConnectionReconfigurationComplete message is used here as message 3 if contention-based random access occurs. In this case, the UE adds the C-RNTI MAC control element to the RRCConnectionReconfigurationComplete message. The C-RNTI

MAC control element contains the C-RNTI of the UE. The UE may also send a Buffer Status Report (BSR) along with the RRCConnectionReconfigurationComplete message to the target eNodeB in order to report pending data in its uplink buffers.

If noncontention-based random access occurs, the target eNodeB sends the PDCCH instance with CRC scrambled by the C-RNTI of the UE. Then the UE considers it as the successful completion of the random access procedure. The PDCCH instance may carry resource allocation for further data transfer.

When the random access procedure has been completed successfully, the UE stops timer T304. From the perspective of the UE, the handover procedure has now been completed successfully. Along with the access attempt at the target eNodeB, the UE also initiates reading the system information from the target cell.

Upon reception of the RRCConnectionReconfigurationComplete message, the target eNodeB verifies the C-RNTI and thus confirms the successful access of the UE.

### 23.3.3.4 Handover Completion

The handover completion phase includes the steps shown below.

*Data Transfer at New Cell*
The target eNodeB has confirmed the successful access of the UE. Thus, the target eNodeB begins the downlink data transfer and it does not wait for the completion of data forwarding from the source eNodeB. The uplink data transfer also begins between the UE and the target eNodeB.

*Path Switch at the Serving GW*
At this juncture, the serving GW switches the downlink data path from the source eNodeB to the target eNodeB. This takes place as follows:

1. The target eNodeB sends a PATH SWITCH REQUEST message to MME notifying that the UE has successfully accessed the target eNodeB and the serving cell has been changed. The PATH SWITCH REQUEST message includes the TAI and ECGI of the target cell. Then the MME determines if the same serving GW can continue serving or not. It is assumed here that the same serving GW can serve. The PATH SWITCH REQUEST message specifies the EPS bearers that need to be switched. It includes the EBI, transport layer address, and GTP tunnel end points for each EPS bearer.
2. The MME sends a MODIFY BEARER REQUEST message to the serving GW. This message includes the EBI, transport address, and DL TEID for the EPS bearers.
3. The serving GW switches the downlink data path from the source eNodeB to the target eNodeB. The serving GW starts sending downlink packets to the target eNodeB using the newly received address and TEIDs.
4. The source eNodeB keeps forwarding data to the target eNodeB as long as it receives packets. The target eNodeB transmits all these forwarded packets to the UE, and thereafter, the target eNodeB transmits the packets received from the serving GW on the new direct path. The EPC attempts

to retain the old GTP tunnel end points for a sufficiently long time in order to minimize the probability of packet losses and avoid any release of the E-RABs.

5. The serving GW sends one or more packets with end marker on the old path to the source eNodeB immediately after switching the path for each E-RAB. The packets with end markers do not contain user data. The serving GW releases the user data path towards the source eNodeB.
6. If forwarding is activated for the particular bearer, the source eNodeB forwards the packets with the end markers to the target eNodeB. The target eNodeB discards the end marker packet and initiates processing to maintain the in-sequence delivery of data received from the source eNodeB over the X2 interface and the data received from the serving GW over S1.
7. The serving GW sends a MODIFY BEARER RESPONSE message to the MME. This message contains the serving GW address and TEID for uplink traffic.
8. The MME sends a PATH SWITCH ACKNOWLEDGE message to the target eNodeB confirming the switching of the data path. This message indicates the EPS bearers that have changed path.. If the UE-AMBR is changed, the updated value of UE-AMBR is included in the PATH SWITCH ACKNOWLEDGE message.

*Resource Release from the Old Cell*

The PATH SWITCH ACKNOWLEDGE message indicates the successful switching of the data path. At this stage, the target eNodeB sends a UE CONTEXT RELEASE message to the source eNodeB indicating that the source eNodeB can release its resources. Then the source eNodeB releases its resources. The whole handover procedure is now complete. However, the source eNodeB can still continue ongoing data forwarding to the target eNodeB, if there is any.

Right after the completion of the whole handover procedure, the UE performs the tracking area update procedure if the tracking area has been changed by the handover action and the new TAI is not one of the TAIs in the TAI list currently possessed by the UE. However, since the MME has been unchanged in the handover procedure described above, the tracking area update procedure is not required here until the periodic tracking area update timer expires.

### 23.3.4 Inter-RAT Handover Procedure

This section explains the general procedure for E-UTRAN to UTRAN inter-RAT handover with a change of the serving GW. The whole handover procedure can be considered as a composition of the following phases:

1. Measurements and reporting;
2. Handover decision;
3. Handover preparation;
4. Handover execution;
5. Routing area update;
6. Handover completion.

The measurements and reporting phase was explained in Section 23.3.2. The handover decision phase is similar with the intra-E-UTRAN handover case as explained in Section 23.3.3.1. The actual RRM algorithm for making this decision depends on the implementation at the eNodeB. Typically, the source eNodeB checks if the signal from the UTRAN cell is sufficiently better than the signal from the serving cell continuously for certain period before a decision for handover is made. The criteria for decision of handover must be consistent with the criteria for the inter-RAT cell reselection. The decision for handover to UTRAN can also be made for the purpose of the CSFB or the reduction of loading of an overloaded E-UTRAN cell.

The handover preparation, handover execution, routing area update, and handover completion phases are explained in the following sections. The exchange of messages is shown in Figure 23.2.

The serving GW acts as the mobility anchor during the handover as mentioned in Section 1.2.2 (i.e., after the handover to UTRAN, the UTRAN forwards uplink data directly to the serving GW and it receives downlink data from the serving GW). Thus, the new data path after the handover consists of the UE, target NodeB, target RNC, target serving GW, and PDN GW if direct tunneling using the S12 interface is available. The new data path consists of the UE, target NodeB, target RNC, target SGSN, target serving GW, and PDN GW if direct tunneling using the S12 interface is not available.

### 23.3.4.1 Handover Preparation

The handover preparation phase consists of two parts, which are explained below.

*Processing Handover Request and Resource Allocation in UTRAN*

1. The source eNodeB initiates the handover preparation by sending the HANDOVER REQUIRED message to the serving MME. This message includes the following major IEs:
   - *Target ID:* The source eNodeB identifies a target RNC based on its stored information. The Target ID IE contains the RNC-ID of the target RNC.
   - *Source to Target Transparent Container:* This IE contains radio-related information, which is transparently passed on to the target RNC.
   - *Handover type:* This IE depicts the type of handover. The value of this IE is set to LTEtoUTRAN.
   - *Cause:* This IE depicts the reason for handover initiation.
2. The serving MME determines the type of handover and target SGSN from the Handover Type IE and the RNC-ID. Then the MME initiates the handover procedure by sending the Forward Relocation Request message to the target SGSN. The Forward Relocation Request message includes the following major information:
   - IMSI of the UE.
   - The RNC-ID of the target RNC.
   - MME TEID and IP address for control plane.
   - *MM context:* It contains UE-AMBR and security-related information (e.g., supported ciphering algorithms).

## 23.3 HANDOVER

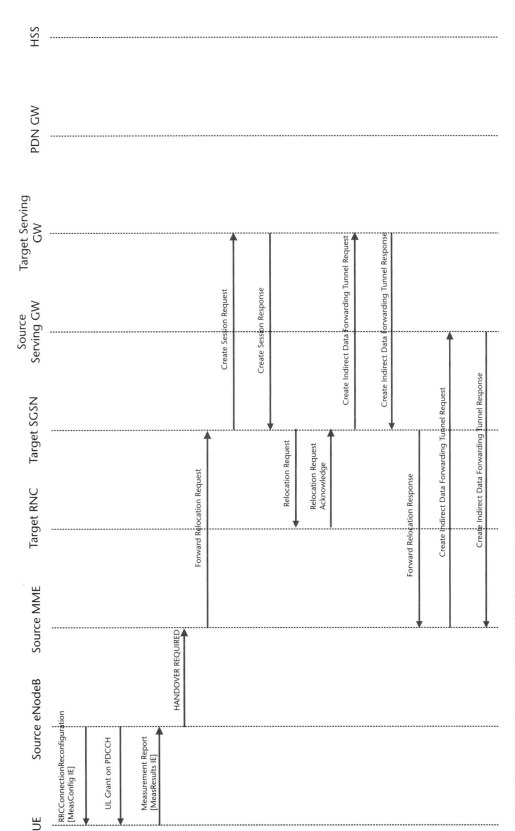

**Figure 23.2** E-UTRAN to UTRAN inter-RAT handover procedure.

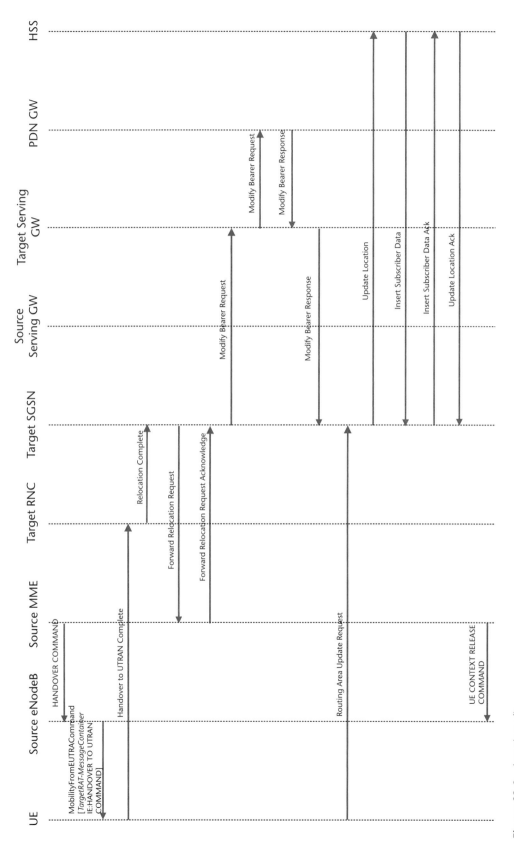

**Figure 23.2** (continued)

## 23.3 HANDOVER

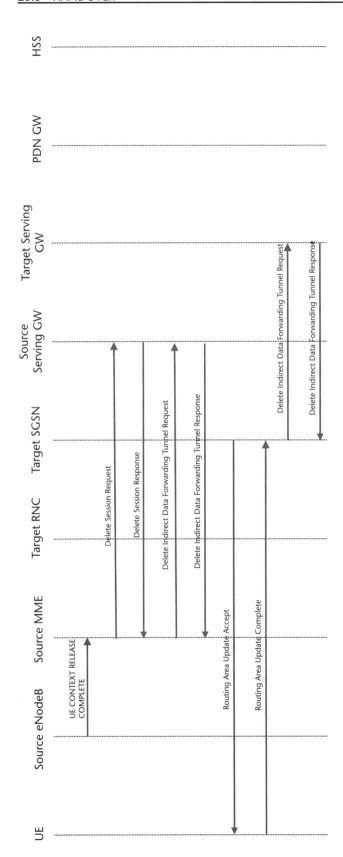

Figure 23.2 (continued)

- *PDN connections:* It contains information related to all active PDN connections. This includes the associated APN, the IP address and the uplink TEID of the serving GW for control plane, and the EPS bearer ID and QCI of the EPS bearers for each PDN connection.
- *Source to Target Transparent Container:* This IE is received from the source eNodeB in the HANDOVER REQUIRED message and is transparently forwarded.
- *RAN cause:* It contains the reason for handover initiation as received from the source eNodeB in the HANDOVER REQUIRED message.
- *ISR supported:* It indicates if the MME and the associated serving GW are capable to activate the ISR for the UE.

3. The target SGSN maps each of the EPS bearers to PDP contexts using one-to-one mapping. Then the target SGSN establishes the EPS bearer contexts and deactivates the EPS bearer contexts that cannot be established. The QoS parameters of the EPS bearer are mapped as follows.

    As shown in Section 2.4.2.1, the Allocation and Retention Priority (ARP) contains the priority level, preemption capability, and preemption vulnerability. The preemption capability and the preemption vulnerability are ignored in the mapping. The priority level is mapped one-to-one to the evolved ARP parameter of a PDP context from Release 9 onwards if supported by the network. The priority level is mapped to the Release 99 bearer parameter ARP as shown in Table 23.9. The values of H (high priority) and M (medium priority) can be configured and modified at the SGSN according to requirements of the operator. The minimum value of H is 1. The minimum value of M is H+1.

    The EPS bearer parameters, the GBR and MBR of a GBR EPS bearer, are mapped one-to-one to the Release 99 bearer parameters, the GBR and MBR of a PDP context. The QCI of the EPS bearer is mapped one-to-one to the Release 99 parameters of Traffic Class, Traffic Handling Priority, Signaling Indication, and Source Statistics Descriptor, as shown in Table 23.10.

4. The target SGSN determines if the serving GW is to be changed. If changed, the target SGSN selects the target serving GW. Then the target SGSN sends a Create Session Request message to the target serving GW for each PDN connection in order to allocate resources at the target serving GW. This message includes the IMSI of the UE, SGSN TEID and IP address for control plane, PDN GW TEID and IP address for control plane, PDN GW IP address and UL TEID for user plane, the protocol type to be used over S5/S8, and EPS bearer contexts.

**Table 23.9** Mapping of the EPS Bearer ARP to the Release 99 Bearer Parameter ARP

| EPS Bearer ARP Priority Value | Release 99 Bearer Parameter ARP Value |
|---|---|
| 1 to H | 1 |
| H+1 to M | 2 |
| M+1 to 15 | 3 |

**Table 23.10** Mapping Between QCIs and Release 99 QoS Parameter Values

| QCI | Traffic Class | Traffic Handling Priority | Signaling Indication | Source Statistics Descriptor | Transfer Delay |
|---|---|---|---|---|---|
| 1 | Conversational | N/A | N/A | Speech | 150 ms |
| 2 | Conversational | N/A | N/A | Unknown | 80 ms or higher |
| 3 | Conversational | N/A | N/A | Unknown | Not specified |
| 4 | Streaming | N/A | N/A | Unknown | Not specified |
| 5 | Interactive | 1 | Yes | N/A | Not specified |
| 6 | Interactive | 1 | No | N/A | Not specified |
| 7 | Interactive | 2 | No | N/A | Not specified |
| 8 | Interactive | 3 | No | N/A | Not specified |
| 9 | Background | N/A | N/A | N/A | Not specified |

The target serving GW allocates resources for the EPS bearers and sends a Create Session Response message to the target SGSN. This message includes serving GW IP address and UL TEID for the user plane and the serving GW TEID and IP address for the control plane.

5. The target SGSN sends a Relocation Request message to the target RNC in order to establish resources at the target RNC. This message includes the IMSI of the UE, a list of RABs to be set up including their QoS information, the Source to Target Transparent Container IE as received from the serving MME in the Forward Relocation Request message, and security-related information (e.g., supported integrity protection algorithms and ciphering algorithms and the CN domain indicator).

The target RNC allocates resources for the RABs and sends a Relocation Request Acknowledge message to the target SGSN. This message includes a list of RABs that have been set up, including the target RNC address and downlink TEID for the RABs. It also includes a list of RABs that could not be set up. The target SGSN later deactivates these EPS bearer contexts using session management (SM) procedures after the routing area update procedure. Furthermore, the Relocation Request Acknowledge message includes the Target RNC to Source RNC Transparent Container IE, which is to be transparently passed on to the source eNodeB.

The target RNC is now ready to receive downlink data for the accepted RABs. It will receive data from the serving GW if direct tunneling using the S12 interface is available. Conversely, it will receive data from the target SGSN if direct tunneling using the S12 interface is not available.

*Preparation for Data Forwarding*
In the case of error-sensitive applications, the handover is required to be lossless. In an attempt for lossless handover, the source eNodeB forwards all yet-to-be-transmitted downlink data to the target serving GW via the source serving GW and this data forwarding requires the preparation explained next. An alternative to this procedure is the direct forwarding of data to the target RNC from the source eNodeB; this is possible if a direct connection between the source eNodeB and the target RNC is available.

1. The target SGSN sends a Create Indirect Data Forwarding Tunnel Request message to the target serving GW. This message includes TEIDs for downlink data forwarding. In addition, this message includes the target RNC address for data forwarding if direct tunneling using the S12 interface is available. Conversely, this message includes target SGSN address for data forwarding if direct tunneling using the S12 interface is not available.

   The target serving GW sends a Create Indirect Data Forwarding Tunnel Response message to the target SGSN. This message includes the target serving GW IP address and DL TEID for data forwarding.

2. The target SGSN sends a Forward Relocation Response message to the source MME. This message includes a serving GW change indication indicating the selection of a new serving GW, SGSN TEID, and IP address for the control plane, RAB setup information, and Target to Source Transparent Container IE. The Target to Source Transparent Container IE contains the value from the Target RNC to Source RNC Transparent Container IE received from the target RNC in the Relocation Request Acknowledge message. The Forward Relocation Response message also includes the target serving GW IP addresses and TEIDs received from the target serving GW in the Create Indirect Data Forwarding Tunnel Response message for data forwarding.

3. The source MME sends a Create Indirect Data Forwarding Tunnel Request message to the source serving GW. This message includes the target serving GW IP addresses and TEIDs received from the target SGSN in the Forward Relocation Response message for data forwarding. Now the source serving GW is capable of forwarding data to the target serving GW.

   The source serving GW sends a Create Indirect Data Forwarding Tunnel Response message to the source MME. This message includes source serving GW IP addresses and TEIDs for data forwarding, provided that the source serving GW supports data forwarding.

### 23.3.4.2 Handover Execution

The handover execution includes the following steps:

1. The source MME sends a Handover Command message to the sourcee NodeB. This message includes the following information:
   - Serving GW IP addresses and TEIDs received from the source serving GW in the Create Indirect Data Forwarding Tunnel Response message for data forwarding;
   - A list of E-RABs to be released;
   - Target to Source Transparent Container IE received originally from the target RNC;
   - Handover Type IE that is set to LTEtoUTRAN.

   The source eNodeB stops downlink data transmission to the UE and it begins to forward the downlink data to the target serving GW via the source serving GW.

2. The source eNodeB sends the MobilityFromEUTRACommand message to the UE as the handover command. This message includes the following major IEs:
   - *TargetRAT-Type:* This IE is set to UTRA.
   - *TargetRAT-MessageContainer:* This IE contains HANDOVER TO UTRAN COMMAND message. This message is defined in UMTS specifications. It contains information regarding the target UTRA cell and relevant radio parameters for access to the UTRA cell. This message is generated based on the Target to Source Transparent Container IE received from the target RNC.
   - *NAS-SecurityParamFromEUTRA:* This IE is contains NAS security-related information.
3. The UE stops reception from the source eNodeB. It generates a mapped security context based on the NAS-SecurityParamFromEUTRA IE in the MobilityFromEUTRACommand message. The UE associates its EPS bearers with the corresponding RABs and accesses the target NodeB based on the information in the TargetRAT-MessageContainer IE of the MobilityFromEUTRACommand message. The UE sends a Handover to UTRAN Complete message to the target RNC. The UE can resume uplink data transmission at the UTRAN.
4. The target RNC sends a Relocation Complete message to the target SGSN. Thus, the SGSN is informed about the successful handover from the E-UTRAN to the RNC and now the target SGSN is ready to receive uplink data from the target RNC. The target SGSN will forward this data directly to the target serving GW.
5. The target SGSN sends a Forward Relocation Complete Notification message to the source MME. This message indicates if the serving GW is changed and if ISR can be activated. The activation of ISR is only possible if the serving GW is not changed. If ISR activation is indicated, then the source MME maintains the UE context if ISR is supported.

    The source MME starts a timer to supervise the release of resources at the source eNodeB and the source serving GW. At the expiry of the timer, the source MME performs the first of the three points from step 1 of the handover completion phase shown in Section 23.3.4.4, to ensure the release of resources in the source eNodeB. Also, at the expiry of the timer, the source MME performs second and third of the three points from step 1 of the handover completion phase shown in Section 23.3.4.4 to ensure the release of resources at the source serving GW.

    The source MME sends Forward Relocation Complete Acknowledge message to the target SGSN. Upon receipt of this message, the target SGSN starts a timer. At the expiry of the timer, the target SGSN performs step 2 of handover completion phase shown in Section 23.3.4.4 to ensure the release of resources at the target serving GW.
6. The target SGSN sends a Modify Bearer Request message to the target serving GW. This message indicates that the target SGSN is now responsible for all the established EPS bearer contexts. This message includes SGSN TEID and IP address for control plane and TEIDs and IP addresses for user

data. The TEIDs and IP addresses are from the target RNC if direct tunneling using the S12 interface is available and they are from the target SGSN if direct tunneling using S12 interface is not available. The target SGSN may activate ISR and then this message also indicates that ISR is activated.

The target serving GW sends a Modify Bearer Request message to the PDN GW to update the PDN GW with the change of RAT type and the change of serving GW. If the policy and charging control (PCC) infrastructure is used, the PDN GW informs the PCRF about the change of the RAT type. The PDN GW sends a Modify Bearer Response message to the target serving GW.

The target serving GW sends a Modify Bearer Response message to the target SGSN. This message acknowledges the user plane switch to the target SGSN. This message includes the target serving GW TEID and IP address for control plane. This completes the establishment of the new data path for all EPS bearer contexts.

### 23.3.4.3 Routing Area Update

As was mentioned in Section 21.3.2, if the ISR is not activated, the routing area update procedure is invoked at this stage. Also, if ISR is activated, the routing area update procedure is invoked if the routing area has been changed after the handover. The UE derives the LAC and the RAC from SIB type 1 broadcast in UTRAN and checks whether the routing area has been changed or not. If the routing area update procedure is invoked, it takes place as described below.

Since the target SGSN is aware of the handover, this routing area update procedure does not require context transfer procedure between the source MME and the target SGSN. The steps in this routing area update procedure are a subset of the typical routing area update procedure:

1. The UE sends a Routing Area Update Request message to the target SGSN. This message includes the routing area identification (RAI), which is MCC + MNC + LAC + RAC. The target SGSN determines the MME from the RAI and this should be the source MME here.
2. The target SGSN updates the HSS as follows:
   - The target SGSN sends an Update Location message to the HSS. This message includes the SGSN number, SGSN address, the IMSI of the UE, and the update type.
   - The HSS sends an Insert Subscriber Data message to the target SGSN. This message includes the IMSI of the UE and the GPRS subscription data.
   - The target SGSN constructs an MM context for the UE and sends an Insert Subscriber Data Ack message to the HSS. This message includes the IMSI of the UE.
   - The HSS sends an Update Location Ack message to the target SGSN. This message includes the IMSI of the UE.
3. The target SGSN sends a Routing Area Update Accept message to the UE. This message assigns a P-TMSI to the UE.
4. The UE sends a Routing Area Update Complete message to the target SGSN. This message acknowledges the assignment of the P-TMSI.

#### 23.3.4.4 Handover Completion

The handover completion phase includes the following steps:

1. When the timer started by the source MME expires, the source MME performs the following tasks as was mentioned in step 5 of the handover execution phase in Section 23.3.4.2:
    - The source MME sends UE CONTEXT RELEASE COMMAND message to the source eNodeB. The source eNodeB releases its resources related to the UE and sends a UE CONTEXT RELEASE COMPLETE message to the source MME.
    - The source MME sends a Delete Session Request message to the source serving GW. This message includes the Operation Indication flag to indicate whether the source serving GW would initiate a delete procedure towards the PDN GW or not. The source serving GW deletes the EPS bearer resources and sends a Delete Session Response message to the source MME.
    - The source MME sends a Delete Indirect Data Forwarding Tunnel Request message to the source serving GW. The source serving GW deletes the temporary resources used for data forwarding and sends a Delete Session Response message to the source MME.
2. When the timer started by the target SGSN expires, as was mentioned in step 5 of the handover execution phase in Section 23.3.4.2, the target SGSN sends a Delete Indirect Data Forwarding Tunnel Request message to the target serving GW. The target serving GW deletes the temporary resources used for data forwarding and sends a Delete Indirect Data Forwarding Tunnel Response to the target SGSN.

# CHAPTER 24

# Multimedia Broadcast Multicast Service

Multimedia Broadcast Multicast Service (MBMS) is introduced in Release 9. As the name suggests, MBMS facilitates the broadcast or multicast of data content to multiple users. The data content may be text, audio, picture, or video via real-time or nonreal-time data transfer. However, MBMS will primarily be used for downlink live video streaming or mobile TV and thus is an alternative solution to technologies like DVB-H, MediaFLO, T-DMB, and ISDB-T. The broadcast service is for any user in the area and the multicast service is only for users who have subscribed to the service. The broadcast and multicast services use different charging rules. MBMS typically uses unidirectional point-to-multipoint (PTM) transmission of data. The UEs can receive MBMS data in both the RRC_CONNECTED state and the RRC_IDLE state. MBMS is not supported in home eNodeBs.

## 24.1 Multicell Transmission

MBMS transmission can be either single-cell transmission or multicell transmission. In case of single-cell transmission, the data content is transmitted over the entire coverage area of a cell, but, more commonly, multicell transmission will be used where the users are supported by multiple cells simultaneously. In this case, the same data content is transmitted from multiple eNodeBs that are synchronized. The signals from different eNodeBs appear as multipath components, instead of interference, to a UE and the UE combines them. Such combination improves the received signal strength greatly especially for cell edge users. A few different types of areas are defined with regard to MBMS transmission:

- *MBSFN area:* The eNodeBs functioning simultaneously for multicell MBMS transmission are referred to as a single frequency network (SFN). The group of cells covered by the SFN is called the *MBSFN area*. The UE feels the same MBMS information transmission available within the cells in the MBSFN area, so the UE does not need to perform handover or RLC reestablishment as it changes cells within the MBSFN area. Each MBSFN area possesses a number as a unique identity in the neighborhood. As the UE moves around within cells, it distinguishes MBSFN areas using this identity. SIB type 13 broadcasts this identity value as shown in Section 24.4.

OFDMA allows the convenient sharing of resources in the MBSFN area. The eNodeBs transmit the same data using the same resources on the OFDMA time-frequency resource grid. A long cyclic prefix allows an easier combination of the signals in the multipath environment. Therefore, MBMS is used with only an extended cyclic prefix for a 15-kHz subcarrier spacing that is 16.67 $\mu$s long. Moreover, for an easier combination of signals from an eNodeB located far away, MBMS can optionally be used with a 33.3-$\mu$s-long cyclic prefix where the subcarrier spacing is 7.5 kHz.

The eNodeBs transmit a time-synchronized common signal called the MBSFN reference signal in the MBSFN area to allow the UE to estimate propagation channel and use coherent detector as shown in Section 5.1.5.3.

- *MBSFN synchronization area:* An MBSFN synchronization area is defined as an area in which all eNodeBs are synchronized. This means that the radio frames with the same system frame number (SFN) are transmitted at the same time from all these eNodeBs. An MBSFN area lies within an MBSFN synchronization area.

- *MBMS service area:* The same geographical area is allowed to be defined for multiple MBSFN areas. This geographical area is regarded as an MBMS service area. In this case, multiple MBSFN reference signals are transmitted with offset among them. There can be eight MBSFN areas maximum in an MBMS service area.

## 24.2 Network Architecture

The network architecture for MBMS operation includes a few entities:

- *Content provider:* This entity provides the data to be broadcast or multicast to the users. This may typically represent the providers of mobile TV channels.

- *Broadcast Multicast Service Center (BM-SC):* BM-SC is the entry point for content providers. This entity is responsible for the authorization and authentication of the content provider, the authorization of the UEs requesting activation of an MBMS service, the charging, scheduling, and announcement of MBMS sessions, and the protection of MBMS data. BM-SC also performs PDCP header compression if it is needed. A SYNC protocol runs between the BM-SC and the eNodeBs, making sure that the same data is transmitted by all the eNodeBs in the MBSFN area. For this purpose, the BM-SC includes a time stamp in the SYNC PDU packets based on a common time reference. The eNodeB first buffers the MBMS packet and then transmits the MBMS packet at the time instant indicated by the time stamp.

- *Multicell/multicast Coordination Entity (MCE):* The MCE controls the allocation of the radio resources used by all eNodeBs in the MBSFN areas and coordinates the synchronous multicell transmission using control plane interface M2 as shown in Figure 24.1. The MCE checks the availability of

## 24.2 Network Architecture

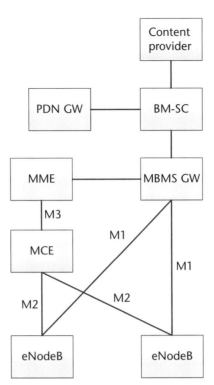

**Figure 24.1** MBMS network architecture.

radio resources for the eNodeBs that the MME is controlling and it initiates MBMS session in the eNodeBs if sufficient radio resources are available. The MCE manages which MBMS service would be assigned to which PMCH. The MCE selects the CSA pattern explained in Section 24.7. The MCE also chooses the order in which the MCHs appear one after another in time in the CSA period. The MCE ensures that all eNodeBs in the MBSFN area use the same configuration for layer 1/layer 2 for each MBMS service. The MCE also makes the decision of the radio resource configuration for MBMS data (e.g., MCS for data transmission). The MCE may be configured as a part of the eNodeB instead of implementing it as a separate entity.

- *MBMS gateway (MBMS GW):* The MBMS GW receives MBMS data from the BM-SC and forwards the MBMS data to eNodeBs with the SYNC protocol using user plane interface M1 as shown in Figure 24.1. The MBMS GW also controls the MBMS session initiation and termination through the MME. The MBMS GW uses IP multicast in order to deliver the MBMS data to the group of eNodeBs in the MBSFN area, and the MBMS GW maintains the IP multicast groups. The MBMS GW allocates the IP multicast group address at the beginning of an MBMS session. When the eNodeB accepts the MBMS Session Start request, it first joins the IP multicast group. Thereafter, the MBMS-GW forwards the MBMS data to the IP multicast address. Similarly, when the MBMS session stops, the MBMS GW releases the IP multicast group address and the eNodeB leaves its multicast group.

## 24.3 Mapping Among Layers

Some particular downlink subframes are defined as the MBSFN subframes. Each MBSFN subframe consists of a non-MBSFN region and an MBSFN region. The non-MBSFN region spans the first one or two OFDM symbols in an MBSFN subframe. The following subframes are used for common purposes and they cannot be defined as MBSFN subframes.

- PSS and SSS are transmitted on subframes 0 and 5 as shown in Section 5.1.3.2.
- MIB is transmitted on subframe 0 as shown in Section 4.3.
- SIB type 1 is transmitted on subframe 5 as shown in Section 4.4.
- Paging occasions are transmitted on subframes 0, 4, 5, and 9 as shown in Section 13.3.1.

Therefore, an MBSFN subframe is allowed to take six different positions in a radio frame with the subframe numbers, 1, 2, 3, 6, 7, and 8. SIB type 2 includes the MBSFN-SubframeConfigList IE, which defines the subframes reserved for the MBSFN in downlink.

The different types of MBMS information are mapped as follows:

1. *MBMS data:* IP packets carrying the MBMS data are mapped on the logical channel called Multicast Traffic Channel (MTCH) and MTCH is mapped on the transport channel called Multicast Channel (MCH). Multiple MTCHs can be mapped on a single MCH. An MCH is mapped on a physical channel called Physical Multicast Channel (PMCH) and there is one-to-one mapping between an MCH and a PMCH. The protocol stack is shown in Figure 24.2.
2. *MBMS-related control information:* MBMS-related RRC PDUs are mapped on the logical channel called Multicast Control Channel (MCCH).

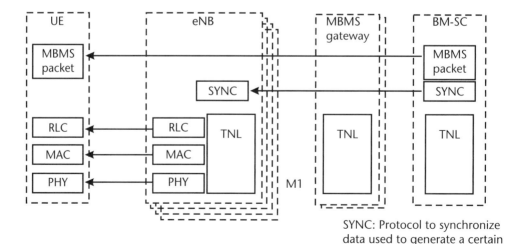

**Figure 24.2** Protocol stacks within the network architecture.

The MCCH is mapped on the MCH transport channel and thereafter is mapped on the PMCH physical channel. A particular MBSFN area has a single MCCH that corresponds to only that MBSFN area. Therefore, if there are multiple MCHs in an MBSFN area, then one of these MCHs contains the MCCH. Thus, one or more MTCHs and, optionally, one MCCH are multiplexed onto one MCH and one PMCH.

In addition to MCCH, some specific MBMS-related control information is transmitted as follows:

- The broadcast control channel (BCCH) logical channel includes SIB type 13, which contains MBMS-related control information.
- SIB type 2 defines the subframes that are reserved for the MBSFN.
- The eNodeB sends the NeighCellConfig IE to inform the MBSFN configuration of the neighbor cells. The NeighCellConfig IE is carried by SIB type 3 for intrafrequency cells and SIB type 5 for interfrequency cells. Also, the MeasObjectEUTRA IE includes the NeighCellConfig IE while configuring measurements of any intrafrequency or interfrequency neighbor cells as shown in Section 23.3.2.2. The NeighCellConfigIE takes on the following values and corresponding indications:
  - *00*: All neighbor cells do not have the same MBSFN subframe allocation as the serving cell.
  - *10*: All neighbor cells have the same MBSFN subframe allocations as the serving cell, or the MBSFN subframe allocations of the neighbor cells are a subset of that of the serving cell.

The PMCH transmission occurs only in the MBSFN region of the MBSFN subframes. The notifications on the PDCCH for a change in the MCCH are also sent only on MBSFN subframes as shown in Section 24.5.3. MBSFN subframes can be used for purposes other than MBMS, (e.g., relaying).

## 24.4 SIB Type 13

SIB type 13 includes MBMS-related control information. The UE can read and update this information in both the RRC_IDLE and RRC_CONNECTED states because the UE can read SI messages in both the states. This helps the MBMS service function in both the RRC_IDLE and RRC_CONNECTED states. The scheduling of SIB type 13 was explained in Chapter 4.

SIB type 13 includes the MBSFN-AreaInfoList IE, which includes the MBSFN-AreaInfo IE for each of the MBSFN areas existing in a particular MBMS service area. The MBSFN-AreaInfo IE includes the MBSFN-AreaId field, which contains the identity of an MBSFN area. This identity is unique for the MBSFN area in its neighborhood and can be any integer up to 255. The bits in the transport block of the PMCH are scrambled prior to modulation. The scrambling sequence generator is initialized for each transport block with a value that is a function of the MBSFN-AreaId.

The MBSFN-AreaInfo IE also includes the non-MBSFNregionLength field, which indicates how many symbols constitute the non-MBSFN region in the beginning of an MBSFN subframe. Since the non-MBSFN region spans the first one or two OFDM symbols in the MBSFN subframe, the value of the non-MBSF-NregionLength can be one or two symbols.

## 24.5 MCCH

A single MCCH is transmitted over an MBSFN area and the MCCH identifies the corresponding MBSFN area. An MCCH instance carries a single RRC message called the MBSFNAreaConfiguration message.

### 24.5.1 Scheduling

The MCCH instances are transmitted periodically. SIB type 13 includes the MBSFN-AreaInfo IE for each of the MBSFN areas in a particular MBMS service area. The MBSFN-AreaInfo IE includes the MCCH-Config IE. The MCCH-Config IE depicts the scheduling of MCCH in the MBSFN area by including the following fields:

- *MCCH-RepetitionPeriod:* This field gives the interval between successive transmissions of the MCCH in the number of radio frames. The interval can be 32, 64, 128, or 256 radio frames. A short interval ensures that if a new UE enters the MBSFN area, the UE does not have to wait long for acquiring MCCH.
- *MCCH-Offset:* This field gives the offset of the radio frame carrying the MCCH from the radio frame with the SFN multiple of the MCCH-RepetitionPeriod. Thus, the SFN of the radio frame carrying MCCH satisfies the relation: SFN mod MCCH-RepetitionPeriod = MCCH-Offset.
- *SF-AllocInfo:* This field indicates the subframe that contains the MCCH while the above two fields indicate the radio frame for the same purpose. Since MCCH can be transmitted only on the MBSFN subframes and there are 6 subframes as such, the SF-AllocInfo field has a 6-bit-long bitmap. The value 1 for a bit in the bitmap indicates the presence of the MCCH in a subframe as shown in Table 24.1.

**Table 24.1** Indications of the Subframe Number for MCCH

| Bit Position | Subframe Number |
|---|---|
| 1 | 1 |
| 2 | 2 |
| 3 | 3 |
| 4 | 6 |
| 5 | 7 |
| 6 | 8 |

### 24.5.2 MCS Configuration

The MCCH-Config IE on SIB type 13 includes the SignalingMCS field, which carries the modulation and coding scheme (MCS) index, $I_{MCS}$. The $I_{MCS}$ value corresponds to a particular MCS according to a specified mapping table. This MCS is used for the MCCH transmission. Additionally, it is used for a specific subframe shown in Section 24.7.

### 24.5.3 Modification

MCCH information can be modified as and when required. MBMS session start and MBMS session stop are example cases for the modification of the MCCH where the MCCH is modified with updated MBMS session information. When there is any change in the MCCH information, the updated MCCH begins to be transmitted in a new period that is regarded as a new modification period. Therefore, modification periods of a fixed length are considered that appear one after another. The eNodeB first notifies the UEs about the change in the MCCH within a modification period. Then, in the next modification period, the eNodeB transmits the updated system information, as shown in Figure 24.3. Such modification period and notification for modification allow the UE to wake up for a short period to check if there is any update in MCCH and thus save battery power. The modification period is the same for all the MCCHs in the MBMS service area.

The MCCH-Config IE on SIB type 13 includes the MCCH-ModificationPeriod field, which gives the length of the modification period in a number of radio frames. The SFN of the radio frames initiating the modification periods are multiples of the MCCH-ModificationPeriod. The value of MCCH-ModificationPeriod can be 512 or 1,024.

Once the UE receives the notification of change in the MCCH, it acquires the updated MCCH from the beginning of the next modification period. The UE uses the old MCCH until it acquires the new MCCH. A similar modification period idea is also used for changes in SI messages as explained in Section 4.5.2. However, in case of a change in the MCCH, a separate notification period is additionally defined. This notification period is a fraction of the modification period (e.g., it may be half or one-fourth of the modification period). The eNodeB sends notifications periodically with an interval of the notification period within the modification period that is located right before the modification period carrying the updated MCCH.

SIB type 13 includes the MBMS-NotificationConfig IE to configure how the notification for a change in the MCCH is scheduled. This configuration applies to

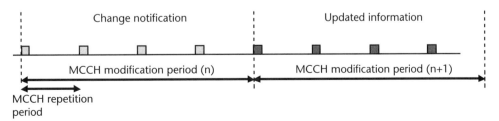

**Figure 24.3** Change of MCCH information.

all MBSFN areas (i.e., all MCCHs in the MBMS service area). The MBMS-NotificationConfig IE includes the following fields:

- *NotificationRepetitionCoeff:* This field indicates the length of the notification repetition period for the indication of a change in the MCCH using: Notification Repetition Period = the shortest MCCH-ModificationPeriod among all MCCHs present)/NotificationRepetitionCoeff. The value of the NotificationRepetitionCoeff can be 2 or 4.
- *NotificationOffset:* This field gives the offset of the radio frame carrying the notification for change in the MCCH from the radio frame with the SFN multiple of the notification repetition period. The SFN of the radio frame carrying notification for MCCH change satisfies the following relation: SFN mod (Notification Repetition Period) = NotificationOffset.
- *NotificationSF-Index:* This field indicates the subframe used to transmit the MCCH change notifications on the PDCCH while the above two fields indicate the radio frame for the same purpose. Since the MCCH change notifications can be transmitted only on the MBSFN subframes and there are 6 subframes as such, the NotificationSF-Index can have a value up to 6 and indicates the presence of notification on the PDCCH in a subframe as shown in Table 24.2.

The eNodeB sends notification for the change in the MCCH using the PDCCH, and the PDCCH instance is sent only on the MBSFN subframes. The PDCCH instance uses DCI format 1C and it is identified by having its CRC scrambled by the MBMS RNTI (M-RNTI). The M-RNTI has a fixed value, FFFD. The DCI format 1C of PDCCH includes 8-bit information because there can be eight MBSFN areas maximum in an MBMS service area. The 8-bit information works as a bitmap and each of these bits indicates if there is any change in the MCCH for a particular MBSFN area. Value 1 of the bit indicates a change in the MCCH and value 0 indicates no change.

The MBSFN-AreaInfo IE that is included in SIB type 13 for each of the MBSFN areas contains the NotificationIndicator field. This field indicates which one out of

**Table 24.2** Indications of the Subframe Number for the Notification for a Change in the MCCH

| Value of NotificationSF-Index | Subframe Number |
|---|---|
| 1 | 1 |
| 2 | 2 |
| 3 | 3 |
| 4 | 6 |
| 5 | 7 |
| 6 | 8 |

the 8 bits on the PDCCH is associated with the particular MBSFN area. The value 0 for this field indicates the least significant bit of the 8 bits.

## 24.6 MBMS Service and MBMS Session

An MBMS service typically represents the service from a particular content provider. Each MCH or PMCH may have one-to-one mapping with an MBMS service. There can be 15 PMCHs maximum in an MBSFN area. A 3-octet-long MBMS service ID identifies a particular MBMS service and it can be used for the group paging of UEs for the service. The MBMS service ID is included in the Temporary Mobile Group Identity (TMGI) IE. The user may need to have a subscription to receive a particular MBMS service and may have to perform authentication with the service provider prior to reception of the service.

An MBMS session may typically represent a mobile TV channel from a service provider. Each PMCH can have multiple MBMS sessions, and there can be 29 MBMS sessions maximum on a PMCH. Every MBMS session is identified by a session ID on the PMCH. Thus, TMGI and session ID together identify a particular MBMS session. Each MBMS session has one-to-one mapping with an MTCH. Each MTCH has a logical channel identity and the MTCH has one-to-one mapping with a radio bearer. Thus, the UE establishes a radio bearer when it wants to initiate an MBMS session. Similarly, the UE releases the radio bearer when it wants to stop an MBMS session.

For every PMCH in the MBSFN area, the PMCH-Info IE on MCCH includes the MBMS-SessionInfoList IE in order to configure MBMS sessions as shown in Section 24.7. The MBMS-SessionInfoList IE includes an individual MBMS-SessionInfo IE for each MBMS session on the PMCH. The MBMS-SessionInfo IE includes the following IEs:

- *TMGI:* This IE has a minimum length of 5 octets and includes the following IEs:
  - *PLMN-Id:* The TMGI IE can optionally include this IE and the TMGI IE becomes 8 octets long when this IE is included. This IE includes the MCC and MNC representing the PLMN.
  - *ServiceId:* This IE contains the MBMS service ID of the MBMS service. The octets 3–5 of the TMGI IE contain the MBMS service ID.
- *SessionId:* This IE contains the session ID of the MBMS session.
- *LogicalChannelIdentity:* This IE contains the logical channel identity of the MTCH associated with the MBMS session.

It may be noted that there can be eight MBSFN areas maximum in an MBMS service area. There can be 15 PMCHs per MBSFN area and 29 MBMS sessions per PMCH at the maximum. Thus, the maximum possible limit in the number of MBMS sessions is $8 \times 15 \times 29 = 3,480$. However, the radio resources cannot support so many TV channels.

The initiation and termination procedures for an MBMS session for broadcast service are shown in the following sections. In the case of multicast services, the users need to be charged specifically, so this procedure is preceded by a subscription process. Also, when the eNodeB notifies the users about the multicast service, the user goes through a joining step before he receives the MBMS session data. Similarly, when the user no longer wants to enjoy the MBMS session, he goes through a leaving step and his reception stops.

### 24.6.1 Session Start

The MBMS session start procedure initiates an MBMS session and notifies the UEs about it. It includes the following steps as shown in Figure 24.4:

1. The MME sends an MBMS Session Start Request message to the MCEs in the MBMS service area. The message includes the IP multicast address, session attributes, and the minimum time to wait before the first data delivery.
2. The MCE sends an MBMS Session Start Response message to the MME as acknowledgment.
3. The MCE checks the availability of radio resources for the eNodeBs that the MME is controlling. If sufficient radio resources are available, the MCE sends the MBMS Session Start message to the eNodeB.
4. The eNodeB sends an MBMS Session Start Response message to the MCE as acknowledgment.
5. The eNodeB notifies the UEs about the MBMS session start. For this purpose, the eNodeB updates the MCCH with the new MBMS session information. Also, the eNodeB sends the MCCH change notification.
6. The eNodeB joins the IP multicast group.
7. The eNodeB receives data for the new MBMS session.
8. The eNodeB establishes an MBMS E-RAB for the new MBMS session and transmits MBMS data at the time instant indicated by the SYNC protocol.

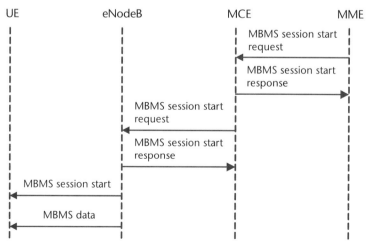

**Figure 24.4** MBMS session start procedure.

### 24.6.2 Session Stop

The MBMS session stop procedure stops an MBMS session and notifies the UEs about it. It includes the following steps as shown in Figure 24.5:

1. The MME sends an MBMS Session Stop Request message to the MCEs in the MBMS service area.
2. The MCE sends an MBMS Session Stop Response message to the MME as acknowledgment.
3. The MCE sends the MBMS Session Stop Request message to the eNodeBs.
4. The eNodeB sends an MBMS Session Stop Response message to the MCE as acknowledgment.
5. The eNodeB notifies the UEs about the MBMS session stop. For this purpose, the eNodeB updates the MCCH and sends the MCCH change notification.
6. The eNodeB stops the transmission of the MBMS data and releases the MBMS E-RAB for the MBMS session.
7. The eNodeB leaves the IP multicast group.

## 24.7 PMCH

MBMS data is transmitted on the PMCH. Up to 15 MCHs or PMCHs are allowed in an MBSFN area. The PMCH uses a single antenna port, antenna port 4. No multiple antenna techniques are supported. The PMCH uses the PTM transmission and does not exploit any feedback from the UEs. Thus, the PMCH uses a fixed MCS level. Also, it does not use the HARQ retransmission or ARQ retransmission; it uses the RLC unacknowledged mode of transmission. However, a new approach called the adaptive PTM transmission that exploits feedback from the UEs is under consideration. In this case, based on feedback, MCS level is selected for groups of UEs and also HARQ retransmissions are performed.

The resource allocation on the PMCH in the time domain involves a few terms:

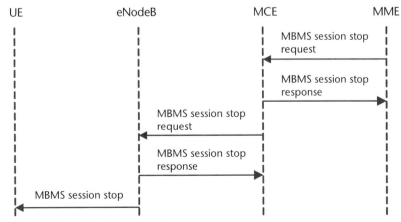

**Figure 24.5** MBMS session stop procedure.

- *Common subframe allocation (CSA) pattern:* In a particular MBSFN area, a common pattern of MBSFN subframes is used for transmission of all the MCHs; it is called the *CSA pattern*. The CSA pattern is repeated periodically; its period is called *CSA period*. The MCHs appear one after another in time within the CSA period (i.e., the MCHs are time-multiplexed within the CSA period).

- *MCH scheduling period (MSP) and MCH subframe allocation (MSA):* An MSP is configured for each MCH. The different MTCHs and the MCCH associated with the MCH or the PMCH are multiplexed during an MSP. The particular sets of subframes used for the PMCH transmission are called the *MSA*. The MSA end indicates the last subframe of the MCH within the CSA period. A single transport block is transmitted in a particular subframe of the MSA. This transport block uses the whole MBSFN region of the subframe. The MSA is indicated using the CSA pattern, the CSA period, and the MSA end for every MCH carrying the MTCH. The MCCH includes the information of these CSA patterns, CSA periods, and MSA ends.

- *MCH scheduling information (MSI) and MCH scheduling information MAC control element:* The eNodeB sends the MSI for each MCH by including the MSI MAC control element in the MAC PDU. Only in the first subframe of each MSA does the MAC PDU include the MSI MAC control element. Thus, the MSI is sent once at the beginning of the MSP. As shown in Section 6.2.1.2, the MCH scheduling information MAC control element indicates the position of each MTCH or MBMS session. The MSI MAC control element includes the logical channel ID for each of the MTCHs and the ordinal number of the subframe within the MSP where the particular MTCH stops. The first MTCH, shown in the MSI MAC control element, starts right after the MSI MAC control element if there is no MCCH and right after the MCCH if there is an MCCH. Other MTCHs in the MSI MAC control element start at the earliest MBSFN subframe after the previous MTCH stops. Thus, the MSI MAC control element indicates the beginning and end points of all MTCHs. The MSI MAC control element includes MBMS sessions in the same order as used in the MBMS-SessionInfoList IE shown in Section 24.6.

The MBSFNAreaConfiguration message on the MCCH configures the resource allocation on the PMCH by including the CommonSF-AllocPatternList IE, which includes the MBSFN-SubframeConfig IE for each of the MBSFN areas. The MBSFN-SubframeConfig IE defines the MBSFN subframes for the CSA pattern pertaining to an MBSFN area. The MBSFN-SubframeConfig IE includes the following fields:

- *RadioframeAllocationPeriod:* The radio frames carrying MBSFN subframes appear periodically with an interval given by this field. This field can represent 1, 2, 4, 8, 16, or 32 radio frames as the length of the period. In every period, one such radio frame may appear or four consecutive radio frames may appear.

- *RadioframeAllocationOffset:* This field gives the offset of the radio frame carrying the MBSFN subframe from the radio frame with the SFN multiple of the RadioframeAllocationPeriod. The SFN of the radio frame carrying the MBSFN subframe satisfies the relation: SFN mod RadioframeAllocationPeriod = RadioframeAllocationOffset. When four consecutive radio frames are scheduled with the MBSFN subframe, this relation is satisfied for the first of the four radio frames. The value of RadioframeAllocationOffset can range between 0 and 7.

- *SubframeAllocation:* While the above two fields indicate the radio frames carrying the MBSFN subframes, this IE indicates the MBSFN subframe within the radio frame. As shown in Section 24.3, the MBSFN subframes can only be subframe 1, 2, 3, 6, 7, and 8.
- *OneFrame:* This field is used when one radio frame carrying MBSFN subframes appear in every RadioframeAllocationPeriod. This field has a 6-bit-long bitmap. The value 1 for a bit in the bitmap indicates the presence of MBSFN subframe in the radio frame as shown in Table 24.3.
- *FourFrames:* This field is used when four consecutive radio frames carrying MBSFN subframes appear in every RadioframeAllocationPeriod. This field has a 24-bit-long bitmap. The value 1 for a bit in the bitmap indicates the presence of MBSFN subframes in the radio frames. Every 6 bits indicate the presence of the MBSFN subframe in a radio frame according to Table 24.3, and this way sequentially provides indications for the four radio frames.

The CommonSF-AllocPeriod field gives the length of the CSA period. The CommonSF-AllocPeriod can be 4, 8, 16, 32, 64, 128, or 256 radio frames.

The MBSFN subframes within the CSA period are time-multiplexed among the PMCHs configured in the MBSFN area. The MBSFN subframes appear repeatedly in the CSA period according to the MBSFN-SubframeConfig IE and they are apportioned to the PMCHs. The period indicated by the CommonSF-AllocPeriod field is repeated one after another.

Since MBSFN subframes can take six different positions maximum in a radio frame, the maximum number of positions available for MBSFN subframes in a CSA period is $256 \times 6 = 1{,}536$.

The PMCH-InfoList IE includes the PMCH-Info IE for each of the PMCHs of the MBSFN area. The PMCH-Info IE configures the PMCH by including the PMCH-Config IE. The PMCH-Config IE includes the following fields:

**Table 24.3** Indications of Position for MBSFN Subframes

| Bit Position | MBSFN Subframe Number |
|---|---|
| 1 | 1 |
| 2 | 2 |
| 3 | 3 |
| 4 | 6 |
| 5 | 7 |
| 6 | 8 |

- *SF-AllocEnd:* This field indicates the last MBSFN subframe of the resources allocated to the PMCH within the period identified by the CommonSF-AllocPeriod. If this PMCH-Info IE takes the $n$th position in the PMCH-InfoList IE, then the time-domain allocation for the PMCH will be as follows.

  The subframes for the PMCH start from the MBSFN subframe right after where the allocation for $(n-1)$th PMCH ends. Then the SF-AllocEnd field gives the position of the last subframe for the PMCH. Since the maximum number of positions available for the MBSFN subframes in the CommonSF-AllocPeriod is 1,536, the value of SF-AllocEnd can range between 0 and 1,535.

  In the case of the first PMCH in the PMCH-InfoList IE, the $(n-1)$th PMCH does not exist and the subframes for the first PMCH start from the first MBSFN subframe in the CommonSF-AllocPeriod.

- *DataMCS:* This field carries the MCS index, $I_{MCS}$, which is applicable for the PMCH transmission. The $I_{MCS}$ value corresponds to a particular MCS according to a specified mapping table. This MCS is not applicable for the first subframe of each MSA which may contain the MCH scheduling information MAC control element. For this particular subframe, the $I_{MCS}$ indicated by the SignalingMCS field on SIB type 13 is rather used. This is mentioned in Section 24.5.2.

- *MCH-SchedulingPeriod:* This field indicates the length of the MSP for the PMCH in number of radio frames. The length of the MSP can be 8, 16, 32, 64, 128, 256, 512, or 1,024 radio frames.

The PMCH-Info IE also includes the MBMS-SessionInfoList IE, which is explained in Section 24.6.

# CHAPTER 25
# Coordination Among Cells

Various types of coordination among cells can be used for different purposes, including the mitigation of Intercell Interference (ICI). The ICI may impact the data rate severely, especially for the cell edge users. The techniques to mitigate ICI are classified as interference randomization, interference cancellation, interference coordination, load balancing, and Self-Optimizing Network (SON). The interference coordination among neighboring cells to mitigate ICI is referred to as Intercell Interference Coordination (ICIC). Apart from interference mitigation, load balancing and SON serve other purposes.

## 25.1 Intercell Interference Coordination (ICIC)

There are different methods for ICIC in uplink and downlink. The ICIC methods are mostly limited to the frequency domain. ICIC controls activities on resource blocks in frequency-domain scheduling with a view to minimize interference among neighboring cells. This may be restricting scheduling resource blocks or controlling the power level in resource blocks. ICIC methods can be applied and adapted statically, semistatically, or dynamically depending on the particular scenario. ICIC does not help improve idle cell interference.

ICIC can be applied in both downlink and uplink. In general, ICIC is found more useful for the uplink transmission as explained here:

1. The UEs transmitting in the uplink are located in various positions in the cell, whereas the eNodeB transmitting in the downlink has a fixed position. Therefore, the interference generated by the UEs in uplink depends largely on the scheduling. The interference in uplink keeps varying more quickly, and frequent adaptation in uplink scheduling is more helpful compared to downlink.
2. The uplink transmission is more likely to be power limited because of lower power availability at the UE compared to the eNodeB. Thus, maintaining good SINR can be more difficult in uplink due to a weak signal and strong cochannel interference compared to downlink.
3. ICIC plays a more important role when the transmission uses a narrow bandwidth. It may be the uplink that uses a narrower bandwidth compared to the downlink because the uplink transmission is more likely to be power limited.

### 25.1.1 Exchange of Information to Support ICIC

There can be exchange of various information among neighboring cells in order to aid ICIC in both downlink and uplink; this is explained in the following sections. The reaction of the neighboring cell that receives such information is not specified and rather depends on implementation. This leads to different types of ICIC implementation techniques depending on traffic, environment, and so forth.

#### 25.1.1.1 ICIC in Downlink

A particular cell can inform its neighboring cells about the level of transmit power that it is going to use in different parts of the downlink bandwidth. This allows the neighboring cells to anticipate which parts of the bandwidth would suffer higher interference. Then a neighboring cell may avoid scheduling its users on the resource blocks with high interference. Those resource blocks can especially be more restricted for the cell edge users who are closer the interfering cell. For this purpose, an eNodeB sends a LOAD INFORMATION message that contains a Relative Narrowband Tx Power (RNTP) IE to its neighboring eNodeB over an X2 interface. An eNodeB, after the reception of the RNTP IE, considersthe IE valid until the reception of a new LOAD INFORMATION message carrying an update. The RNTP IE includes the following IEs:

- *RNTP per PRB:* This IE gives a bitmap. Each bit in the bitmap represents one PRB in the frequency domain. The bit string can be 110 bits long at the maximum since the maximum number of PRBs is 110. The value 0 of the bit indicates that the transmission power on the corresponding PRB is not going to exceed the value of the RNTP threshold. The value 1 of the bit indicates no promise about the transmit power on the corresponding PRB. Thus, this IE depicts whether the eNodeB is planning to keep the transmit power for the PRB below a certain limit or not.
- *RNTP threshold:* This IE gives the value of the RNTP threshold indicating the maximum intended EPRE. A small cell uses a low transmit power at the eNodeB, whereas a large cell uses a higher transmit power at the eNodeB. Thus, the value of the RNTP threshold should vary according to the transmit power. The RNTP threshold is expressed, normalizing it with the maximum transmit power per hertz. The RNTP Threshold IE gives the maximum intended EPRE of a resource element normalized with the value Maximum transmit power at eNodeB/total effective bandwidth [i.e., Maximum transmit power at eNodeB/($N_{RB}^{DL} N_{SC}^{RB} \times \Delta f$)]. The value of the RNTP Threshold IE can be $-\infty$, $-11$, $-10$, $-9$, $-8$, $-7$, $-6$, $-5$, $-4$, $-3$, $-2$, $-1$, 0, 1, 2, or 3 dB.

#### 25.1.1.2 ICIC in Uplink

There can be two types of exchange of information among neighboring cells in order to aid ICIC in uplink:

1. *Overload indication (OI):* An eNodeB sends a LOAD INFORMATION message that contains a UL Interference Overload Indication (OI) IE to its

neighboring eNodeB over an X2 interface to inform the level of interference that is being experienced in different parts of the uplink bandwidth currently. This is typically used by a cell to inform neighboring cells that there is a currently unacceptably high level of interference on the indicated resource blocks. The message is sent to a cell that is potentially causing the high interference, although it may sometimes be difficult to detect the interfering cell. Then the interfering cell may take some action to cause less interference to the reporting cell. For example, it may reduce the uplink transmit power on the resource blocks that are indicated to have high interference. Alternatively, it may not schedule those resource blocks to its cell edge users, or it may not even schedule those resource blocks at all.

The UL Interference OI List in the LOAD INFORMATION message contains UL Interference OI IEs for each Physical Resource Block (PRB) in the uplink. There can be 110 UL Interference OI IEs at the maximum since the maximum number of PRBs is 110. The UL Interference OI IE indicates if high, medium, or low interference exists for the respective PRB.

The eNodeB, after the reception of the UL Interference OI IE, considers the IE valid until the reception of a new LOAD INFORMATION message carrying an update.

2. *High interference indication (HII):* An eNodeB sends a LOAD INFORMATION message that contains a UL High Interference Indicator (HII) IE to its neighboring eNodeB over the X2 interface to inform the level of interference that it is going to create in different parts of the bandwidth. This is typically used by a cell to inform the neighboring cells that it intends to schedule cell edge users on the indicated resource blocks, so high interference may be created on those resource blocks. The cell edge users may be identified based on their measurement reports that the UEs send to help handover decision. After receiving the message, a neighboring cell may not schedule its cell edge users on the same resource blocks in order to avoid interference or it may not even schedule those resource blocks at all.

The HII IE in the LOAD INFORMATION message contains a bitmap indicating interference sensitivity for each PRB in the uplink. The bit string can be 110 bits long at the maximum since the maximum number of PRBs is 110. The values 0 and 1 of a bit in the bitmap indicate low and high interference sensitivity on the respective resource block, respectively.

The eNodeB, after the reception of the UL HII IE, considers the IE valid until the reception of a new LOAD INFORMATION message carrying an update.

### 25.1.2 Fractional Frequency Reuse (FFR)

An important type of ICIC technique is FFR. FFR is the opposite of hard frequency reuse (HFR), which splits the system bandwidth into a number of distinct subbands based on a chosen frequency reuse factor and where the neighboring cells in a cluster transmit on different subbands. The user interference at the cell edge is maximally reduced in HFR, but the spectrum efficiency drops by a factor equal to the reuse factor. On the other hand, FFR allows the use of orthogonal frequency resources among neighboring cells around the cell borders. FFR allows the UEs near the base station to reuse the same subcarriers in neighboring cells, which makes the frequency reuse factor 1. However, for UEs closer to the cell edge, the subcarriers are allocated in a coordinated manner. FFR provides a good performance gain with low complexity. The cell edge users may be identified based on their CQI reports. The following variants of FFR are promising:

1. *Partial Frequency Reuse (PFR):* The PFR splits the system bandwidth into two parts to serve the cell center users and cell edge users in each cell. The part for the cell center users is completely reused by all cells. Then the part of the bandwidth for the cell edge users is further divided into parts, and each cell in a neighborhood uses only one of these parts. As shown in Figure 25.1, the three cells transmit only on their own parts of the spectrum in the case of cell edge users. Since the cell center users have better channel SINR compared to the cell edge users, they need less power to achieve a similar link performance. Thus, the part of the system bandwidth for the cell center users is loaded with less power than the part for the cell edge users. The effective reuse factor of this scheme is always greater than 1.
2. *Soft Frequency Reuse (SFR):* The overall bandwidth is shared by all cells, but different parts of the bandwidth apply a certain power limit among cells in the neighborhood, as shown in Figure 25.2. The users are allocated resources from the entire bandwidth with different power limits. Since the cell center users have a better channel SINR compared to the cell edge users, they need less power to achieve a similar link performance and they have a lower power limit. However, the cell edge users are allowed a higher power limit if they use a particular part of the bandwidth. The cell edge users in different cells in the neighborhood can be allocated different parts of the bandwidth with a higher power limit. The frequency reuse factor is 1 for cell center users and it is equal or greater than 1 for cell edge users.

FFR can be applied based on static planning. In addition, the FFR techniques can be adapted semistatically or even fully dynamically. The adaptation can depend

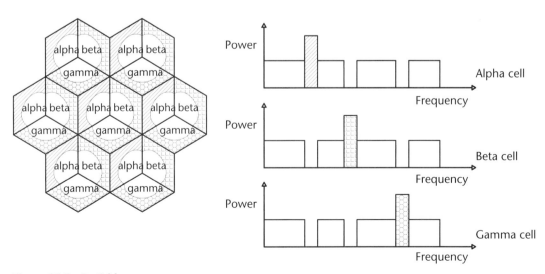

**Figure 25.1** Partial frequency reuse.

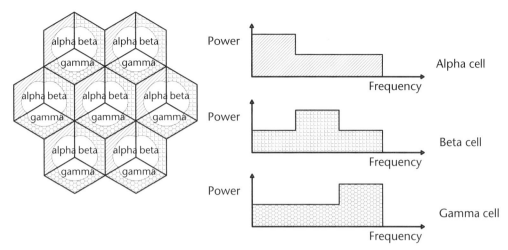

**Figure 25.2**  Soft frequency reuse.

on the radio link condition and the traffic condition at both cell level and individual user levels. Such adaptation can be assisted by the exchange of information between cells. For example, the UL HII IE can be used to identify the part of bandwidth that is used at cell edge by a neighboring cell.

## 25.2  Load Balancing Among Cells

It is not unlikely that the user distribution becomes nonuniform in an area and the cells in the area become unequally loaded. Such a traffic load imbalance may cause some cells to be overloaded while leaving some neighboring cells lightly loaded. The load balancing functionality stops allocating resources to some users in the overloaded cells and schedules those users in lightly loaded neighboring cells. The cell edge users are preferably moved to a new cell. The load balancing helps minimize the traffic load imbalance among neighboring cells and improves the system capacity. The load balancing also helps avoid too much interference from an overloaded cell and thus helps mitigate ICI. The load balancing requires that the neighboring eNodeBs notify each other about their respective loading information. For this purpose, an eNodeB sends a request to perform load measurements to its neighboring eNodeB, as shown in Figure 25.3. Then the requested eNodeB performs measurements and sends measurement results periodically to the first eNodeB. This is explained next assuming eNodeB 1 and eNodeB 2 neighboring eNodeBs.

1. *Resource status request:* The eNodeB 1 sends RESOURCE STATUS REQUEST message to its neighboring eNodeB 2 to request the performance and report of certain load measurements. The RESOURCE STATUS REQUEST message contains the following IEs:
    - *Registration request:* This IE contains the value of start or stop, which indicates if the eNodeB 2 would start or stop the measurements.
    - *Report characteristics:* This IE indicates what the eNodeB 2 would measure. It can be the usage of resource blocks, the load on the transport network layer of the S1 interface, and/or the hardware load.

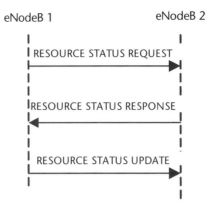

**Figure 25.3** Exchange of resource status between eNodeBs.

- *Cell ID list:* This IE contains the ECGI of the cells for which the eNodeB 2 would perform the measurement.
- *Reporting periodicity:* The measurement reporting is performed periodically. This IE gives the time interval that the eNodeB 2 would allow between two measurement reports. It can be 1,000 ms, 2,000 ms, 5,000 ms, and so forth.

2. *Resource status response:* The eNodeB 2 sends a RESOURCE STATUS RESPONSE message to eNodeB 1 to indicate that the requested measurements have been initiated.
3. *Resource status update:* After the completion of the measurements, the eNodeB 2 sends a RESOURCE STATUS UPDATE message to eNodeB 1 to report the results of the measurements. This message contains the measurement results for each of the cells reported and it contains the following IEs for a particular cell:
    - *Radio resource status:* This IE is included and contains the measurements results if the Report Characteristics IE of the RESOURCE STATUS REQUEST message indicated the measurement of the usage of resource blocks. The eNodeB 2 checks the usage separately for the GBR resource blocks and the non-GBR resource blocks for both uplink and downlink. This IE includes the usage report for GBR resource blocks and non-GBR resource blocks and the average usage of total resource blocks in percentage. The average usage of total resource blocks is calculated as follows for both uplink and downlink over a certain time period:

(All PRBs used for transmission/Total number of PRBs available) × 100

The PRB usage per traffic class instead of all PRBs if calculation of resource blocks usage is made per traffic class.
    - *TNL load indicator:* This IE is included and contains measurement results if the Report Characteristics IE indicated the measurement of the load in the S1 interface. This IE contains the S1 TNL Load Indicator IE for the downlink and uplink to indicate the status of the load in the S1

interface of eNodeB 2 for the downlink and uplink. It can indicate the status of the load as the low load, medium load, high load, and overload.
- *Hardware load indicator:* This IE is included and contains measurements results if the Report Characteristics IE indicated the measurement of the hardware load. This IE contains the Hardware Load Indicator IE for the downlink and uplink to indicate the status of the hardware load for downlink and uplink. It can indicate the status of load as low load, medium load, high load, and overload.

## 25.3 Load Balancing and Rebalancing Among MMEs

A number of MMEs may be connected by a mesh network to a set of eNodeBs. Then the eNodeB serving the UE can be associated with multiple MMEs (i.e., multiple MMEs are able to serve the particular UE). An imbalance in load of MMEs in an MME pool area may cause some MMEs to be overloaded while leaving some MMEs lightly loaded. The MME load balancing and rebalancing functionalities attempt to achieve a proper load balancing among MMEs within the MME pool. For this purpose, each MME in the MME pool is assigned a weight factor. The weight factor is an integer between 0 and 255. The weight factor is typically set according to the relative processing capacities of the MMEs within the MME pool. The MME sends the MME CONFIGURATION UPDATE message to the eNodeB. This message includes the Relative MME Capacity IE, which contains the weight factor of the MME. The eNodeB sends the MME CONFIGURATION UPDATE ACKNOWLEDGE message to the MME to acknowledge the updated information.

When a UE enters an MME pool area, the UE is associated with an appropriate MME from the MME pool. The NAS node selection function (NNSF) in the eNodeB selects the MME for association with the particular UE. The MME load balancing functionality facilitates that the eNodeB prioritizes an MME higher in the selection if it has a higher weight factor.

If there is a change in the MME capacities within the MME pool, then the weight factor of the MME can be updated. However, the weight factors are not expected to be changed frequently in a mature network. It may be changed, for example, when an MME is added to the MME pool. When a new MME is installed, it may be assigned a high weight factor for an initial period in order to allow it to take over its intended load quickly.

In order to allow a proper load balancing among MMEs, a UE that is already registered with the network needs to be moved from one MME to another MME within the MME pool. This is performed using the MME load rebalancing functionality. The MME load rebalancing functionality is not intended to be used when the MME becomes overloaded. It is rather the MME load balancing functionality that attempts to maintain a similar MME loading within the MME pool by associating the UE with an appropriate MME when the UE enters the MME pool area. The MME load rebalancing functionality may be used, for example, when an MME is removed from the MME pool.

If the UE and the MME are in the ECM-CONNECTED state, then the MME load rebalancing functionality requires off-loading the UE from the current MME. Thus, the current MME initiates the S1 Release procedure and the ReleaseCause field in the RRCConnectionRelease message takes on the valueLoadBalancing-TAURequired. This releases NAS signaling connection. Then the UE performs the tracking area update procedure. During this procedure, when the UE sends an RRCConnectionSetupComplete message to the eNodeB as a part of the RRC connection establishment, the UE does not include the GUMMEI of the previously registered MME in the RegisteredMME field. Then the eNodeB selects a new MME based on the weight factor of the MME in the MME pool and the UE hooks up with the new MME. On the other hand, if the UE is in the ECM-IDLE state, then the MME first pages the UE, brings it to the ECM-CONNECTED state, and then performs the above steps.

## 25.4 Self-Optimizing Networks (SON)

LTE supports SON in order to minimize manual involvement in the planning, deployment, operation, and optimization of the network. SON allows self-configuration and self-optimization of the network.

### 25.4.1 Self-Configuration (SC)

SC allows the automatic configuration of a newly added infrastructure or service. Thus, when a network node is added, an automatic installation takes place and basic configuration parameters are loaded, making the new node operational. This requires an initial configuration of a number of parameters or algorithms. Different types of SC techniques are considered, including the following:

- *Automatic eNodeB installation:* When a new eNodeB is hooked up, the automatic configuration of the IP address and the detection of OAM take place. In addition, the authentication of the eNodeB and its association with gateways are performed automatically. Then the necessary configuration parameters are loaded to make the eNodeB operational. This includes the configuration of neighboring cells and coverage- and capacity-related parameters. Moreover, the automatic selection of PCI is made for the cells of the eNodeB, ensuring that the neighboring cells do not have identical PCIs. For this purpose, the eNodeB can obtain the neighboring PCIs via reports from UEs, reports from neighboring eNodeBs over the X2 interface, and/or reports received using any implementation-specific technique.

- *Dynamic configuration of the S1 interface:* The automatic configuration of the S1 interface connecting an eNodeB and an MME is made through the exchange of information between the eNodeB and the MME. The eNodeB sends its name, eNodeB global ID, and supported TAIs to the MME. The MME sends information like its name, GUMMEI, serving PLMNs, and relative MME capacity to the eNodeB. This was shown in Section 14.3.1.

- *Dynamic configuration of the X2 interface:* The automatic configuration of the X2 interface connecting two eNodeBs is made through exchange of information between eNodeBs. The two eNodeBs exchange information such as eNodeB global ID, served cell information, and globally unique group ID list.
- *Neighbor cell list configuration:* The automatic neighbor relation function (ANRF) allows the eNodeB to identify neighbor cells and set up neighbor relation automatically based on reports from the UEs. Thus, when a new eNodeB is added, it finds out the existing neighbor cells around itself. This relieves manual configuration and allows better optimization. The identified neighbors can be intrafrequency, interfrequency, or intra-RAT cells. To discover the neighbor cells, the eNodeB may ask a UE to measure the neighbor cells. The UE measures and derives the global cell identity from system information for the neighbor cells. Finally, the UE reports back to the serving eNodeB with the global cell identity allowing the serving eNodeB to populate its neighbor cell list. This measurement of the UE is not for the purpose of handover decision but to generate a neighbor cell list.

As shown in Section 23.3.2, for handover decision, the UE finds out and measures E-UTRAN neighbor cells without relying on any neighbor list. Thus, for example, a UE may detect a new cell with sufficient signal quality and then report it to the serving eNodeB. In this case, the serving eNodeB may want to include this cell in its neighbor relation list assuming that this cell was unknown before. The received measurement report contains PCI of the cell but not ECGI, so the newly detected cell is not fully known. Then the eNodeB asks the UE to read the ECGI and TAI from SIB type 1 and report to the serving eNodeB. The eNodeB can check on the transport layer address of the detected eNodeB and include the cell in its neighbor relation list. The serving eNodeB may even trigger the X2 interface setup procedure with the detected eNodeB.

### 25.4.2 Self-Optimization (SO)

SO provides for continuous automatic tuning of the existing network. Different types of SO techniques are considered, including the following:

- *Exchange of load information:* The eNodeBs exchange load information over X2 interface for ICIC and appropriate load balancing between eNodeBs. The exchange of information was explained in Section 25.1.1. Based on this information, the eNodeBs perform resource scheduling, power control, and decision for handover between two eNodeBs.
- *Coverage/capacity optimization:* Different parameters are automatically optimized for better system coverage and capacity.
- *Mobility parameters optimization:* Cell reselection and handover parameters are automatically optimized to attempt for appropriate cell reselection and handover and minimization of radio link failures.

- *Random Access Channel (RACH) optimization:* RACH and other common channels are configured based on measurements. The following RACH procedures may be optimized:
  - Resource unit allocation for RACH;
  - Preamble allocations for dedicated random access, nondedicated random access, Group A, and nondedicated random access, Group B;
  - RACH persistence level and back-off control;
  - RACH transmission power control.
- *Neighbor cell list optimization:* The eNodeBs updates its list of neighbor cells using ANRF as shown for SC.

# CHAPTER 26

# Public Warning System

The Public Warning System (PWS) feature allows broadcasting alerts, warnings, and critical information regarding disasters and other emergencies to public. This can effectively help people to take appropriate action in order to save themselves from injury or death and damage of property. Thus, damage or loss due to threats from diseases, terrorism, or toxic spills or due to disasters such as earthquakes, tsunamis, hurricanes, flood, or wildfires can be reduced dramatically. The warning notifications should be delivered to people very quickly so people can take necessary actions, such as evacuation. In addition to warning people of impending danger, instructions can also be provided with a view to guide people to safety. The PWS includes two types of warning notifications:

1. Earthquake and Tsunami Warning System (ETWS);
2. Commercial Mobile Alert Service (CMAS).

If a UE has the capability of receiving PWS notifications within notification areas, then it is called a *PWS-UE*. In order to notify the user of the warning, the PWS-UE offers a dedicated alerting indication that is distinct from other alerts of the UE. This can be a dedicated audio attention signal and vibration cadence. The reception and presentation of the warning notifications does not preempt the current data transfer sessions.

## 26.1 Earthquake and Tsunami Warning System (ETWS)

The ETWS delivers warning notifications specific to earthquakes and tsunamis. The network receives the notifications from warning notification providers and delivers them to the UEs within the notification areas. The warning notification provider can be national governments, local governments, public service organizations, and so forth.

Three forms of waves produced in the case of earthquakes are as follows.

1. *Primary waves (P-waves):* These waves have the least destructive force and travel at about 6 km/sec speed.
2. *Secondary wave (S-waves):* These waves have more destructive force and travel at about 3.5 km/sec speed.

3. *Surface waves:* These waves cause most of the destruction and travel at about 2 km/sec speed.

Due to the much faster speed of the primary waves, a helpful warning can be broadcast for earthquakes based on the detection of primary waves before the arrival of other waves. If the location and magnitude of an earthquake meet the known criteria for the generation of a tsunami, then a tsunami warning is issued to warn of an imminent tsunami hazard.

The ETWS warning notification is classified into two types depending on the urgency.

1. *ETWS primary notification:* This is used to notify users about the most urgent event, for example, the imminent occurrence of earthquake. For a quick delivery, the amount of data sent should be a few bytes only. The eNodeB sends the ETWS primary notification on SIB type 10.
2. *ETWS secondary notification:* This is used to notify users of supplementary information that is of lesser urgency, for example, instructions on necessary actions, information about the location of assistance, a map indicating the route to an evacuation site, a timetable of food distribution, and so forth. The amount of data sent on secondary notification is large. The eNodeB sends the ETWS secondary notification on SIB type 11.

Depending on the policy of the warning notification provider, the notification can be generated in any of the following ways:

- Only primary notification;
- Only secondary notification;
- Both primary and secondary notifications.

## 26.2 Commercial Mobile Alert Service (CMAS)

The CMAS was introduced in Release 9. The CMAS facilitates sending a text message to a large number of users and thus helps in warning the public. The CMAS supports messages of up to 90 characters in English text. The eNodeB sends CMAS notifications on SIB type 12. The CMAS notifications carry three different classes of warning:

1. Presidential;
2. Imminent threat;
3. Child abduction emergency.

## 26.3 PWS Notification

The eNodeB can send ETWS primary notifications on SIB type 10, ETWS secondary notifications on SIB type 11, and CMAS notifications on SIB type 12 at any point in time. These notifications involve a modification in the SIB types. Thus, the modification in SIB types 10, 11, and 12 can occur at any time regardless of any modification period, whereas any change in other SIB types of SI messages must occur only in a new modification period as explained in Section 4.5.2. The warning notification takes place as follows:

1. The eNodeB sends a paging message to inform the PWS-UEs that are either in the RRC_IDLE state or in the RRC_CONNECTED state. In the RRC_CONNECTED state, the PWS-UEs attempt to read paging message at least once in every DefaultPagingCycle in order to check on PWS notifications.

   If the parameter ETWS-Indication is present in the paging message, then it indicates the transmission of the ETWS primary notification on SIB type 10 and/or the ETWS secondary notification on SIB type 11. Similarly, if the parameter CMAS-Indication is present in the paging message, then it indicates the transmission of one or more CMAS notifications on SIB type 12.

2. If the UE is ETWS- and/or CMAS-capable and it receives a paging message including the ETWS-Indication and/or CMAS-Indication, then it needs to start reading the ETWS primary notification, the ETWS secondary notification, and/or the CMAS notification, respectively. To do so, it first reads SIB type 1 immediately without waiting for the next system information modification period boundary. The UE checks on the SchedulingInfoList IE, which includes the SchedulingInfo IE for each of the SI messages. The SchedulingInfo IE includes the parameter SIB-MappingInfo, which depicts which SIB types are transmitted.

3. The UE reads SIB types 10, 11, and/or 12 depending on whether the parameter SIB-MappingInfo includes SIB types 10, 11, and/or 12.

## 26.4 SIB Content for PWS

The SIB types 10, 11, and 12 contain the following common IEs:

1. *MessageIdentifier:* This is a 16-bit-long IE identifying the source and type of the PWS notification. A particular value of this IE represents a combination of source and type. The EPC sets the value of this IE.
2. *SerialNumber:* This is a 16-bit-long IE to distinguish the variations of broadcast messages from the same combination of source and type. Thus, the value of this IE is changed every time the broadcast message from the same source and type is changed (i.e., the broadcast message with the same message identifier is changed).

The SIB type 10 contains the ETWS primary notification. It includes the following additional major IEs:

1. *WarningType:* This IE indicates that a primary notification is included and also indicates the type of the disaster. This IE includes three fields as follows:

- *Warning type value:* This field gives the type of warning. The type of warning can be any of the following: earthquake, tsunami, earthquake and tsunami, testing, or other.
- *Emergency user alert:* It is a binary bit that, if set, activates emergency user alert on the UE upon the reception of ETWS primary notification. This can lead to the generation of an alerting tone, vibration, and so forth according to the UE's capability.
- *Pop-up indication:* It is a binary bit that, if set, activates the message pop-up on the display of the UE in order to alert the user upon the reception of the ETWS primary notification.

2. *WarningSecurityInfo:* This field is used only when ETWS primary notification is sent with security. This field provides the information needed for securing the primary notification. This field contains a 7-byte-long timestamp and a 43-byte-long digital signature.

The SIB type 11 contains the ETWS secondary notification. The SIB type 12 contains the CMAS notification. Both SIB types 11 and 12 include the following additional IEs in order to carry their notifications:

1. *WarningMessageSegment:* This IE carries a segment of the Warning Message Contents IE. The Warning Message Contents IE contains a warning message as a secondary notification for the user. Segmentation can be applied for the delivery of the Warning Message Contents IE.
2. *WarningMessageSegmentNumber:* This IE gives the segment number of the segment of the Warning Message Contents IE included in this SIB. For the first segment, the segment number is 0. For the second segment, the segment number is 1 and so on.
3. *WarningMessageSegmentType:* This IE indicates whether the segment of the Warning Message Contents IE included in this SIB is the last segment or not.

# Appendix A:
# Major Information Elements (IEs) on RRC Messages

The RRC messages contain various IEs in the form of trees (i.e., one IE includes several IEs and each of them may contain other IEs). The positions of most significant IEs in these trees are shown in the following sections. In Figures A.1, A.2, and A.3, the message or IE at the beginning of an arrow includes the IEs at the tip of the arrow.

## A.1  IEs on SIB Type 2

The eNodeB broadcasts system information block (SIB) type 2 in the cell. The UEs must read SIB type 2 in both the RRC_IDLE state and the RRC_CONNECTED state. SIB type 2 includes various IEs as shown in Figure A.1.

## A.2  IEs on RadioResourceConfigDedicated IE

The eNodeB sends an RRCConnectionReconfiguration message to the UE in order to set up or modify DRBs or SRB2. In addition, the eNodeB sends the RRCConnectionSetup message to the UE to establish SRB1 and sends RRCConnectionReestablishment message to the UE to reestablish SRB1 resolving contention. These messages include the RadioResourceConfigDedicated IE for the configuration of radio resources for the particular UE. The RadioResourceConfigDedicated IE includes various IEs as shown in Figure A.2.

## A.3  IEs Used During Handover

When handover takes place, the source eNodeB sends the RRCConnectionReconfiguration message to the UE as an instruction for handover action. The RRCConnectionReconfiguration message includes the MobilityControlInfo IE, which specifies how the UE would act toward the target eNodeB. The MobilityControlInfo IE includes various IEs as shown in Figure A.3.

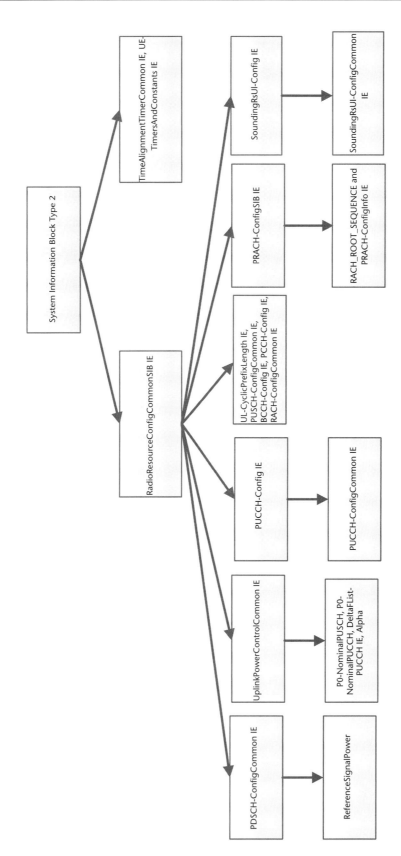

**Figure A.1** IEs on SIB Type 2.

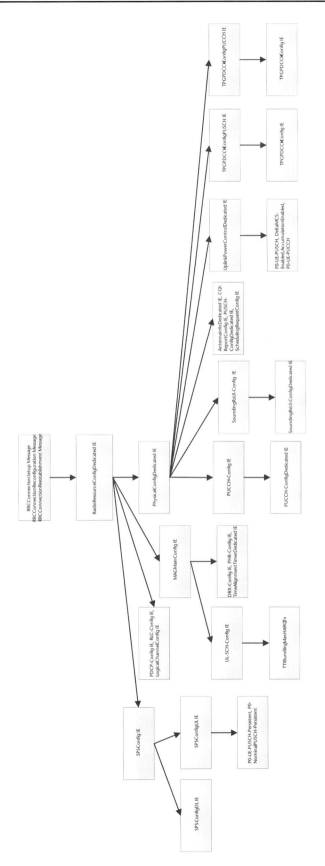

**Figure A.2** IEs on RadioResourceConfigDedicated IE.

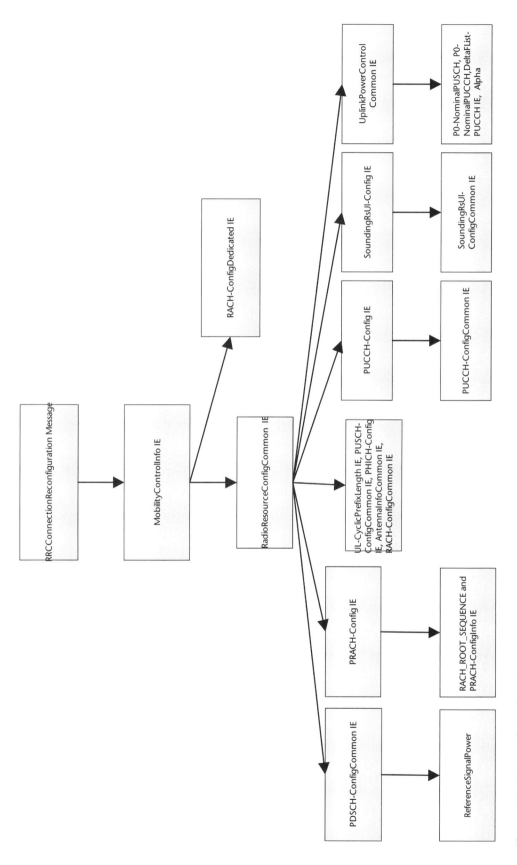

**Figure A.3** IEs used during handover.

# Acronyms

| | |
|---|---|
| A-GNSS | Assisted-GNSS |
| AM | Acknowledged mode |
| AMC | Adaptive modulation and coding |
| AMD PDU | Acknowledged mode data PDU |
| ANRF | Automatic neighbor relation function |
| AoA | Angle of arrival |
| APN | Access point name packet data network PDN; access point name |
| APN-AMBR | APN aggregate maximum bit rate |
| ARP | Allocation and retention priority |
| ARQ | Automatic repeat request |
| AS | Access stratum |
| AuC | Authentication center |
| BCCH | Broadcast control channel |
| BCH | Broadcast channel |
| BM-SC | Broadcast multicast service center |
| BSIC | Base station identity code |
| BSR | Buffer status report |
| CB | Cell broadcast |
| CC | Chase combining |
| CCCH | Paging control channel |
| CCE | Control channel element |
| CDD | Cyclic delay diversity |
| CDS | Channel-dependent scheduling |
| CFI | Control format indicator |
| CGI | Cell global identification |
| CID | Context identifier |
| CM | Cubic metric |
| CMAS | Commercial Mobile Alert Service |
| Co-MIMO | Cooperative MIMO |
| CP | Cyclic prefix |
| CPICH | Common pilot channel |
| CPT | Control PDU type |
| CQI | Channel quality indicator |
| C-RNTI | Cell radio network temporary identifier |
| CS | Circuit switching |

| | |
|---|---|
| CSA | Common subframe allocation |
| CSG | Closed subscriber group |
| DCCH | Dedicated control channel |
| DCI | Downlink control information |
| DFT | Discrete Fourier transform |
| DM RS | Demodulation reference signal |
| DOA | Direction of arrival |
| DRB | Data radio bearer |
| DRX | Discontinuous reception |
| DTCH | Dedicated traffic channel |
| DTX | Discontinuous transmission |
| EARFCN | E-UTRA absolute radio frequency channel number |
| EBI | EPS bearer identity |
| ECGI | E-UTRAN cell global identifier |
| ECI | E-UTRAN cell identifier |
| E-CID | Enhanced Cell-ID |
| ECM | EPS connection management |
| EHPLMN | Equivalent home PLMN |
| EIR | Equipment identity register |
| EMM | EPS mobility management |
| EPC | Evolved packet core |
| EPRE | Energy per resource element |
| EPS | Evolved packet system |
| EPS AKA | EPS authentication and key agreement |
| E-RAB | E-UTRAN radio access bearer |
| ESM | EPS session management |
| E-SMLC | Enhanced serving mobile location center |
| ETWS | Earthquake and Tsunami Warning System |
| E-UTRAN | Evolved UMTS Terrestrial Radio Access Network |
| FDD | Frequency division duplex |
| FDE | Frequency domain equalization |
| FDS | Frequency diverse scheduling |
| FEC | Forward error correction |
| FFR | Fractional frequency reuse |
| FFT | Fast Fourier transform |
| FMS | First missing PDCP SN |
| FO state | First-order state |
| FSS | Frequency selective scheduling |
| FSTD | Frequency-switched transmit diversity |
| GBR | Guaranteed bit rate |
| GNSS | Global Navigation Satellite System |
| GTP | GPRS Tunneling Protocol |
| GUMMEI | Globally unique MME identity |
| GUTI | Globally unique temporary UE identity |
| HARQ | Hybrid automatic repeat request |
| HARQ ID | HARQ process identifier |
| HeNB GW | Home eNodeB gateway |
| HFN | Hyper frame number |

| | |
|---|---|
| HFR | Hard frequency reuse |
| HPLMN | Home public land mobile network |
| HSS | Home subscription server |
| ICI | Intercell interference |
| ICIC | Intercell interference coordination |
| IE | Information element |
| IFFT | Inverse fast Fourier transform |
| IMEI | International mobile station equipment identity |
| IMS | IP multimedia subsystem |
| IR | Incremental redundancy |
| IR state | Initialization and refresh state |
| ISI | Intersymbol interference |
| KSI | Key set identifier |
| LAC | Location area code |
| LB | Long block |
| LBI | Linked bearer identity |
| LBS | Location-based service |
| LCG | Logical channel group |
| LI | Length indicator |
| LSF | Last segment flag |
| LTE | Long-term evolution |
| MAC | Medium access control |
| MAC RAR | MAC random access response |
| MAG | Mobile access gateway |
| MBMS | Multimedia broadcast multicast service |
| MBR | Maximum bit rate |
| MCC | Mobile country code |
| MCCH | Multicast control channel |
| MCE | Multicell/multicast coordination entity |
| MCH | Multicast channel |
| MCS | Modulation and coding scheme |
| MGL | Measurement gap length |
| MGRP | Measurement gap repetition period |
| MIB | Master information block |
| MIMO | Multiple input multiple output |
| MISO | Multiple input single output |
| MME | Mobility management entity |
| MMEC | MME code |
| MMEGI | MME group ID |
| MMEI | MME identifier |
| MNC | Mobile network code |
| M-RNTI | MBMS radio network temporary identifier |
| MSA | MCH subframe allocation |
| MSI | MCH scheduling information |
| MSP | MCH scheduling period |
| MTCH | Multicast traffic channel |
| MU-MIMO | Multiple users MIMO |
| NACK_SN | Negative acknowledgment SN |

| | |
|---|---|
| NAS | Nonaccess stratum |
| NDI | New data indicator |
| NNSF | NAS node selection function |
| NLOS | Nonline of site |
| NSF | Node selection function |
| OFDM | Orthogonal frequency division multiplex |
| OFDMA | Orthogonal frequency division multiple access |
| OOB | Out-of-band |
| OTDOA | Observed time difference of arrival |
| PAPR | Peak-to-average power ratio |
| PBCH | Physical broadcast channel |
| PCC | Policy and charging control |
| PCCH | Paging control channel |
| PCEF | Policy control enforcement function |
| PCFICH | Physical control format indicator channel |
| PCH | Paging channel |
| PCI | Physical cell identifier |
| PCO | Protocol configuration option |
| PCRF | Policy control and charging rules function |
| PDB | Packet delay budget |
| PDCCH | Physical downlink control channel |
| PDCP | Packet Data Convergence Protocol |
| PDN | Packet data network |
| PDSCH | Physical downlink shared channel |
| PDU | Protocol data unit |
| PF | Paging frame; proportional fair |
| PFR | Partial frequency reuse |
| P-GW | PDN gateway |
| PHICH | Physical hybrid ARQ indicator channel |
| PHR | Power headroom report |
| PL | Path loss |
| PMCH | Physical multicast channel |
| PMI | Precoding matrix indicator |
| PMIP | Proxy mobile IP |
| PO | Paging occasion |
| PRACH | Physical random access channel |
| PRB | Physical resource block |
| P-RNTI | Paging radio network temporary identifier |
| PRS | Positioning reference signal |
| P/S | Parallel-to-serial |
| PS | Packet-switched |
| PSC | Primary scrambling code |
| PSS | Primary synchronization signal |
| PTI | Procedure transaction ID |
| PTM | Point to multipoint |
| PUCCH | Physical uplink control channel |
| PUSCH | Physical uplink shared channel |
| P-wave | Primary wave |

| | |
|---|---|
| PWS | Public Warning System |
| QCI | QoS class identifier |
| QoS | Quality of service |
| QPP | Quadratic permutation polynomial |
| QZSS | Quasi-Zenith Satellite System |
| RAC | Routing area code |
| RACH | Random access channel |
| RAPID | Random access preamble identifier |
| RAR | Random access response |
| RA-RNTI | Random access radio network temporary identifier |
| RB | Radio bearer; resource block |
| RB ID | Radio bearer identity |
| RBG | Resource block group |
| RE | Resource element |
| REG | Resource element group |
| RF | Resegmentation flag |
| RI | Rank indication |
| RIV | Resource indication value |
| RLC | Radio link control |
| RNL | Radio network layer |
| RR | Round robin |
| RRC | Radio resource control |
| RSRP | Reference signal received power |
| RSRQ | Reference signal received quality |
| RTT | Round-trip time |
| RV | Redundancy version |
| SAP | Service access point |
| SAW | Stop-and-wait |
| SB | Short block |
| SBAS | Satellite-Based Augmentation Systems |
| SC-FDMA | Single-carrier FDMA |
| SCTP | Stream Control Transmission Protocol |
| SDF | Service data flow |
| SDMA | Space-division multiple access |
| SDU | Service data unit |
| SFBC | Space-frequency block coding |
| SFN | Single frequency network; system frame number |
| SFR | Soft frequency reuse |
| S-GW | Serving gateway |
| SI | System information |
| SIB | System information block |
| SIMO | Single input multiple output |
| SI-RNTI | System information radio network temporary identifier |
| SLP | SUPL location platform |
| SM | Spatial multiplexing |
| SN | Sequence number |
| SN id | Serving network's identity |
| SO | Segmentation offset |

| | |
|---|---|
| SO state | Second-order state |
| SON | Self-optimizing networks |
| SPS C-RNTI | Semipersistent scheduling C-RNTI |
| SR | Scheduling request |
| SRB | Signaling radio bearer |
| SRS | Sounding reference signal |
| SSS | Secondary synchronization signal |
| STBC | Space-time block code |
| S-TMSI | SAE TMSI |
| SU-MIMO | Single-user MIMO |
| SUPL | Secure user plane location |
| S-wave | Secondary wave |
| TAC | Tracking area code |
| TAI | Tracking area identity |
| TAU | Tracking area update |
| TB | Transport block |
| TBS | Transport block size |
| TDD | Time division duplex |
| TEID | Tunnel end point identifier |
| TFT | Traffic flow template |
| TM | Transparent mode |
| TMD PDU | Transparent mode data PDU |
| TMGI | Temporary mobile group identity |
| TNL | Transport network layer |
| TPC | Transmit power control |
| TPMI | Transmitted precoding matrix indicator |
| TRI | Transmitted rank indication |
| TSTD | Time-switched transmit diversity |
| TTI | Transmission time interval |
| UE | User equipment |
| UE-AMBR | UE aggregate maximum bit rate |
| UM | Unacknowledged node |
| UICC | Universal integrated circuit card |
| USIM | Universal subscriber identity module |
| VPLMN | Visited public land mobile network |
| VRB | Virtual resource block |
| V-TSTD | Virtual time switched transmit diversity |
| ZC | Zadoff-Chu |

# Bibliography

3GPP Tdoc, REV-090005, LTE RAN Architecture Aspects, 2009.

3GPP Technical Specification 22.011, "Service Accessibility," Release 9.

3GPP Technical Specification 22.016, "International Mobile Station Equipment Identities (IMEI)," Release 9.

3GPP Technical Specification 22.168, "Earthquake and Tsunami Warning System (ETWS) Requirements; Stage 1," Release 9.

3GPP Technical Specification 22.220, "Service Requirements for Home NodeBs and Home eNodeBs," Release 9.

3GPP Technical Specification 22.268, "Universal Mobile Telecommunications System (UMTS); Public Warning System (PWS) Requirements," Release 9.

3GPP Technical Specification 23.003, "Technical Specification Group Core Network and Terminals; Numbering, Addressing and Identification," Release 9.

3GPP Technical Specification 23.032, "Universal Mobile Telecommunications System (UMTS); Universal Geographical Area Description (GAD)," Release 9.

3GPP Technical Specification 23.041, "Technical Specification Group Core Network and Terminals; Technical Realization of Cell Broadcast Service (CBS)," Release 9.

3GPP Technical Specification 23.122, "Universal Mobile Telecommunications System (UMTS); LTE; Non-Access-Stratum (NAS) Functions Related to Mobile Station (MS) in Idle Mode," Release 8.

3GPP Technical Specification 23.203, "Policy and Charging Control Architecture," Release 8.

3GPP Technical Specification 23.272, "Universal Mobile Telecommunications System (UMTS); LTE; Circuit Switched (CS) Fallback in Evolved Packet System (EPS); Stage 2," Release 9.

3GPP Technical Specification 23.401, "LTE; General Packet Radio Service (GPRS) Enhancements for Evolved Universal Terrestrial Radio Access Network (E-UTRAN) Access," Release 10.

3GPP Technical Specification 24.007, "Technical Specification Group Core Network and Terminals; Mobile Radio Interface Signalling Layer 3; General Aspects," Release 8.

3GPP Technical Specification 24.008, "Technical Specification Group Core Network and Terminals; Mobile Radio Interface Layer 3 Specification; Core Network Protocols; Stage 3," Release 10.

3GPP Technical Specification 24.301, "Technical Specification Group Core Network and Terminals; Non-Access-Stratum (NAS) Protocol for Evolved Packet System (EPS); Stage 3," Release 10.

3GPP Technical Specification 25.133, "Universal Mobile Telecommunications System (UMTS); Requirements for Support of Radio Resource Management (FDD)," Release 9.

3GPP Technical Specification 25.214, "Technical Specification Group Radio Access Network; Physical Layer Procedures (FDD)," Release 10.

3GPP Technical Specification 25.302, "Technical Specification Group Radio Access Network; Services Provided by the Physical Layer," Release 10.

3GPP Technical Specification 25.304, "User Equipment (UE) Procedures in Idle Mode and Procedures for Cell Reselection in Connected Mode," Release 9.

3GPP Technical Specification 25.331, "Technical Specification Group Radio Access Network; Radio Resource Control (RRC); Protocol Specification," Release 10.

3GPP Technical Specification 33.401, "Technical Specification Group Services and System Aspects; 3GPP System Architecture Evolution (SAE): Security Architecture," Release 8.

3GPP Technical Specification 36.101, "Technical Specification Group Radio Access Network; Evolved Universal Terrestrial Radio Access (E-UTRA); User Equipment (UE) Radio Transmission and Reception," Release 10.

3GPP Technical Specification 36.104, "LTE; Evolved Universal Terrestrial Radio Access (E-UTRA); Base Station (BS) Radio Transmission and Reception," Release 9.

3GPP Technical Specification 36.133, "LTE; Evolved Universal Terrestrial Radio Access (E-UTRA); Requirements for Support of Radio Resource Management," Release 9.

3GPP Technical Specification 36.201, "LTE; Evolved Universal Terrestrial Radio Access (E-UTRA); LTE Physical Layer; General description," Release 10.

3GPP Technical Specification 36.211, "LTE; Evolved Universal Terrestrial Radio Access (E-UTRA); Physical Channels and Modulation," Release 10.

3GPP Technical Specification 36.212, "LTE; Evolved Universal Terrestrial Radio Access (E-UTRA); Multiplexing and Channel Coding," Release 10.

3GPP Technical Specification 36.213, "LTE; Evolved Universal Terrestrial Radio Access (E-UTRA); Physical Layer Procedures," Release 1.

3GPP Technical Specification 36.214, "Technical Specification Group Radio Access Network; Evolved Universal Terrestrial Radio Access (E-UTRA); Physical Layer—Measurements," Release 8.

3GPP Technical Specification 36.300, "Technical Specification Group Radio Access Network; Evolved Universal Terrestrial Radio Access (E-UTRA) and Evolved Universal Terrestrial Radio Access Network (E-UTRAN); Overall Description; Stage 2," Release 10.

3GPP Technical Specification 36.302, "Technical Specification Group Radio Access Network; Evolved Universal Terrestrial Radio Access (E-UTRA); Services Provided by the Physical Layer," Release 9.

3GPP Technical Specification 36.304, "Technical Specification Group Radio Access Network; Evolved Universal Terrestrial Radio Access (E-UTRA); User Equipment (UE) Procedures in Idle Mode," Release 10.

3GPP Technical Specification 36.305, "LTE; Evolved Universal Terrestrial Radio Access Network (E-UTRAN); Stage 2 Functional Specification of User Equipment (UE) Positioning in E-UTRAN," Release 9.

3GPP Technical Specification 36.306, "Technical Specification Group Radio Access Network; Evolved Universal Terrestrial Radio Access (E-UTRA) User Equipment (UE) Radio Access Capabilities," Release 10.

3GPP Technical Specification 36.321, "Technical Specification Group Radio Access Network; Evolved Universal Terrestrial Radio Access (E-UTRA); Medium Access Control (MAC) Protocol Specification," Release 9.

3GPP Technical Specification 36.322, "Technical Specification Group Radio Access Network; Evolved Universal Terrestrial Radio Access (E-UTRA); Radio Link Control (RLC) Protocol Specification," Release 9.

3GPP Technical Specification 36.323, "LTE; Evolved Universal Terrestrial Radio Access (E-UTRA); Packet Data Convergence Protocol (PDCP) Specification," Release 9.

3GPP Technical Specification 36.331, "Technical Specification Group Radio Access Network; Evolved Universal Terrestrial Radio Access (E-UTRA); Radio Resource Control (RRC); Protocol Specification," Release 10.

3GPP Technical Specification 36.355, "LTE; Evolved Universal Terrestrial Radio Access (E-UTRA); LTE Positioning Protocol (LPP)," Release 9.

3GPP Technical Specification 36.401, "LTE; Evolved Universal Terrestrial Radio Access Network (E-UTRAN); Architecture description," Release 10.

3GPP Technical Specification 36.413, "LTE; Evolved Universal Terrestrial Radio Access Network (E-UTRAN); S1 Application Protocol (S1AP)," Release 9.

3GPP Technical Specification 36.423, "Technical Specification Group Radio Access Network; Evolved Universal Terrestrial Radio Access Network (E-UTRAN); X2 Application Protocol (X2AP)," Release 9.

3GPP Technical Specification 36.440, "LTE; Evolved Universal Terrestrial Radio Access Network (E-UTRAN); General Aspects and Principles for Interfaces Supporting Multimedia Broadcast Multicast Service (MBMS) Within E-UTRAN," Release 9.

3GPP Technical Specification 36.902, "Technical Specification Group Radio Access Network; Evolved Universal Terrestrial Radio Access Network (E-UTRAN); Self-Configuring and Self-Optimizing Network (SON) Use Cases and Solutions," Release 9.

3GPP Technical Specification 44.031, "Technical Specification Group GSM/EDGE Radio Access Network; Location Services (LCS); Mobile Station (MS)–Serving Mobile Location Centre (SMLC) Radio Resource LCS Protocol (RRLP)," Release 10.

3GPP Technical Specification 45.008, "Digital Cellular Telecommunications System (Phase 2+); Radio Subsystem Link Control," Release 9.

Bohge, M., J. Grossy, and A. Wolisz, "Optimal Power Masking in Soft Frequency Reuse Based OFDMA Networks," *European Wireless Conference*, Denmark, May 2009, pp. 162–166.

Holma, H., and A. Toskala, LTE for UMTS: OFDMA and SC-FDMA Based Radio Access, New York: John Wiley & Sons, 2009.

Khan, F., *LTE for 4G Mobile Broadband: Air Interface Technologies and Performance,* New York: Cambridge University Press, 2009.

Leitao, F. A., S. S. Freire, and S. R. Lima, "SMS over LTE: Interoperability Between Legacy and Next Generation Networks," *IEEE Symposium on Computers and Communications,* 2010, pp. 634–639.

# About the Author

Mohammad T. Kawser is an assistant professor at the Electrical and Electronic Engineering Department at the Islamic University of Technology, Bangladesh. Previously, he served as a senior RF and tools engineer at Accuver Americas (formerly WirelessLogix, Inc.), Texas. He received an M.S. in electrical engineering from Virginia Tech. He is a member of the editorial boards for the *International Journal of Computer and Electrical Engineering* (IJCEE) and the *International Journal of Computer Theory and Engineering* (IJCTE).

# Index

## A

Access Point Names (APNs)
   defined, 21
   DNS naming conventions, 21–22
Access Stratum (AS), 11–13
   ciphering, 175, 181
   COUNT, 180
   integrity protection, 175, 181
   security activation, 179
   security in, 179–81
Acknowledged mode (AM), 125–35
   actions after acknowledgment, 127–28
   defined, 125
   functions, 125–30
   initial transmission, 126–27
   PDU format, 130–35
   retransmission of data, 129–30
   RLC Control PDU, 132–35
   RLC Data PDU, 130–32
   status report, 128–29
   use cases, 125
   *See also* Radio Link Control (RLC)
Active time, 234
Adaptive HARQ, 297–98
Adaptive modulation and coding (AMC), 291–92
Allocation and Retention Priority (ARP), 36–37
   defined, 36
   information, 36–37
AM Control PDU, 130, 132–35
AM Data PDU, 130–32
   extension part of header, 132
   fixed part of header, 130–31
   segment, 132, 133–35
Aperiodic CQI/PMI/RI reporting, 286
   defined, 286
   number of subbands, 288
   reporting modes, 287
   subband size, 288
Aperiodic power control, 237
Assisted-GNSS (A-GNSS), 20
Asynchronous HARQ retransmissions, 298
Attach procedure
   authentication, 211–12
   changes caused by, 205
   illustrated, 207–8
   initial, 205–23
   location update, 212–13
   ME identity check, 212
   NAS security activation, 211–12
   NAS signaling connection, 206–11
   old EPS bearer context delete, 212
   use of, 205
   user identification, 211
Automatic repeat request (ARQ)
   defined, 292
   HARQ relationship, 296–97

## B

Battery life, 237
Beamforming, 262–64
   with precoding, 263
   transmission mode 6, 263–64
   transmission mode 7, 263–64
   use of, 262–63
Bidirectional optimistic mode (O-mode), 141
Bidirectional reliable mode (R-mode), 141
Blind handover
   inter-RAT, 346
   intra-E-UTRAN, 345
Broadcast Multicast Service Center (MB-SC), 382

BSR classifications, 107
    buffer size levels, 108
    formats, 107
    long BSR, 107, 109
    padding BSR, 109–10
    periodic BSR, 109
    regular BSR, 108–9
    short BSR, 107, 109
    truncated BSR, 109–10
Bucket Size Duration (BSD), 281
Buffer Status Report (BSR), 276
Buffer Status Report (BSR) MAC control element, 106–10

## C

Camping, 153
Cell coordination, 395–404
Cell Radio Network Temporary Identifier (C-RNTI), 23
Cells
    coordination among, 395–404
    GERAN, 338
    interfrequency, 336
    intrafrequency, 336
    load balancing among, 399–401
    UTRAN, 337–38
Cell search, 149–50
Cell selection, 149–53, 334–43
    criteria, 339–43
    initial, 149–53
    measurement of neighboring cells, 336–39
    methods, 149
    priority 1, 340
    priority 2, 340–41
    priority 3, 341–42
    stored information, 153
Cell-specific power control, 236
Cell-specific reference signals, 80, 81–82
Change of cells, 333–79
    cell reselection, 334–43
    handover, 343–79
    intrafrequency cell scenarios, 334
    neighbor cells, 333
Channel-dependent scheduling (CDS), 266–67, 291
Channel Quality Indicator (CQI), 87
    index, 285
    reports, 285
    value to offset level, 288
Channels
    logical, 27–28, 34, 279–81
    mapping among bearers and, 32–34
    physical, 28–29, 34, 67–80
    transport, 28
    *See also* Specific channels
Channel-state information reference signals, 80, 87–89
Chase combining (CC), 293
Ciphering, 175, 181
Closed-loop power control, 236
Closed-loop spatial multiplexing, 258–60
Code blocks (CBs), 53, 54
Coding, 53
Commercial Mobile Alert Service (CMAS), 406
Compressor states, 142
Contention-based random access, 164–71
    cases, 164
    contention resolution, 169–71
    handover, 164
    initiation of downlink data transfer, 164
    message 3, 168–69
    RAR, 167–68
    RRC connection, 191, 196–97
    RRC connection establishment, 164
    RRC connection reestablishment, 164
    successful RAR, 167
    transmission of random access preamble, 165–67
    unsuccessful RAR, 168
Contention resolution, 169–71
    successful, 170–71
    timer, 170
    unsuccessful, 171
Control Channel Elements (CCEs), 67
    defined, 67
    for PDCCH, 76
Control messages, piggybacking, 39–40
Control plane, 11–13
    Access Stratum (AS), 11–13
    Non Access Stratum (NAS), 13
Control regions, 67
CQI/PMI/RI reports
    aperiodic, 286
    aperiodic reporting modes, 287

low-speed users, 285
periodic, 286–89
representation, 285–86
subband size, 288
time frequency, 286
CRC insertion, 53
C-RNTI MAC control element, 110
CS Fallback (CSFB), 312–18
call establishment methods, 312
defined, 311
MO call setup, 313–15
MT call setup, 315–18
Cyclic delay diversity (CDD), 252
large delay, 260
multilayer operation, 261
Cyclic prefix (CP), 44–45
advantages, 44–45
disadvantages, 45
length of, 50
uplink, 57

## D

Data forwarding, 375
Data radio bearers (DRBs), 31–32
defined, 34
of RLC acknowledged mode, 366
of RLC unacknowledged mode, 367
Data transfer
activities, 283–300
downlink steps, 283–84
feedback from UE, 283, 284–89
in handover execution, 366–67
HARQ, 283–84, 292–300
link adaptation, 283, 284, 289–92
process, 283, 284
resource allocation, 283, 284
seeking resource allocation, 284
uplink steps, 284
Data transfer session setup, 301–22
dedicated EPS bearer setup, 301–2, 306–9
service request procedure, 301, 302–6
SMS, 318–22
voice services, 310–18
DCI formats
format 1, 72
format 1A, 72–73
format 1B, 73

format 1C, 73–74, 77
format 1D, 74
format 2, 71, 74–75
format 2A, 75
format 2B, 75
format 2C, 75
format 3, 75
format 3A, 75
format 4, 76
Decompressor states, 142–43
Dedicated EPS bearer setup
classification, 301
network-initiated, 306–8
UE-requested, 309
Demodulation reference signals
(DM RSs), 88–90
position for normal cyclic prefix, 89
transmission of symbols, 88
Detach procedure, 224
classification, 224–25
explicit detach, 224
MME initiated, 227–29
UE-initiated, 225
use of, 205, 224
Discontinuous Reception (DRX), 231–34
continuous reception and
resumption of, 233
cycle, 231
cycle, configuration in, 232–33
HARQ during, 233–34
in RRC_CONNECTED state, 231–34
in RRC_IDLE state, 231
Distributed scheduling, 269
Downlink
CDS, 267
data transfer steps, 283–84
DCI formats 1, 2, 2A, and 2B, 273–74
DCI formats 1A, 1B, 1C, and 1D, 275
HARQ for, 294, 298
ICIC in, 396
logical channels, 27
multiple antenna techniques in, 249–64
OFDMA for, 43–48
physical channels, 28, 67–80
physical layer process, 53–54
physical signals, 80–87
time-frequency response grid, 48–53

DCI formats 1A, 1B, 1C, and 1D (continued)
    transport channels, 28
Downlink allocation, 272–75
Downlink synchronization, 150, 155–58
    Physical Cell Identifiers (PCIs), 156
    PSS sequence, 156–57
    scheduling PSS and SSS, 155
    SSS sequence, 157–58
    with target eNodeB, 367
    uplink transmission timing, 160
Dynamic power control, 236
Dynamic scheduling, 237, 268, 271

## E

Earthquake and Tsunami Warning System (ETWS), 405–6
    defined, 405
    urgency classification, 406
    wave types, 405–6
    *See also* Public Warning System (PWS)
Energy Per Resource Element (EPRE), 21
Enhanced Cell-ID (E-CID), 20
ENodeB identifiers (eNB IDs), 22
ENodeBs
    base station use, 13
    defined, 6
    exchange of resource status between, 400
    functions, 6–7
    Home, 14, 15–17
    interfaces, 7–8
EPS Authentication and Key Agreement (EPS AKA) procedure, 211–12
EPS Bearer Identities (EBIs), 35
EPS bearers, 34–39
    Allocation and Retention Priority (ARP), 36–37
    defined, 33
    establishment of, 34
    GBR, 37
    identities, 34–35
    non-GBR, 37–38
    QCI, 38–39
    QoS profile, 36
    service data flows (SDF), 35
    traffic flow templates (TFT), 35
EPS Mobility Management (EMM)
    connection-related services, 13
    defined, 13
    states, 41–42
EPS Session Management (ESM)
    defined, 13
    states, 42
Equipment Identity Register (EIR)
    defined, 9
    mobile equipment records, 212
E-UTRA absolute radio frequency channel numbers (EARFCNs), 4
E-UTRA measurement reporting
    conditions for, 353
    defined, 350–52
    event-triggered, 351–52
    parameters, 350–51
    periodical, 352
E-UTRAN cell global identifiers (ECGIs), 22
E-UTRAN cell identifiers (ECIs), 22
E-UTRAN Radio Access Bearers (E-RABs), 34
Event-triggered measurement reporting, 349–50
    E-UTRA, 351–52
    inter-RAT, 354–55
    parameters, 350–51, 352–53
    periodical, 355–57
Evolved packet core (EPC), 5
    components, 8–9
    EIR, 9
    HSS, 8–9
    interfaces, 9–10
    MME, 8
    node selection, 215–16
    PCRF, 9
    P-GW, 8
    S-GW, 8
Evolved Packet System (EPS), 176
    Authentication and Key Agreement (EPS AKA), 176–77
    defined, 5
    GTP, 6
    network architecture for, 5
    SCTP, 6
    security context, 176
Evolved UMTS Terrestrial Radio Access (E-UTRA), 1
Evolved UMTS Terrestrial Radio Access Network (E-UTRAN), 6–8
    architecture illustration, 7

defined, 6
eNodeB functions, 6–7
eNodeB interfaces, 7–8
Explicit detach, 224

## F

Feedback from UE, 284–89
    aperiodic CQI/PMI/RI reporting, 286
    CQI index, 285
    defined, 283
    mapping differential, 288
    mapping subband differential, 288
    periodic CQI/PMI/RI reporting, 286–89
    periodic reporting modes, 290
    reporting modes, 287
    subband size for reports, 288
Forward error correction (FEC), 292
Four-antenna codebook, 258
Fractional frequency reuse (FFR), 397–99
    defined, 397–98
    performance gain, 398
    PFR, 398
    SFR, 398, 399
Frequency division duplex (FDD), 2
Frequency-switched transmit diversity (FSTD), 252

## G

GERAN cells, 338
Global eNodeB identifiers (Global eNB IDs), 22
Globally unique MME identity, 22
Globally unique temporary UE identity (GUTI)
    defined, 23
    independent reallocation, 223
    reallocation procedure, 223
Global Navigation Satellite System (GNSS), 18
GPRS Tunneling Protocol (GTP), 6
Guaranteed bit rate (GBR) bearers, 37

## H

Handover, 343–79
    blind, 345, 346
    complete, 368–69
    completion, 379
    configuration procedure, 346–47
    in contention-based random access, 164
    decision, 362–64
    execution, 365–68, 376–78
    hard, 343
    IEs used during, 409, 412
    instruction, 365
    inter-RAT, 344, 345–46, 369–79
    intra-E-UTRAN, 344–45, 362–69
    measurement and reporting, 346–62
    measurement configuration UE update, 359
    measurement gap, 357–58
    measurement object, 346, 347–49
    measurement reporting, 347, 349–57
    measurement triggering configuration, 358–59
    in noncontention-based random access, 171, 173
    performing measurements, 360
    preparation, 364–65, 370–76
    random access configuration, 163
    request, 364
    request acknowledge, 364–65
    RLF, 345, 346
    S1-based, 344
    sending measurement reports, 360–62
    types of, 344
    UE-assisted, 344–46
    X2-based, 344
    *See also* Change of cells
Hard frequency reuse (HFR), 397
Header compression, 139–43
    basic ROHC procedure, 140–41
    bidirectional optimistic mode (O-mode), 141
    bidirectional reliable mode (R-mode), 141
    compressor states, 142
    decompressor states, 142–43
    modes of operation, 141
    ROHC profiles, 139
    typical header size, 140
    unidirectional mode (U-mode), 141
Heterogeneous network, 13
High interference indication (HII), 397
Home eNodeB, 14, 15–17
    access mode, 15–17
    gateway, 15
    network architecture, 16
Home Subscription Server (HSS) functions, 8–9
Hybrid automatic repeat request (HARQ), 292–300
    adaptive, 297–98

Hybrid automatic repeat request (HARQ)
(continued)
  ARQ relationship, 296–97
  asynchronous, 298
  chase combining (CC), 293
  data transfer, 283, 284
  defined, 292
  for downlink transmission, 294, 298
  IR, 293
  in MAC layer, 293–96
  nonadaptive, 298
  on PDCCH, 297
  process, 292–96
  retransmissions, 297–98
  round trip time (RTT), 293
  stop-and-wait (SAW) operation, 292
  synchronous, 298
  for uplink transmission, 295, 299–300

# I

Implicit detach, 224–25
Information bearers, 27–42
  channels and signals, 27–29
  EPS bearer, 34–39
  mapping among channels and, 32–34
  NAS layer states, 41–42
  piggybacking control messages, 39–40
  radio bearer, 29–32
Information Elements (IEs)
  during handover, 409, 412
  on RadioResourceConfigDedicated IE, 409, 411
  on RRC messages, 409–12
  on SIB type 2, 409, 410
Initial cell selection, 149–53
  camping, 153
  cell search, 149–50
  PLMN selection, 151–52
  reading MIB, 150
  reading other SIB types, 152
  reading SIB type 1, 150–51
  suitability criteria, 152
  *See also* Cell selection
Initial context setup, 218–22
  context setup complete, 222
  context setup request, 218
  radio bearer configuration, 220–22

  radio bearer establishment and attach accept, 220
  radio bearer setup complete, 222
  AS security activation, 218
  UE capability transfer, 218
Integrity check failure, 196
Integrity protection, 175, 181
Intercarrier interference (ICI), 43
  OFDMA, 44
  subcarrier spacing and, 45
Intercell interference (ICI), 237
Intercell interference coordination (ICIC), 395–99
  in downlink, 396
  exchange of information to support, 396–97
  fractional frequency reuse (FFR), 397–99
  methods, 395
  in uplink, 395, 396–97
Interfrequencies and inter-RAT measurement conditions, 338–39
International Mobile Station Equipment Identity (IMEI), 22–26
International Mobile Subscriber Identity (IMSI)
  defined, 22–23
  as permanent identifier, 23
Inter-RAT handover, 345–46
  blind, 346
  data forwarding preparation, 375–76
  defined, 344
  handover completion, 379
  handover execution, 376–78
  handover request processing, 370–75
  phases, 369
  preparation, 370–76
  procedure, 369–79
  RLF, 346
  routing area update, 378
  UE-assisted, 345–46
  *See also* Handover
Inter-RAT measurement reporting
  conditions for, 355
  defined, 352
  event-triggered, 354–55
  parameters, 352–53
  periodical, 355–57
Intra-E-UTRAN handover, 344–45

Index 431

blind, 345
defined, 344
downlink synchronization, 367
handover complete indication, 367–68
handover completion, 368–69
handover decision, 362–64
handover execution, 365–68
handover instruction, 365
handover preparation, 364–65
handover request, 364
handover request acknowledge, 364–65
phases, 362
procedure, 362–69
procedure over X2 interface, 363
proper delivery of packets, 366–67
random access response (RAR), 367
RLF, 345
S1-based, 344
transmission of random access preamble, 367
UE-assisted, 344–45
X2-based, 344
*See also* Handover
Intrafrequency cells, 334
IP address allocation, 213–15
entities, 213
IPv4 address allocation, 214–15
IPv6 address allocation, 215
PDN connectivity request, 214
IP Multimedia System (IMS), 310

## L

Limited service, 154
Link adaptation, 289–92
adaptive modulation and coding (AMC), 291–92
channel-dependent scheduling (CDS), 291
downlink data transfer, 283
multiple antennas schemes, 292
transmission power control, 292
uplink data transfer, 284
Linked Bearer Identities (LBIs), 35, 307
Load balancing
among cells, 399–401
among MMEs, 324, 401–2
scheduling decision, 271
Localized scheduling, 268–69

Location-Based Services (LBSs), 18
Location update
attach procedure, 212–13
PDN connection establishment, 213–23
Logical channels, 27–28
mapping, 34
resource apportionment among, 279–81
Long Term Evolution (LTE)
Advanced, 1
coverage issue, 1
defined, 1
end-to-end QoS support, 1
FDD, 2, 3
heterogeneous network, 13–18
high performance support, 1
network architecture, 5–11
operating bands, 2–4
optimized signaling, 1
protocol stack, 11–13
TDD, 2
UE positioning, 18–20
UE positioning techniques, 19
LTE. *See* Long Term Evolution
LTE-Advanced, 1
LTE Positioning Protocol (LPP), 18

## M

MAC control elements, 105–11
BSR, 106–10
categories, 105
C-RNTI, 110
defined, 102
on DL-SCH, 105–6
DRX command, 105
example illustration, 102
on MCH, 106
power headroom, 110–11
timing advance command, 105, 106
UE contention resolution identity, 105, 106
on UL-SCH, 106–10
MAC headers
defined, 102
example illustration, 102
fields, 104
LCID values, 104
MAC RAR, 112–13
subheaders, 103

MAC RAR (continued)
    variable size, 103
MAC PDU, 102–15
    format, 102–11
MAC control element, 102, 105–11
MAC header, 102, 103–5
MAC SDU, 102
MCH, 102
    padding, 103
    for random access response, 111–15
    special format, 111–15
    for transparent MAC, 111
MAC RAR, 111–15
    back-off indicator, 112, 113
    defined, 111
    fields, 113–15
    illustrated, 112
MAC header, 112–13
    random access preamble identifier (RAPID), 112, 114
MAC SDU, 102
Maximum allowable path loss (MAPL), 1
Maximum C/I, 268
MBMS Radio Network Temporary Identifier (M-RNTI), 25
MBMS Service ID and Temporary Mobile Group Identity (TMGI), 25
MBSFN reference signals, 80, 84
Measurement gap, 357–58
Measurement identity, 347
Measurement object, 347–49
    defined, 346
MeasObjectEUTRA, 348
MeasObjectGERAN, 349
MeasObjectUTRA, 348–49
    *See also* Handover
Measurement reporting, 349–57
    defined, 347
    E-UTRA, 350–52
    event-triggered, 349, 350–55
    periodical, 350–51, 352, 355–57
    reports, sending, 360–62
    *See also* Handover
Measurements, performing, 360
Measurement triggering configuration, 358–59
Medium Access Control (MAC) layer, 13, 101–15

defined, 101
functions, 101
PDU, 102–15
*See also* MAC control elements; MAC headers
Message 3, 168–69
    on CCCH, 169
    on DCCH, 169
    defined, 168
    MIB messages, 61
    mapping, 59–60
    reading, 150
MME codes (MMECs), 22
MME group ID (MMEGI), 22
MME identifiers (MMEIs), 22
MME initiated detach procedure, 227–29
    detach accept, 229
    detach request, 229
    EPS bearer context delete, 229
    illustrated, 228
    NAS connection, 227–28
    paging, 227
    S1 release, 229
    service request, 227
    steps, 227–29
Mobility Management Entities (MMEs)
    capacities, 401
    functions, 8
    load balancing, 324, 401–2
    load rebalancing, 401–2
    mesh network connection, 11, 401
    pool area, 10–11, 199
    S1 connection, 199–201
    selection, 198–99, 210–11
    UE-associated logical S1-connection setup, 210–11
    unable to accept S1 setup request, 200
Modulation, 54
Modulation and Coding Scheme (MCS), 268
Multicast control channel (MCCH), 386–89
    change of information, 387
    configuration, 387
    modification, 387–89
    scheduling, 386
    transmission, 386
Multicell/multicast Coordination Entity (MCE), 382–83

Multicell transmission, 381–82
Multimedia Broadcast Multicast Service
        (MBMS), 381–94
    BM-SC, 382
    content provider, 382
    control information, 384–85
    data, 384
    defined, 381
    gateway (MBMS GW), 383
    mapping among layers, 384–85
    MCCH, 386–89
    MCE, 382–83
    multicell transmission, 381–82
    network architecture, 382–84
    PMCH, 391–94
    protocol stacks, 384–85
    service request procedure, 389
    sessions, 389
    session start, 390
    session stop, 391
    SIB type 13, 385–86
Multiple antennas techniques, 262
    beamforming, 262
    in channels, 250–51
    in downlink, 249–64
    MIMO, 249–50, 251–52
    MISO, 249
    SIMO, 249
    spatial multiplexing, 255–62
    transmission mode 2, 252–55
    transmission mode 3, 260–62
    transmission mode 4, 259
    transmission mode 5, 262
    transmission mode 6, 259–60, 264
    transmission mode 7, 263–64
    transmit diversity, 252–55
Multiple input multiple output (MIMO), 262
    configurations, 251
    defined, 249–50
    feedback to eNodeB, 252
    illustrated, 250
    layers, 251
    multiple-user (MU-MIMO), 255–56, 262
    single-user (SU-MIMO), 255, 256–62
Multiple input single output (MISO)
    defined, 249
    illustrated, 250

Multiple-user MIMO (MU-MIMO)
    configuration, 262
    defined, 255–56
    SU-MIMO comparison, 256
    transmission mode 5, 262

# N

NAS Node Selection Function (NNSF), 199
NAS signaling connection, 189–203
    attach request and, 206–11
    MME initiated detach procedure, 227–28
    MME selection, 198–99
    release, 201–3
    RRC connection, 189–98
    S1-MME connection, 199–201
    TAU, 324, 326–27
    UE-initiated detach procedure, 225–26
    UE-triggered service requests, 302
    *See also* Non Access Stratum (NAS)
Neighbor cells, 333
    intrafrequency measurement conditions, 338
    list configuration, 403
    list optimization, 404
    measurement of, 336–39
Network architecture, 5–11
    EPC, 8–10
    for EPS, 5
    E-UTRAN, 6–8
    Home eNodeB, 16
    Multimedia Broadcast Multicast Service
        (MBMS), 382–84
    for relay, 17
    tracking area and MME pool area, 10–11
Network-initiated dedicated EPS bearer setup,
        306–8
Network services, 153–54
    limited, 154
    normal, 153–54
    operator, 154
Network-triggered service requests, 305–6
    configuring, 306–7
    defined, 301
    illustrated, 305, 307
    radio bearer establishment, 307–8
    setup accept, 308
Non Access Stratum (NAS), 13
    ciphering, 175

Non Access Stratum (NAS) (continued)
    integrity protection, 175
    layer states, 41–42
    messages, piggybacking, 39–40
    security activation, 178
    security in, 178–79
    *See also* NAS signaling connection
Nonadaptive HARQ, 297–98
Noncontention-based random access, 171–74
    defined, 171
    handover, 171, 173
    random access preamble assignment, 172–73
    random access preamble transmission, 173
    RAR, 173–74
Nondedicated random access preamble, 165
Non-GBR bearers, 37–38
Nonline-of-site (NLOS) links, 4
Normal service, 153–54

## O

Observed Time Difference of Arrival (OTDOA), 19–20
OFDMA
    cyclic prefix (CP), 44–45
    defined, 44
    for downlink, 43–48
    implications of, 44–45
    intercarrier interference (ICI), 44
    reception, 47–48
    subcarrier spacing, 45
    time-dispersive radio channel, 44
    transmission, 46–47
Open-loop power control, 236
Open-loop spatial multiplexing, 260–62
Operator service, 154
Orthogonal frequency division multiplex (OFDM), 43, 54
Over-the-top voice, 311

## P

Packet Data Convergence Protocol (PDCP), 13, 137–47
    ciphering, 181
    control PDU, 146–47
    data packet discard, 143
    data packet processing, 137–43
    data PDU, 145–46
    defined, 137
    entities, 137
    functional view, 138
    header compression, 139–43
    integrity protection, 181
    layer functions, 137–44
    packet delivery after handover, 144
    PDU, 145–47
    radio bearers, 138
    reestablishment of, 144
    security function, 138–39
    status report, 147
    structural view, 139
    transmission side, 143
Padding bits, 103
Padding BSR, 109–10
Paging, 183–87
    CMAS-Indication, 187
    cycle, 185
    default cycle, 185–86
    defined, 183
    ETWS-Indication, 187
    frames, 185
    frames, allocation of, 186
    mapping among layers, 184
    messages, 183, 186–87
    messages, indication of, 184–85
    MME initiated detach procedure, 227
    occasions, allocation of, 186
PagingRecordList, 187
    SystemInfoModifications, 187
Paging Radio Network Temporary Identifier (P-RNTI), 24
Partial frequency reuse (PFR), 398
PDCP control PDU, 146–47
    illustrated, 147
    interspersed ROHC feedback packet, 146
PDCP status report, 147
PDCP data PDU, 145–46
    control plane, 146
    defined, 145
    illustrated, 145
    user plane, 146
PDN connection establishment
    attach complete/begin uplink transmission, 222

Index                                                                                     435

bearer modify/begin downlink transmission, 222–23
EPC node selection, 215–16
initial context setup, 218–22
IP address allocation, 213–15
location update, 213–23
session setup, 216–17
PDN gateway (P-GW), 9
Periodical measurement reporting, 350
  E-UTRA, 352
  parameters, 350–51, 352–53
Periodic BSR, 109
Periodic CQI/PMI/RI reporting, 286–89
  bandwidth parts, 289
  defined, 286–89
  reporting modes, 290
  as semistatically configured, 289
  subband size, 289
Periodic power control, 237
Periodic update, 323–24
Physical Broadcast Channel (PBCH), 68
Physical cell identifiers (PCIs), 22, 156
Physical channels, 28–29
  downlink, 67–80
  mapping, 34
Physical Broadcast Channel (PBCH), 68
Physical Control Format Indicator Channel (PCFICH), 68–69
Physical Downlink Control Channel (PDCCH), 69–77
Physical Hybrid ARQ Indicator Channel (PHICH), 77–80
Physical Random Access Channel (PRACH), 94–99
Physical Uplink Control Channel (PUCCH), 91–94
Physical Control Format Indicator Channel (PCFICH), 68–69
Physical Downlink Control Channel (PDCCH), 69–77
  allocation of, 76–77
  CCEs for, 76
  control information, 69–70
  DCI formats, 70–76
  defined, 69
  format size and scope, 77
  HARQ information on, 297

instance identification, 70
instances, 69, 70
search spaces, 78
UE searches for candidates, 76
Physical Hybrid ARQ Indicator Channel (PHICH), 77–80
  allocation, 79
  HARQ ACK/NACK, 77
  HARQ indicators, 78
  instances, 78
  number of groups, 79
  orthogonal sequence for, 79
  transmission, 79
Physical layer properties, 43–58
  downlink physical layer process, 53–54
  downlink time-frequency resource grid, 48–53
  OFDMA for downlink, 43–48
  SC-FDMA for uplink, 54–57
  uplink transmission features, 57–58
Physical multicast channel (PMCH), 391–94
  antenna port, 391
PMCH-Config IE, 393–94
PMCH-Info IE, 393, 394
  resource allocation, 391–92
Physical Random Access Channel (PRACH), 94–99
  defined, 94
  format types, 98
  preamble formats, 95, 96
  search spaces, 98
  subcarrier spacing, 96
  transmission, 95
  ZC sequences, 97
Physical resource blocks (PRBs), 51–52
  allocated, 275
  defined, 51
Physical signals, 29, 80–87
  downlink, 80–87
  uplink, 87–91
Physical Uplink Control Channel (PUCCH), 91–94
  defined, 91
  dynamically configurable terms, 245–47
  format 1, 93
  format 1a, 93
  format 1b, 93

Physical Uplink Control Channel (PUCCH) (continued)
    format 2, 94
    format 2a, 94
    format 2b, 94
    format types, 92
    power control on, 244–47
    region, 91
    semistatically configurable terms, 244
    UE transmission, 91
Pico cells, 14
Piggybacking
    control messages, 39–40
    NAS messages, 39–40
    RRC messages, 40
    S1AP messages, 39
Policy Control and Charging Rules Function (PCRF), 9
Positioning reference signals, 80, 84–86
    defined, 84
    fields, 86
    mapping of, 86
Power capability, 237
Power control, 235–47
Power headroom MAC control element, 110–11
Precoding, 54
Precoding Matrix Index (PMI), 87, 252
Primary Synchronization Signals (PSSs), 155, 358
    scheduling, 155
    sequence, 156
Prioritized bit rate (PBR), 280
Priority aware scheduling, 268
Proportional fair (PF), 268
Protocol data units (PDUs), 32
    MAC, 102–15
    PDCP, 145–47
    TMD, 119
    UMD, 122–25
Protocol stack
    control plane, 11–13
    illustrated, 12
    types, 11
    user plane, 13
Public Land Mobile Network (PLMN) codes
    defined, 21
    selection, 151–52
Public Warning System (PWS), 405–8
    Commercial Mobile Alert Service (CMAS), 406
    defined, 405
    Earthquake and Tsunami Warning System (ETWS), 405–6
    notification, 406–7
    SIB content for, 407–8
    types of, 405
PUSCH power control, 237–44
    dynamically configurable terms, 239–44
    semistatically configurable terms, 238
    uplink transmit power computation, 237

## Q

QoS Class Identifiers (QCIs), 38–39
Quadratic permutation polynomial (QPP) interleaver, 53

## R

Radio bearers, 29–32
    configuration, 220–21
    data (DRB), 31–32
    establishment, 220
    Packet Data Convergence Protocol (PDCP), 138
    setup complete, 222
    signaling (SRB), 29
Radio frame structures, 48–51
    cyclic prefix length, 50
    radio frame, 49–50
    supported, 49
    symbol period, 50–51
Radio Link Control (RLC), 13, 117–35
    acknowledged mode (AM), 125–35
    defined, 117
    modes, 117
    transparent mode (TM), 117–19
    unacknowledged mode (UM), 119–25
Radio link failure (RLF), 193–95
    actions to overcome, 195
    defined, 193
    detection of, 193–95
    *See also* RRC connection
Radio link failure (RLF) handover
    inter-RAT, 346

intra-E-UTRAN, 345
Radio Resource Control (RRC) layer, 11–13
    messages, IEs on, 409–12
    messages, piggybacking, 40
    *See also* RRC connection
Radio signal measurement, 21
Random access, 163–74
    configuration of, 163
    contention-based, 164–71
    noncontention-based, 171–74
    procedure, 277
Random Access Channel (RACH), 404
Random access preambles
    assignment of, 172–73
    contention-based random access, 165–67
    division of, 166
    nondedicated, 165
    transmission of, 165–67, 173
Random Access Radio Network Temporary Identifier (RA-RNTI), 25
Random access response (RAR)
    contention-based random access, 167–68
    in handover execution, 367
    MAC PDU for, 111–15
    message transport, 111
    noncontention-based random access, 173–74
    successful RAR, 167
    unsuccessful RAR, 168
Rank Indicator (RI), 87, 252
Rate matching, 53–54
Reception
    OFDMA, 47–48
    single-carrier FDMA (SC-FDMA), 56–57
Reference signal (RS), 29
Reference Signal Received Power (RSRP), 21
Reference Signal Received Quality (RSRQ), 21
Reference signals (RSs), 29, 80–87
    cell-specific, 80, 81–82
    channel-state information, 80, 87
    demodulation, 88
    downlink, 80–87
    MBSFN, 80, 84
    positioning, 80, 84–86
    sounding (SRSs), 90–91
    UE-specific, 80, 82–84
    uplink, 87–88
Regular BSR, 108–9

Relay, 17
Resource allocation
    apportionment among logical channels, 279–81
    downlink, 272–75
    in downlink data transfer, 283
    dynamic scheduling, 271
    with frequency hopping, 277–79
    inter-RAT handover, 370–75
    physical multicast channel (PMCH), 391–92
    procedure, 271–79
    semipersistent scheduling, 272
    type 0, 273–74
    type 1, 274
    type 2, 275
    uplink, 275–79
    in uplink data transfer, 284
    without frequency hopping, 277
Resource blocks (RBs), 51–53
    physical (PRBs), 51–52
    resource elements, 58
    uplink, 57
    virtual (VRBs), 52–53
Resource Element Groups (REGs), 67
Resource indication values (RIVs), 275
RLC Control PDU, STATUS PDU, 132–35
Round robin (RR), 267
Routing area update, inter-RAT handover, 378
RRC_CONNECTED state, 231–34
RRC_IDLE state, 231
RRC connection, 189–98
    contention-based random access, 191, 196–97
    defined, 189
    establishment, 190–92, 206–10
    integrity check failure, 196
    radio link failure (RLF), 193–95
    reconfiguration, 198
    reconfiguration failure, 196
    reestablishment, 195–98
    reestablishment complete, 197
    reestablishment failure, 197–98
    reject, 192
    release, 193
RRCConnectionRequest message, 206
RRCConnectionSetupComplete message, 206–10

RRCConnectionSetup message, 206
    setup, 191, 197
    setup configuration, 192
    setup failure, 192
    *See also* Radio Resource Control (RRC) layer

## S

S1
    bearers, 34
    release procedure, 202
    setup procedure, 199–200
S1-based handover, 344
S1-MME connection, 199–201
    setup, 199–200
    UE-associated logical connection setup, 200–201
S5/S8 bearers, 34
SAE TMSI (S-TMSI), 23
Scheduling
    channel-dependent (CDS), 266–67, 291
    for delay-limited capacity, 268
    distributed, 269
    dynamic, 268, 271
    limitation in TTI, 270–71
    localized, 268–69
    multicast control channel (MCCH), 386
    queue and priority aware, 268
    semipersistent, 268, 271
Scheduling decision, 265–71
    antenna configuration, 271
    apportionment among UEs, 267–68
    backhaul support, 271
    channel dependent scheduling (CDS), 266–67
    considerations in, 266–71
    dynamic/semipersistent scheduling, 268
    frequency hopping in uplink, 269
    ICI coordination, 271
    information for, 266
    limitation of scheduling in TTI, 270–71
    load balancing, 271
    localized or distributed scheduling, 268–69
    measurement gaps, 270
    modification, 265
    Modulation and Coding Scheme (MCS), 268
    power or bandwidth limitation, 269–70
    queue/priority aware scheduling, 268
    TTI bundling in uplink, 270
    UE capabilities, 270
Scheduling Request (SR), 276
Scrambling, 54
Search spaces
    common, 76
    PDCCH format types and, 78
    purposes, 78
    UE-specific, 76–77
Secondary Synchronization Signals (SSSs), 155
    scheduling, 155
    sequence, 157–58
Security
    in AS layer, 179–81
    in NAS layer, 178–79
    PDCP, 138–39
Self-configuration (SC), 402–3
    automatic eNodeB installation, 402
    dynamic configuration of S1 interface, 402
    dynamic configuration of X2 interface, 403
    neighbor cell list configuration, 403
Self-optimization (SO), 403–4
Self-optimizing networks (SONs), 402–4
    defined, 402
    self-configuration (SC), 402–3
    self-optimization (SO), 403–4
Semipersistent scheduling, 237, 268, 271
Semipersistent Scheduling Cell Radio Network Temporary Identifier (SPS C-RNTI), 24
Semistatic power control, 236
Service data flows (SDF), 35
Service data units (SDUs), 32
Service request procedure
    classification, 301
    network-triggered, 301, 305–6
    UE-triggered, 301, 302
Serving cells, 153
Serving Gateway (S-GW), 9
Session setup, 216–17
Short message service (SMS), 318–22
    IMS, 318
    interim voice solutions, 318
    MO, 319–20
    MT, 320–22

over SGs, 318–22
as revenue source, 318
solutions, 318
SIB messages, 59–60, 61
IEs on, 409, 410
for PWS, 407–8
reading, 150–51
type 13, 385–86
Signaling radio bearers (SRBs), 29–31
defined, 29
in handover execution, 367
SRB0, 30
SRB1, 30–31
SRB2, 31
Signals
physical, 29, 80–87
reference, 29, 80–87
synchronization, 29
SI messages, 62–65
modification, 63
notification, 64–65
scheduling, 62–63
Single-carrier FDMA (SC-FDMA)
defined, 54
reception, 56–57
subcarriers, 54
transmission, 55–56
for uplink, 54–57
Single input multiple output (SIMO)
defined, 249
illustrated, 250
Single-user MIMO (SU-MIMO)
adaptive control, 256
closed-loop spatial multiplexing, 258–60
defined, 255
MU-MIMO comparison, 256
open-loop spatial multiplexing, 260–62
precoding in LTE, 257
Rank 1, 256
Rank 2/3/4, 257
transmission mode 3, 260–62
transmission mode 4, 259
transmission mode 6, 259–60
Soft frequency reuse (SFR), 398, 399
Sounding reference signals (SRSs), 90–91
Space-frequency block coding (SFBC), 252
Space-time block codes (STBCs), 252

Spatial multiplexing, 255–62
closed-loop, 258–60
function of, 255
MU-MIMO, 255–56, 262
open-loop, 260–62
SU-MIMO, 255, 256–62
STATUS PDU, 132–35
defined, 132–34
fields, 134–35
Stored information cell selection, 153
Stream Control Transmission Protocol (SCTP), 6
Subcarrier spacing, 45
choice of, 45
PRACH, 96
uplink transmission, 57
Symbol period, 50–51
Synchronization, 155–62
downlink, 150, 155–58
uplink, 158–62
Synchronization signal, 29
Synchronous HARQ retransmissions, 298
System architecture evolution (SAE), 5
System information, 59–65
mapping among layers, 59–60
MIB messages, 61
required, 60–61
SIB messages, 61
SI messages, 62–65
System Information Radio Network Temporary Identifier (SI-RNTI), 25

## T

Temporary C-RNTI, 24
Time division duplex (TDD), 2
Time-frequency response grid
downlink, 48–53
radio frame structure, 48–51
resource block (RB), 51–53
uplink, 57
Timing alignment
advance command, 159
establishing, 159–61
factors affecting, 158
items, 159
maintenance, 161–62
offset, 159

TPC-PUCCH-RNTI, 24
TPC-PUSCH-RNTI, 24
Tracking area, 10–11
Tracking area code (TAC), 22
Tracking area identity (TAI), 22
Tracking area update (TAU)
    accepted, 330–31
    bearer update, 328–29
    change in UE properties, 324
    HSS update, 329
    MME load rebalancing, 324
    NAS signaling connection establishment, 326–27
    NAS signaling connection recovery, 324
    periodic update, 323–24
    procedure, 323–25
    procedure illustration, 325
    S1 release, 331
    UE context update, 327–28
    UE mobility, 323
Traffic flow templates (TFT), 35
Transmission
    OFDMA, 46–47
    single-carrier FDMA (SC-FDMA), 55–56
Transmission Time Interval (TTI)
    scheduling decision and, 265
    *See also* TTI bundling
Transmit diversity, 252–55
    application with SFBC and FSTD, 252
    configuration, 252
    for two/four transmit antennas, 252, 254
Transparent mode (TM)
    functions, 118–19
    PDU format, 119
    receiving RLC entity, 119
    transmitting RLC entity, 118
    use cases, 117–18
    *See also* Radio Link Control (RLC)
Transport blocks (TBs), 58
Transport channels
    downlink, 28
    mapping, 34
    uplink, 28
TTI bundling
    defined, 279
    scheduling decision, 270
    uplink allocation, 279
    *See also* Transmission Time Interval (TTI)
Tunnel end point identifiers (TEIDs), 6
Two-antenna codebook, 258

## U

UE-assisted handover
    inter-RAT, 345–46
    intra-E-UTRAN, 344–45
UE-initiated detach procedure, 225
    detach accept, 227
    EPS bearer context delete, 226–27
    NAS signaling connection, 225–26
    S1 release, 227
UE positioning
    determination, 18
    E-CID, 20
    A-GNSS, 20
    network architecture, 18–19
    OTDOA, 19–20
    techniques, 19–20
    *See also* User equipment
UE-requested dedicated EPS bearer setup, 309
UE-specific power control, 236
UE-specific reference signals, 80, 82–84
UE-triggered service requests, 302–5
    authentication and security activation, 303
    bearer modify, 305
    defined, 301
    illustrated, 303
    initial context setup, 303–4
    NAS signaling connection, 302
Unacknowledged mode (UM), 119–25
    concatenation, 121
    defined, 119
    discarding multiple PDUs, 122
    extension part of header, 125
    fixed part of header, 122–23
    functions, 119–22
    PDU format, 122–25
    receiving RLC entity, 120
    reordering, 121–22
    segmentation, 121
    transmitting RLC entity, 119–20
    *See also* Radio Link Control (RLC)
Unidirectional mode (U-mode), 141
Universal Subscriber Identity Module (USIM), 22

Uplink
  bandwidth, 237
  CDS, 267
  data transfer steps, 284
  frequency hopping in, 269
  HARQ for, 295, 299–300
  ICIC in, 395, 396–97
  physical channels, 91–99
  physical signals, 87–91
  resource blocks (RBs), 57
  SC-FDMA for, 54–57
  subcarrier spacing, 57
  time-frequency response grid, 57
  transmission features, 57–58
  TTI bundling in, 270
Uplink allocation, 275–79
  with frequency hopping, 277–79
  seeking resources, 276–77
  TTI bundling, 279
  without frequency hopping, 277
  *See also* Resource allocation
Uplink Control Information (UCI), 87
Uplink logical channels, 28
Uplink physical channels, 29
Uplink power control, 235–47
  aperiodic, 237
  battery life, 237
  cell-specific, 236
  closed-loop, 236
  considerations, 235–37
  control and data transmit power, 237
  dynamic, 236
  dynamic scheduling, 237
  intercell interference (ICI), 237
  link adaptation techniques, 236
  objectives, 235
  open-loop, 236
  periodic, 237
  power capability, 237
  on PUCCH, 244–47
  on PUSCH, 237–44
  radio link characteristics, 235–36
  semipersistent scheduling, 237
  semistatic, 236
  UE-specific, 236
  uplink bandwidth, 237
  use of, 235

Uplink synchronization, 158–62
  downlink transmission timing, 160
  time alignment, 158
  timing advance command, 159
  timing alignment, 159
  timing alignment establishment, 159–61
  timing alignment maintenance, 161–62
  timing control, 158
  timing offset, 159
Uplink transport channels, 28
User equipment (UE)
  apportionment among, 267–68
  capabilities, informing, 219–20
  capability transfer, 218
  categories, 25
  category capabilities, 26
  cell selection, 149–53
  feedback from, 283, 284–89
  identities, 22–26
  identities construction, 24
  logical S1-connection setup, 200–201
  mobility, 323
  network services, 153–54
  operating band support, 2
  power class, 25
  powering up, 149–54
  updating with measurement configurations, 359
  *See also* UE positioning
User identification, attach procedure, 211
User plane, 13
UTRAN cells, 337–38

## V

Virtual resource blocks (VRBs), 52–53
Visited public land mobile network (VPLMN), 23
Voice over LTE via Generic Access (VoLGA), 310–11
Voice services, 310–18
  call classification, 310
  CS Fallback (CSFB), 311, 312–18
  implementation problems, 310
  IP Multimedia System (IMS), 310
  over-the-top voice, 311
  solution comparison, 311
  solutions, 310–11

Voice over LTE via Generic Access (VoLGA), 310–11

**X**

X2-based handover, 344